高等院校"十二五"信息技术规划教材

Android
游戏开发案例教程

张 辉 编著

清华大学出版社
北京

内容简介

本书主要内容包括 Android 常用游戏类型的视角与内容设计、项目结构、资源管理、生命周期、布局与基础组件、事件处理、多线程与消息处理、游戏视图、图形特效、游戏动画、多媒体与传感器、游戏中的数学与物理学知识、碰撞检测及 Android 平台下常用的游戏物理引擎。

本书采用以"案例驱动"为主线的"基础—实践—综合—训练"这一循序渐进的学习体系,理论知识及实验内容立足于教学实际,案例选择来源于商业实际应用;结合"理论—实践"一体化及"工学结合"的教学理念,突出 CDIO 教学理念的实用性、灵活性、先进性和技巧性;力求"素质、能力、知识"合一和"教、学、做"合一;各章知识点整体以从易到难、由浅入深的形式呈现,通过验证实例、基础实例、综合实例让读者循序渐进地学习和掌握 Android 游戏开发的相关知识与技巧。

本书兼具技术手册和教材的特点,适合作为高等学校数字媒体技术、软件工程、计算机科学与技术等专业和各类培训机构相关课程的教材,也可供移动游戏开发从业人员参考。

本书封面贴有清华大学出版社防伪标签,无标签者不得销售。
版权所有,侵权必究。举报: 010-62782989, beiqinquan@tup.tsinghua.edu.cn。

图书在版编目(CIP)数据

Android 游戏开发案例教程/张辉编著. —北京: 清华大学出版社,2015(2024.1重印)
高等院校"十二五"信息技术规划教材
ISBN 978-7-302-39985-8

Ⅰ. ①A… Ⅱ. ①张… Ⅲ. ①移动电话机-游戏程序-程序设计-教材 Ⅳ. ①TN929.53 ②TP311.5

中国版本图书馆 CIP 数据核字(2015)第 077241 号

责任编辑: 焦 虹 战晓雷
封面设计: 常雪影
责任校对: 白 蕾
责任印制: 沈 露

出版发行: 清华大学出版社
    网　　址: https://www.tup.com.cn, https://www.wqxuetang.com
    地　　址: 北京清华大学学研大厦 A 座　　邮　编: 100084
    社 总 机: 010-83470000　　邮　购: 010-62786544
    投稿与读者服务: 010-62776969, c-service@tup.tsinghua.edu.cn
    质量反馈: 010-62772015, zhiliang@tup.tsinghua.edu.cn
    课件下载: https://www.tup.com.cn, 010-83470236
印 装 者: 三河市龙大印装有限公司
经　　销: 全国新华书店
开　　本: 185mm×260mm　　印 张: 23.5　　字 数: 540 千字
版　　次: 2015 年 6 月第 1 版　　印 次: 2024 年 1 月第 11 次印刷
定　　价: 59.80 元

产品编号: 060492-03

前言

1. 本书主要内容

如今的 Android 系统市场份额节节攀升,势不可挡,越来越多的开发者加入到 Android 应用开发的行列。2013 年 Android 市场应用相比 2012 年增长了 9 倍之多;而这些与日俱增的 Android 应用程序中,无论是按使用量还是按总收入排名,位居榜首的应用 80% 都是游戏。

全书总共 9 章,内容概括如下。

第 1 章:介绍 Android 平台下常用游戏类型在游戏视角、游戏内容两个方面的设计规范。

第 2 章:讲解 Android 环境配置、项目结构、生命周期、资源管理、国际化、消息和对话框等知识。

第 3 章:讲解 Android 游戏开发中常用的界面布局、基础组件、程序单元和事件处理机制。

第 4 章:讲解线程与消息处理、Android 游戏视图框架、绘图类、图像特效和游戏动画制作。

第 5 章:讲解 Camera 图像采集、游戏音乐与音效、视频播放和传感器框架。

第 6 章:讲解 Android 数据存储、网络编程和网页浏览组件的使用。

第 7 章:讲解游戏开发中的数学和物理学知识、二维游戏中的碰撞检测和模拟粒子系统的方法。

第 8 章:讲解飞行射击类游戏案例的设计与开发,对前面各章知识进行综合演练。

第 9 章:介绍 Android 平台下常用的二维和三维游戏物理引擎。

2. 课程教学方法

本书采用理论与实践相结合,运用任务驱动法、启发式教学法及引导式教学法,在灵活、直观的示范操作中,讲授课程内容及实际运用。

（1）构建应用情境，"教、学、做"一体化的教学模式（做中学、学中做）。

注重与学生的互动，师生一体，共同实现"网状"知识运用模型；注重创新思维的引导和演示操作有机结合，让学生在"思"与"学"、"做"与"用"的过程中掌握课程知识和实践应用技能。

（2）基于行动导向的"六步法"情境化教学过程。

在每个学习任务的教学实施过程中，按照基于行动导向的"资讯、决策、计划、实施、检查、评价"六步法，以"任务描述→任务资讯→任务分析（决策、计划）→任务实施→任务检查→任务评价与总结→拓展训练"的过程实施教学。

本书强调重点知识讲解的深入浅出和案例选择的合理性、时效性、实用性和科学性。从一线教学和实际研发需要出发，立足于项目实施和未来行业发展，重点培养学生自主学习能力和实践能力，强化工程意识与创新思维。

由于作者水平有限，书中难免有不足之处，恳请广大读者和同行批评指正。

张　辉

2015 年 3 月

目录

第1章 Android 常见游戏类型 ... 1

1.1 射击类游戏 ... 1
1.1.1 游戏视角 ... 1
1.1.2 游戏内容设计 ... 2

1.2 竞速类游戏 ... 2
1.2.1 游戏视角 ... 3
1.2.2 游戏内容设计 ... 3

1.3 益智类游戏 ... 3
1.3.1 游戏视角 ... 4
1.3.2 游戏内容设计 ... 4

1.4 角色扮演类游戏 ... 5
1.4.1 游戏视角 ... 5
1.4.2 游戏内容设计 ... 6

1.5 闯关动作类游戏 ... 6
1.5.1 游戏视角 ... 7
1.5.2 游戏内容设计 ... 7

1.6 冒险类游戏 ... 7
1.6.1 游戏视角 ... 8
1.6.2 游戏内容设计 ... 8

1.7 策略类游戏 ... 8
1.7.1 游戏视角 ... 9
1.7.2 游戏内容设计 ... 9

1.8 养成类游戏 ... 10
1.8.1 游戏视角 ... 10
1.8.2 游戏内容设计 ... 10

1.9 经营类游戏 ... 11
1.9.1 游戏视角 ... 11

  1.9.2 游戏内容设计 ·················· 11
1.10 体育类游戏 ························ 12
  1.10.1 游戏视角 ··················· 12
  1.10.2 游戏内容设计 ················ 13
1.11 本章小结 ·························· 13
1.12 思考与练习 ······················· 13

第 2 章 Android 基础知识 ················ 14

2.1 Android 平台简介 ················ 14
2.2 搭建 Android 开发环境 ············ 15
2.3 Eclipse Debug 调试程序 ············ 16
  2.3.1 Eclipse 调试器 ················ 16
  2.3.2 Logcat ······················· 17
2.4 Android 系统架构 ················· 18
2.5 创建第一个 Android 项目 ·········· 20
  2.5.1 使用 Eclipse 创建项目 ·········· 20
  2.5.2 使用命令行创建项目 ··········· 20
2.6 Android Project 项目结构 ··········· 21
2.7 Android 资源使用 ················· 25
  2.7.1 字符串资源 ·················· 26
  2.7.2 数组资源 ···················· 26
  2.7.3 颜色资源 ···················· 27
  2.7.4 尺寸资源 ···················· 27
  2.7.5 Drawable 资源 ················· 27
  2.7.6 样式和主题资源 ·············· 28
  2.7.7 布局资源 ···················· 29
  2.7.8 原始资源 ···················· 31
  2.7.9 原始资产 ···················· 31
  2.7.10 其他 XML 文件 ··············· 32
2.8 屏幕方向改变的应对策略 ·········· 32
2.9 Android 中常用的计量单位 ········· 33
2.10 Android 中的国际化 ·············· 33
2.11 消息提示与对话框 ··············· 34
  2.11.1 用 Toast 类显示消息 ··········· 34
  2.11.2 用 AlertDialog 类实现对话框 ···· 35
  2.11.3 基础实例：自定义视图对话框 ··· 38
2.12 本章小结 ························ 40
2.13 思考与练习 ····················· 40

第 3 章　Android 游戏开发之视图界面 ·· 41

3.1　界面布局 ·· 41
- 3.1.1　线性布局 ·· 41
- 3.1.2　表格布局 ·· 42
- 3.1.3　相对布局 ·· 45
- 3.1.4　帧布局 ··· 46
- 3.1.5　绝对布局 ·· 47

3.2　游戏开发常用组件 ·· 47
- 3.2.1　按钮类组件 ··· 47
- 3.2.2　文本类组件 ··· 53
- 3.2.3　进度条类组件 ·· 60
- 3.2.4　选项卡组件 ··· 64
- 3.2.5　列表类组件 ··· 66
- 3.2.6　日期、时间类组件 ·· 71

3.3　基本程序单元——活动 ·· 73
- 3.3.1　Android 生命周期 ··· 74
- 3.3.2　用 Intent 切换页面 ·· 79
- 3.3.3　用 Intent 实现活动间简单参数传递 ··· 81
- 3.3.4　Bundle 类在活动传值中的使用 ·· 83
- 3.3.5　用 Intent 实现活动间传递对象参数 ··· 87

3.4　Android 事件处理 ··· 91
- 3.4.1　处理键盘事件 ·· 91
- 3.4.2　处理触摸事件 ·· 92

3.5　综合实例一：游戏菜单及选项设置界面 ······································ 93
- 3.5.1　功能描述 ·· 93
- 3.5.2　关键技术 ·· 93
- 3.5.3　实现过程 ·· 94

3.6　综合实例二：BMI 计算器 ··· 103
- 3.6.1　功能描述 ··· 103
- 3.6.2　关键技术 ··· 103
- 3.6.3　准备知识 ··· 103
- 3.6.4　实现过程 ··· 104
- 3.6.5　实例扩展 ··· 107

3.7　综合实例三：猜猜看 ·· 112
- 3.7.1　功能描述 ··· 112
- 3.7.2　关键技术 ··· 113
- 3.7.3　实现过程 ··· 113

3.8 本章小结 ······ 117
3.9 思考与练习 ······ 117

第4章 Android 游戏开发之图形界面 ······ 118

4.1 线程与消息处理 ······ 118
    4.1.1 循环者类 Looper ······ 118
    4.1.2 Handler 消息传递机制 ······ 119
    4.1.3 消息类 Message ······ 119
    4.1.4 基础实例：快乐舞者 ······ 120
    4.1.5 基础实例：风中的气球 ······ 122
4.2 Android 二维游戏开发视图 ······ 125
    4.2.1 View 框架 ······ 126
    4.2.2 SurfaceView 框架 ······ 128
4.3 常用绘图类 ······ 133
    4.3.1 Paint 类 ······ 133
    4.3.2 Canvas 类 ······ 134
    4.3.3 Bitmap 类 ······ 135
    4.3.4 BitmapFactory 类 ······ 135
    4.3.5 基础实例：游戏角色行走控制 ······ 136
4.4 绘制 2D 图像 ······ 139
    4.4.1 绘制文本 ······ 139
    4.4.2 绘制几何图形 ······ 141
    4.4.3 绘制路径 ······ 143
    4.4.4 绘制图片 ······ 144
4.5 图像特效 ······ 146
    4.5.1 旋转图像 ······ 146
    4.5.2 缩放图像 ······ 147
    4.5.3 倾斜图像 ······ 147
    4.5.4 平移图像 ······ 148
    4.5.5 渲染图像 ······ 149
4.6 剪切区域 ······ 150
    4.6.1 剪切区域原理 ······ 150
    4.6.2 基础实例：RPG 游戏地图生成 ······ 150
    4.6.3 基础实例：游戏中的自动滚屏 ······ 153
4.7 游戏动画 ······ 156
    4.7.1 逐帧动画 ······ 156
    4.7.2 补间动画 ······ 158
    4.7.3 自定义动画 ······ 164

4.8 综合实例一：小小弹球 167
　　4.8.1 功能描述 167
　　4.8.2 关键技术 167
　　4.8.3 实现过程 168
　　4.8.4 实例拓展 173
4.9 综合实例二：动态游戏导航界面 175
　　4.9.1 功能描述 175
　　4.9.2 关键技术 175
　　4.9.3 实现过程 176
　　4.9.4 实例拓展 180
4.10 综合实例三：打地鼠 184
　　4.10.1 功能描述 184
　　4.10.2 关键技术 184
　　4.10.3 实现过程 185
4.11 综合实例四：游戏中的瞄准镜 188
　　4.11.1 功能描述 188
　　4.11.2 关键技术 188
　　4.11.3 实现过程 188
4.12 综合实例五：发疯的小猪 190
　　4.12.1 功能描述 190
　　4.12.2 关键技术 191
　　4.12.3 实现过程 191
4.13 综合实例六：开心涂鸦 194
　　4.13.1 功能描述 194
　　4.13.2 关键技术 194
　　4.13.3 实现过程 195
4.14 本章小结 199
4.15 思考与练习 200

第5章 Android 多媒体与传感器 201

5.1 Camera 图像采集 201
5.2 游戏音乐与音效 204
　　5.2.1 MediaPlayer 类 205
　　5.2.2 SoundPool 类 208
　　5.2.3 基础实例：游戏音效 210
　　5.2.4 基础实例：游戏开场动画 214
5.3 播放视频 215
5.4 传感器 219

5.4.1　传感器介绍 ··· 219
　　5.4.2　传感器框架 ··· 221
　　5.4.3　基础实例：战机飞行 ·· 223
5.5　综合实例一：控制相机拍照 ·· 227
　　5.5.1　功能描述 ·· 227
　　5.5.2　关键技术 ·· 227
　　5.5.3　实现过程 ·· 227
5.6　综合实例二：游戏导航摇杆 ·· 231
　　5.6.1　功能描述 ·· 231
　　5.6.2　关键技术 ·· 231
　　5.6.3　实现过程 ·· 232
5.7　综合实例三：多点触屏缩放 ·· 234
　　5.7.1　功能描述 ·· 234
　　5.7.2　关键技术 ·· 234
　　5.7.3　实现过程 ·· 234
5.8　本章小结 ·· 236
5.9　思考与练习 ··· 236

第 6 章　Android 数据存储与网络编程 ·· 237

6.1　游戏数据存储 ·· 237
　　6.1.1　SharedPreferences ·· 238
　　6.1.2　使用 Files 对象存储数据 ··· 241
　　6.1.3　SQLite 数据库应用 ·· 245
6.2　基于 Socket 的网络编程 ··· 260
6.3　基于 HTTP 的网络编程 ··· 264
　　6.3.1　使用 HttpURLConnection 类访问网络 ··································· 264
　　6.3.2　使用 HttpClient 类访问网络 ··· 272
6.4　用 WebView 组件显示网页 ·· 276
6.5　本章小结 ·· 279
6.6　思考与练习 ··· 279

第 7 章　游戏中的数学与物理学 ·· 280

7.1　游戏中常用的数学知识 ··· 280
7.2　游戏中常用的物理学知识 ·· 284
7.3　碰撞检测 ·· 287
　　7.3.1　矩形碰撞检测 ··· 288
　　7.3.2　圆形碰撞检测 ··· 291

7.3.3 像素碰撞检测 ··· 293
7.4 游戏中的粒子系统 ··· 297
7.5 本章小结 ··· 303
7.6 思考与练习 ··· 303

第 8 章 案例演练——疯狂战机 ··· 304

8.1 游戏背景及功能概述 ··· 304
   8.1.1 游戏类型 ··· 304
   8.1.2 功能简介 ··· 304
8.2 游戏的策划及准备工作 ··· 304
   8.2.1 游戏的策划 ··· 305
   8.2.2 Android 平台下游戏的准备工作 ······························ 305
8.3 游戏的架构 ··· 307
   8.3.1 游戏中各个类的简介 ··· 307
   8.3.2 游戏运行界面 ··· 308
8.4 游戏中的实体相关类 ··· 309
   8.4.1 主战飞机类 Plane ··· 309
   8.4.2 敌机类 Enemy ·· 315
   8.4.3 子弹类 Bullet ·· 325
   8.4.4 道具类 Property ··· 336
8.5 游戏中的界面相关类 ··· 341
   8.5.1 游戏显示类 PlaneGameActivity ······························ 341
   8.5.2 游戏主界面类 GameView ·· 341
   8.5.3 游戏界面绘制类 GameScreen ································· 344
   8.5.4 菜单界面类 MenuScreen ··· 348
   8.5.5 数据存储类 GameStore ·· 352
8.6 游戏中的辅助类 ··· 354
   8.6.1 Tools 类 ··· 354
   8.6.2 GameMusic 类 ·· 355
8.7 本章小结 ··· 356
8.8 思考与练习 ··· 356

第 9 章 Android 游戏物理引擎 ··· 357

9.1 常用 2D 物理引擎 ·· 357
9.2 常用 3D 物理引擎 ·· 359
9.3 本章小结 ··· 359
9.4 思考与练习 ··· 360

参考文献 ··· 361

第1章

Android 常见游戏类型

学习目标：
- 了解目前 Android 平台常见的游戏类型。
- 了解不同类型手机游戏的视角及内容设计方法。

本章导读：
当今流行的手机游戏类型繁多，不同类型的游戏都会有其独特的设计方式以吸引玩家。本章结合目前手机游戏产业的现状，介绍 Android 平台常见的 10 种游戏类型的视角及内容设计方法。

1.1 射击类游戏

射击类游戏(shooting game)是手机游戏中常见的游戏类型，目前市面上比较著名的射击类游戏有空战系列、坦克大战等。

手机中的射击类游戏通常为单人游戏，游戏节奏较快，要求玩家通过快速反应与游戏进行交互。操作方式单一，主要是控制游戏角色的行走方向以及向目标开火（手动和自动）或施放特殊技能。在游戏开始时会为玩家分配若干条生命用以进行后续的游戏，当耗费光所给的生命数目后游戏就会结束。射击类游戏多数为关卡类游戏，即玩家用有限的生命值挑战难度不断提升的关卡（一般也是有限的）。有些游戏的关卡可以由程序生成，为无限关卡模式。

1.1.1 游戏视角

射击类游戏的界面背景一般是根据相对运动原理，采用卷轴式的自动滚屏方法。就是使用一幅比游戏屏幕长的图片首尾相接作为游戏界面背景，在游戏的运行过程中不断循环滚动来达到背景变换的效果。另外，也可以用树木、建筑等一个一个的小图片拼接成游戏背景，这些小图片称为图元，采用图元技术可以方便地搭建出 2D、斜 45°角的 2.5D、90°角的 2.5D 的游戏场景。

飞行射击类游戏大多采用 2D 视角，例如著名的"雷电"系列（如图 1-1 所示）。还有一些采用 2.5D 视角，例如"坦克大战"系列（如图 1-2 所示）。

图 1-1 "雷电"游戏　　　　　　　图 1-2 "坦克大战"游戏

还有一些采用桌面电脑中比较著名的第一人称视角模式的射击类游戏,例如经典的"反恐精英"、"使命召唤"、"荣誉勋章"等都有了手机端版本,极大地提高了手机射击类游戏的可玩性。

1.1.2　游戏内容设计

射击类游戏的发展节奏较快,要求的是玩家的快速反应,虽然有着炫目的爆炸和声音效果,但单一的玩法不会吸引玩家的长期关注。因此,设计合理的关卡剧情是非常重要的,例如增加一段背景故事,塑造一个游戏主角,在游戏中适当出现人物的对话等,也可以把相互独立的关卡用背景故事串联起来。而针对相对简单的游戏规则,除了限制玩家的生命数目之外,还可以在游戏过程中不断出现一些增加生命数、加血量、增加控制角色的伤害输出以及设置积分机制等奖励措施。

1.2　竞速类游戏

竞速类游戏就是模拟驾驶交通工具进行比赛,通过高速移动时所带来的视觉和听觉上的体验以及冲破各种障碍到达终点、获得好名次的成就感来吸引玩家。这类游戏通常分为以下 3 种模式:

(1) 夹杂打斗模式。游戏在进行中允许玩家和其他选手进行简单的战斗,使游戏更加紧张刺激。例如"暴力摩托"就采用这种模式(如图 1-3 所示)。

(2) 职业联赛模式。玩家每赢得一场比赛,其等级就会提升,可以参加更高级的比赛来获取更多的等级,例如"世界摩托大奖赛"就采用这种模式(如图 1-4 所示)。

(3) 任务驱动模式。由任务系统控制游戏流程,由于任务之间的前后关系,一般情况下玩家不能任意选择比赛。

图 1-3 "暴力摩托"游戏　　　　图 1-4 "世界摩托大奖赛"游戏

1.2.1 游戏视角

目前，大部分的竞速类游戏都采用第一人称视角，这样容易给玩家一种身临其境的感觉。游戏画面一般分为主画面、模拟特定交通工具的操纵界面、任务列表和小地图4个部分。

1.2.2 游戏内容设计

可以自由选择交通工具决定了竞速类游戏的发展方向，目前以赛车为题材的较多。游戏的策划人员不应该只看到一种交通工具，其他交通工具可能会更加具有可玩性。

（1）水上交通工具：这类交通工具有机动艇、帆船、摩托艇等。赛道的设计可以添加一些有趣的障碍，如旋涡、台风甚至是水怪等。

（2）空中交通工具：主要是飞机和飞船两种。近地的空中赛道的设计可以添加飞鸟、云层等元素，太空场景可添加小行星群、陨石等。另外，这两种交通工具的驾驶方式（尤其是飞船）比较复杂，可以将操作方式做一些简化。

竞速类游戏大部分时间都让玩家处于高度紧张的状态，在游戏中适当穿插转场动画和剧情会加强游戏的氛围，特别是对于任务驱动模式的竞速游戏。

1.3 益智类游戏

益智类游戏（puzzle game）是另外一种深受玩家欢迎的游戏类型，其特色就是游戏中会更多地依靠智力去解决问题，例如纸牌类游戏、棋类游戏等都属于益智类游戏。这类游戏取胜的条件一般很简单，通常都会有限时功能，或者把消耗的时间作为计算积分时考量的因素。也有的取胜条件虽然简单，但很难实现，如"大富翁"系列游戏很难在短时间取得游戏的胜利。

很多玩家把益智类游戏称作休闲游戏，但实际上很多益智类游戏玩起来并不轻松，例如一些需要频繁思考的游戏(如数独等)。休闲游戏中很大一部分并不属于"益智"范畴,有些养成类游戏一般也划为休闲游戏。

不同的益智类游戏由于设计的内容相差很多,各有不同的玩法。但一般来讲,解谜类的游戏大都为单人游戏,例如经典游戏"吃豆子"(如图 1-5 所示)、"推箱子"、"拼图"、"走迷宫"等,都是以关卡作为提升难度的手段。关卡可以是有限的,也可以是由程序自动生成的,例如迷宫游戏中的地图就有很多成熟的算法。还有一些益智类游戏是多人对战模式,例如各种棋牌类游戏,经典的"大富翁"系列(如图 1-6 所示)就是多人联网游戏的典型,这里的"人"也可以是人工智能(AI)。

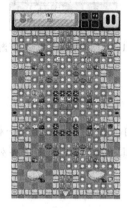
图 1-5 "吃豆子"游戏

图 1-6 "大富翁"游戏

1.3.1 游戏视角

游戏视角的改变可以提高玩家的体验,相比于其他吸引玩家的手段而言更直观且更容易实现。一般益智类游戏的视角可以看到整个游戏场景,场景不会很大,基本不需要滚屏,都是平面游戏。近年来很多平面游戏界面都被改造成 3D 效果,例如 3D 版的"推箱子"。当然也有例外,比如曾经很流行的"掘金者"游戏每一关的场景都需要滚屏,这也跟游戏节奏的快慢有关。

1.3.2 游戏内容设计

益智类游戏最大的魅力在于对智力的挑战,所以玩家往往对游戏的剧情并不是非常感兴趣。难度等级必须是可调节的,对于不设关卡的益智类游戏来说,通常在游戏开始时提示玩家选择一个难度等级。

游戏规则设计需要把握好 3 个方面:一是入门难易程度,游戏开始时不能让玩家感觉太简单,也不能让玩家太沮丧;二是趣味性,益智类游戏吸引人的地方不仅是对智力的挑战,还必须生动有趣,尽量营造一种轻松的环境;三是耐玩度,怎样让玩家愿意再玩一次,也是需要在设计的时候多加考虑的。

1.4 角色扮演类游戏

角色扮演类游戏(role playing game)一般要求玩家投入较多的注意力和较长的关注时间,同时一款优秀的角色扮演游戏开发成本也相对较高,例如网络游戏"梦想海贼王"与单机回合游戏"神雕侠侣"。

手机平台的角色扮演类游戏有单人模式的,也有网络模式的,但是论数量,还是单机模式的占多数。单机版的主线比较明朗,往往会把玩家控制的角色定义为"英雄"的形象,整个游戏都会围绕这个角色展开,来串接故事情节并影响游戏的发展方向,其取胜条件是由剧情决定的。网络版的角色扮演游戏一般对单个玩家没有很高的定位,所以对于玩家来说,游戏主线在于控制自己的角色进行各种探险、战斗并以此来提升自己的属性,也可以通过复杂的任务系统让玩家体会游戏剧情的发展,一般没有取胜条件。

★注意:任务系统在角色扮演类游戏中比较常见,好的任务系统对游戏剧情起到推动作用。

1.4.1 游戏视角

角色扮演类游戏基本上不会以2D的视角来呈现,通常都是2.5D或3D视角,而对于2.5D又有斜45°俯视和正90°俯视两种。例如,一款移植于电脑游戏的"仙剑奇侠传"的视角采用的就是斜45°俯视(如图1-7所示),而手机游戏"仙侣情缘之麒麟劫"的视角是正90°俯视(如图1-8所示)。

图 1-7 "仙剑奇侠传"游戏

图 1-8 "仙侣情缘之麒麟劫"游戏

目前这类游戏以正90°视角居多,其实现方法也比斜45°视角简单些。但不管是斜45°还是正90°,都是采用图元技术加上多个图层叠加实现的,所以这类角色扮演类游戏中地图设计将会是一个非常重要的环节。出于剧情和玩家需要,还必须为游戏创建不同用途的界面,例如角色属性面板、物品及装备面板、技能面板、战斗系统面板等。

1.4.2 游戏内容设计

对于角色扮演类游戏,故事情节的好坏在很大程度上影响了游戏带给玩家的体验,所以在游戏设计初期必须选好一个题材,这是决定游戏成败的关键要素之一。通常游戏的题材背景会选择一个不同于普通人生活的世界,例如武侠文学、西方魔幻文学、科幻小说或电影等。确定了题材,还要有丰富的剧情,一般来说,角色扮演类的游戏方式主要包括探险、接受任务以及战斗,合理分配这 3 种游戏方式才能使游戏的可玩性达到最高。

玩家控制的角色在游戏中不断成长可以体现游戏的趣味性,同时也是游戏情节发展的主线。所以在设计游戏时需要根据故事情节让主角不断成长,这种成长包括个人属性(技能、血量、等级、法力等)的提升以及游戏剧情的逐步铺开,主角的成长方向同时也是吸引玩家坚持玩到底的原因之一。

对于一般玩家,角色扮演类游戏很少能够在短时内通关,所以必须为游戏增加存储功能。游戏中可以采用到达指定地方才可以存储的模式,也可以做成菜单选项让玩家随时存储。

1.5 闯关动作类游戏

闯关动作类游戏的节奏比较轻快,玩家的成就感主要来源于胜利完成对一个个关卡的挑战。因此设计的重点不在战斗,而是在闯关,这样适应的玩家人群会更广。经典的如"超级玛丽"(如图 1-9 所示)和"冒险岛"(如图 1-10 所示)等。

图 1-9 "超级玛丽"游戏

图 1-10 "冒险岛"游戏

闯关动作类游戏的主要目标一般都在于过关斩将,并不十分需要别的玩家参与或协助,所以通常都为单机模式。其操作的方式也不会太复杂,只是简单地控制游戏角色的移动和释放技能即可。一般是在最后的关卡会出现游戏的 BOSS(关底,一般指游戏中最强大的敌人),将其打败就宣告玩家成功通关。

1.5.1 游戏视角

最为经典的闯关动作类游戏画面是横向滚屏，采用这种游戏视角可以让玩家能够较早地看到前面可能要遇到的障碍和挑战，适应游戏的节奏变化；也有的采用纵向滚屏，这样通关的难度就会有所提高。

1.5.2 游戏内容设计

闯关动作类游戏一般不会以复杂的故事背景去吸引玩家，只是在简单的故事背景中向玩家介绍闯关的动机和玩家所要追求的终极目标。然后应该用层出不穷的关卡来吸引玩家的注意力，而不是对剧情念念不忘。因此，要考虑关卡难易度的把握。增加难度的方式有多种，除了重新设计高难度的通关条件外，对游戏进行小范围的调整也可以实现难易度的改变，例如给玩家更短的时间、将游戏中的奖励放在更危险的地方等。另外，还可以根据关卡的不同制定不同的游戏规则，也可以设计出不同的场景，这样不容易使玩家产生游戏疲劳感。

1.6 冒险类游戏

冒险类游戏（adventure game）也是一种需要故事情节的游戏，与角色扮演类游戏类似，不同的是冒险类游戏是在故事中添加了游戏元素，而角色扮演是在游戏中穿插故事。冒险类游戏一般不需要什么策略或技巧，玩此类游戏就像是读历险记和讲故事一样。除非是为了给结局留悬念，一般最后都是好的结局。

不管是解谜类的冒险游戏，例如"寂静岭"（如图 1-11 所示），还是逃脱类的冒险游戏，例如"越狱 24 小时"（如图 1-12 所示），游戏中遇到的困难基本上类似。例如打开一扇需要钥匙的门、找齐全套的物品进行组合等。在游戏中玩家通常不是靠快速反应或长时间思考来解决问题，而是依靠已获得的游戏线索或物品道具甚至技能排除阻碍。

图 1-11 "寂静岭"游戏

图 1-12 "越狱 24 小时"游戏

★**注意**：冒险类游戏很难让玩过一次的玩家再去经历一次冒险。这也体现了故事情节对于冒险游戏的重要性，一旦熟悉了故事情节，就再也无法给人以刺激和惊奇了。

1.6.1 游戏视角

冒险类游戏的视角很多情况下与角色扮演类游戏类似，不过也有很多以第一人称视角来呈现，以增强玩家身临其境的感觉。为了更好地讲述一个历险故事，除了文字，要尽量提高画面的空间感。另外，在游戏中增加一个缩略地图是一种很好的选择，这样玩家不至于迷路，也不会走重复路。

1.6.2 游戏内容设计

冒险类游戏如果没有跌宕起伏的故事内容，就基本失去了可玩性；而如果只像故事书那样讲故事，也会使游戏变成一种负担。游戏中展示的是场景而不是故事，所以在将故事改编成游戏或为游戏增加故事情节时，要注意规划故事结构，把各种剧情插入到不同的场景之中，场景和场景之间又相互关联，最后形成相互串连的游戏场景。

为了适应冒险类游戏漫长的故事情节，通常要将故事切成若干个关卡，这些关卡的连接方式有多种，例如，单线索方式是将所有关卡连成一条线索循序渐进地向玩家铺开，而多线索方式则是在进入新关卡前让玩家选择，或者根据玩家现在的游戏状态进入相应的关卡。

为了增加游戏的趣味性，在游戏中可以加入对话，对象则是NPC（Non-Player-Controlled Character，非玩家控制角色），既可以使玩家体会到游戏的互动性，也可以为玩家提供重要的游戏线索，所以在设计冒险类游戏时应该合理地设计对话。

1.7 策略类游戏

手机平台下的策略类游戏基本上是从PC平台移植过来的，最初是模拟类游戏的一个分支。随着策略类游戏的不断发展，也衍生出了很多不同的形式，例如回合制策略游戏和即时策略游戏。在策略类游戏中，玩家常常没有具体的角色，或者说玩家控制不止一个角色，玩家扮演的角色是统筹各个方面的总管，这在一定程度上也增加了游戏的复杂程度。与PC平台下的策略游戏不同的是：PC平台下一般不会限制玩家的个数，玩家可以同电脑对战，也可以和其他玩家一起游戏，或者合作，或者对抗；而在手机平台下，受设备功能、网络环境等多方面因素限制，大部分都是单人模式的，玩家主要和电脑中的敌人或朋友一起游戏。

策略类游戏更偏重于思考和谋划，因此策略类游戏所消耗的时间有可能很长，要显示的信息量也很大。同时有些即时策略游戏包含任务系统，这些连续的任务也不是短时间能够完成的。按照策略游戏的原则，取胜条件在于征服，即完全消灭游戏中的敌人或者被敌人消灭才宣告游戏结束。

1.7.1 游戏视角

手机平台下的策略游戏由于不容易变换视角,通常采用 2.5D 俯视视角,这样不仅玩家的视野会比较开阔,也比较容易操作游戏中的不同对象。例如手机版的"帝国时代"(如图 1-13 和图 1-14 所示)采用的就是斜 45°俯视视角。

图 1-13 "帝国时代"的生产场景　　　　图 1-14 "帝国时代"的战斗场景

★注意:策略类游戏中的地图都非常大,这样能够保证开展游戏的多方阵营在初期平衡地生存。在这种情况下缩略地图对于玩家来说就是一个非常有用的工具了。利用缩略地图,玩家可以迅速切换镜头和了解游戏局势。

1.7.2 游戏内容设计

策略类游戏的题材形式并不多,一般都与战争有关,只是选择的故事背景不同,有历史题材的,也有魔幻题材的。一个成功的策略类游戏,其引人入胜的背景故事起着至关重要的作用,一些策略类游戏甚至是从电影或文学作品改编而来的。

策略类游戏中往往会有许多阵营,如手机版的"文明"(如图 1-15 和图 1-16 所示),不同阵营之间所具有的游戏角色和发展路线也不一样,否则会降低游戏的可玩性。而如何让不同阵营在游戏的进行中保持平衡就需要好好考虑了,一款失衡的策略游戏将不会有生存的可能。

图 1-15 "文明"的菜单界面　　　　图 1-16 "文明"的战斗场景

1.8 养成类游戏

养成类游戏来自电子宠物，强调主人（即玩家自己）同被养者之间的亲密程度，所以一般这类游戏都是单人模式。不过有些养成类游戏为了给玩家一个展示成果的机会，增加了联网的模块来让不同的玩家带着自己的宠物进行展示。与其他游戏不同的是，养成类游戏没有胜利的概念，不过还是会有失败的情况，例如培养过程中一时疏忽使宠物死掉了。此类游戏通常不需要玩家费过多的脑筋，游戏节奏也很慢，因为一个优秀的宠物不是一朝一夕就能培养出来的。这类游戏主要的乐趣来源于宠物在自己的照料下一点一点地成长，接受训练并能和玩家进行一些简单的情感交流等。

1.8.1 游戏视角

养成类游戏的显示元素比较单一，主要是被养的对象和生活场景。所以为了提高玩家的视觉体验，应该对游戏的画面做好规划，一般采用 2D 平面视角，如"口袋妖怪"（如图 1-17 所示）。游戏中通常还需要设计管理面板等界面，如宠物属性面板等（如图 1-18 所示）。

图 1-17 "口袋妖怪"游戏

图 1-18 "口袋妖怪"属性面板

1.8.2 游戏内容设计

养成类游戏中的玩家基本上不会出现在游戏中，主要显示的内容就是宠物，因此宠物的造型设计是非常重要的，相当于整个游戏的招牌。养成类游戏中的宠物往往不是现实生活中的宠物，可以设计成奇怪的生物或科幻的产物，有些养成类游戏中被养的对象还可能是人。设计宠物除了外在形象，还需要设计其各种属性，如宠物所具有的技能和成长的路线等。

养成类游戏的培养方式设计也很重要,否则玩家就会发现其宠物过于迟钝或是太聪明,无法从中体会到成就感。另外,游戏中的人工智能(AI)也是不得不考虑的设计内容,宠物如何在玩家不干涉的情况下自动衰减自身的属性,例如饥饿、心情等,如何响应主人(玩家)的问候之类的交互等,这些都需要开发相应的人工智能算法来实现。

1.9 经营类游戏

经营类游戏一般模拟现实世界中的某种行业,例如手机游戏"地产大亨"(如图 1-19 和图 1-20 所示)就是让玩家通过对房屋土地的买卖或出租成为富翁。玩家的游戏目标在于超越游戏中的其他对手,获得自己的发展,并从经营和管理自己在现实中很少接触的事物获得乐趣。

图 1-19 "地产大亨"出售界面

图 1-20 "地产大亨"工作界面

手机平台下的经营类游戏一般为单机模式,这类游戏中玩家主要关心的是如何管理好自己的资源,没有一个确定的取胜条件,也有些经营类游戏需要像策略类游戏那样不断发展自己的势力,最后打败其他竞争对手赢得胜利。因此,游戏中往往会为玩家提供一些竞争对手来增加挑战性。

1.9.1 游戏视角

经营类游戏中,玩家关注的地方很多,所以在手机平台下一般都采用 45°(或 90°)的 2D 俯视视角来呈现游戏的画面。同时还要为玩家提供一个管理面板来完成对其所掌握的资源进行分配等操作。

1.9.2 游戏内容设计

不同于策略类游戏,经营类游戏中玩家对于整个游戏进程的影响不是特别大(除非到了后期玩家已经很强大的时候),所以游戏中大部分时间需要在程序中控制游戏的发展流程，例如调动 AI（人工智能）去和玩家竞争或随机产生突发事件等。

经营类游戏从本质上讲，就是玩家不断收集资源发展自己的过程，所以每个经营类游戏内部都会有一个资源系统。有些资源会在游戏中被玩家消耗掉，然后产生新的，有些则是在不同人的手中来回交换或交易，还有的则是自动产生消耗。所以在设计游戏的时候需要对涉及的资源进行详细分类，并在游戏过程中进行科学的管理。

1.10 体育类游戏

体育类游戏是面向体育爱好者的一类游戏，虽然玩家群体不如角色扮演或益智类游戏多，但是体育类游戏还是在众多的手机游戏种类中因独特的内容题材占有一席之地。由于手机平台的局限性，此类游戏多为单机模式，这时游戏的可玩性很大程度上取决于 AI（人工智能）的真实程度。

体育类游戏主要是模仿现实中体育竞技运动，所以取胜方式就是赢得比赛，或根据剧情赢得一系列的比赛，例如"NBA 职业篮球"（如图 1-21 所示），玩家主要操控的对象是一个或多个运动员。也有的体育类游戏融合了经营类游戏的元素，使得游戏的乐趣不在于取得竞技上的胜利，而是把一个俱乐部经营好，玩家扮演的角色不是运动员，而是教练或经理之类的管理职务，例如"实况足球经理"（如图 1-22 所示）。

图 1-21 "NBA 职业篮球"游戏

图 1-22 "实况足球经理"游戏

1.10.1 游戏视角

体育类游戏的视角取决于竞技项目。如果是一对一的比赛，如网球或摔跤等，那么既可以采用第一人称视角，也可以采用其他视角。对于团队竞技项目，例如篮球、足球等，一般不会只采用第一人称一种视角，因为玩家需要实时掌握场上的局面。对于一些竞速性质或带有跑道的项目，例如滑雪、游泳等，一般采用背后视角来设计。还有一些会采用闯关类动作游戏的滚屏式设计。

★注意：体育类游戏中可以提供多种视角的切换功能以增强用户的体验和可操作性。

1.10.2 游戏内容设计

对于需要模拟真实竞技场景的体育类游戏，人物造型、人物动作和人工智能的设计是很重要的，否则玩家会因为生硬的线条、扭曲的人物动作和不合理的人物行为提前放弃游戏。

由于竞技方式的多样化，往往要向所控制的角色下达很多命令，而现在智能手机的操作接口是有限的，所以需要设计游戏操作的软接口和触屏控制、重力感应控制等，确保玩家能够感觉到完整的运动体验。

1.11 本章小结

本章针对当今游戏市场现状，对目前流行的手机游戏类型进行介绍，主要说明了游戏视角和游戏内容设计方法，以便在开发相应类型的手机游戏时能够对相关环节有所了解。但随着手机产业及游戏开发技术的不断发展，手机游戏类型的分类也越来越模糊，很多手机游戏中融合了不同游戏类型的风格，还有一些新的游戏类型正在出现，在实际开发中需要注意这些问题。

1.12 思考与练习

(1) 以自己喜欢的手机游戏为例，说说该类游戏具有哪些特色。
(2) 试着撰写一款 Android 手机游戏的策划简案，游戏类型不限。
(3) 思考提升手机游戏可玩性和用户体验的方法。

第 2 章

Android 基础知识

学习目标：
- 了解 Android 的体系结构、特性及版本。
- 掌握 Android 开发环境搭建。
- 掌握 Android 的生命周期。
- 掌握 Android 应用的国际化方法。
- 了解 Android 中的计量单位。
- 学会 Android 项目的运行和调试。
- 掌握 Android 项目资源的创建与使用。
- 掌握消息提示与对话框的使用。

本章导读：

本章主要介绍 Android 开发环境的搭建和程序调试方法，另外重点讲解 Android 的项目结构、生命周期、资源组织、计量单位、国际化、消息提示及对话框等。本章是进行 Android 开发的基础，无论是对应用系统开发还是游戏开发，本章介绍的内容都至关重要。

2.1 Android 平台简介

Android 是一种以 Linux 为基础的开源操作系统，主要用于移动设备。Android 最初由 Andy Rubin 开发，主要支持手机，2005 年 8 月由 Google 收购。2007 年 11 月，Google 与 84 家硬件制造商、软件开发商及电信营运商组建开放手机联盟共同研发改良 Android 系统，逐渐扩展到平板电脑及其他领域。

2011 年第 1 季度，Android 在全球的市场份额首次超过 Symbian 系统跃居全球第一。2012 年 2 月的统计数据表明，Android 占据全球智能手机操作系统市场 59% 的份额，在中国市场的占有率为 68.4%，Android 智能手机的全球销量为 4.815 亿部。在 2013 年，这个数字上升到了 7.812 亿部、78.9%。2013 年 5 月，Google I/O 大会上公布数据，Android 设备激活总量已超过 9 亿，这与 Android 低价平板的增多不无关系。Android 在 2014 年、2015 年仍将保持这一优势。

Android 操作系统具有开放性，是可定制的，允许被用于其他电子产品，包括笔记本电脑和上网本、智能本、电子书阅读器和智能电视(谷歌电视)。此外,Android 已经可以应用到手表、耳机、车载 CD 和 DVD 播放机、智能眼镜、冰箱、车载卫星导航系统、家庭自动化系统、游戏机、镜子、摄像头、便携式媒体播放器、固定电话、跑步机等终端设备。

2.2 搭建 Android 开发环境

首先要下载所需要的开发工具,相关软件的名称及下载网址如表 2-1 所示。

表 2-1　Android 开发所需的软件及下载地址

软件名称	下载地址	本书使用的版本
JDK	http://www.oracle.com	JDK 7
Eclipse	http://www.eclipse.org	Eclipse IDE for Developers(4.2)
Android SDK	http://www.android.com	Android 4.2 SDK
ADT	http://dl-ssl.google.com/android/eclipse https://dl-ssl.google.com/android/eclipse/	ADT 21.0.1

JDK 是 Java 的核心,包括 Java 的运行环境(Java Runtime Environment)、类库以及 Java 开发工具等。

Eclipse 是一个 IDE 集成开发环境,有两个版本:Eclipse Classic x.x.x 版本和 Eclipse IDE for Java EE 版本。这两个版本没有本质上的区别,前者是后者的子集,后者只是包含较多的包。

Andriod SDK 是 Andriod 开发工具包,内含 Andriod 虚拟设备(模拟器)。

★注意:可以从 http://developer.android.com/sdk/index.html 网址下载 ADT Bundle for Windows 绿色压缩包,有 32 位和 64 位两个版本(本书使用的是 adt-bundle-windows-x86_64-20140702 版本)。

ADT 是 Google 研发的一个插件,此插件集成在 Eclipse 中,可为 Andriod 提供专属开发环境,其中包括创建实例、运行和除错等功能。

选择 Eclipse 主菜单,执行 Window→Perferences 命令进入配置窗口。如果 ADT 正确安装了,就会在 Preferences 界面左侧出现 Android 一栏。单击 Android 选项,然后在右侧单击 Browse,选择 Android SDK 解压后的路径。

选择"窗口"→Android Virtual Device Manager 命令或选择工具栏上的图标按钮,打开管理对话框,单击右边的"新建"按钮打开创建 AVD(Android 模拟器)对话框(如图 2-1 所示)。

另外,如果 CPU 是 Intel 芯片(支持 Intel VT-x 加速技术),可以使用 Intel 的 X86 镜像,结合 Intel 的硬件加速执行管理器(HAXM)驱动,完成 Android 模拟器的加速配置(如图 2-2 所示)。

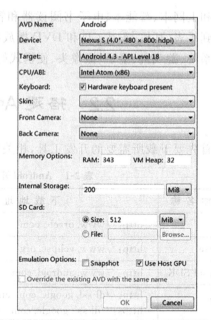

图 2-1　创建模拟器对话框　　　　图 2-2　创建 Intel 硬件加速模拟器对话框

★**注意**：对于非 Intel 芯片的计算机，可以使用 Genymotion 工具，它提供了 Android 虚拟环境。支持 Windows、Linux 和 Mac OS 等操作系统，容易安装和使用。

2.3　Eclipse Debug 调试程序

2.3.1　Eclipse 调试器

Android SDK 提供了从 Dalvik 字节码到 Java 源代码的映射，可以直接使用 Eclipse 功能强大的调试器调试 Android 应用程序。

在代码中设置断点是常用的一种调试手段，在 Eclipse 中可以通过以下 3 种方法设置断点：

- 使用菜单命令。首先将光标放置到想要设置断点的行，然后执行菜单命令 Run→Toggle Breakpoint。
- 使用键盘。选择想要设置断点的行，在键盘上按下快捷键 Ctrl＋Shift＋B。
- 在编辑器中直接双击想要设置断点的行左边的空白处。

执行菜单命令 Run→Debug→Android Application，开始对程序进行调试。初始化过程与正常运行程序一样，如果需要会对项目进行重新构建，然后启动模拟器，加载程序。程序正常启动后，在模拟器上就会出现 DebugTest 的用户界面。程序将会在断点位置停止执行，Eclipse 会自动切换到 Debug 布局。在 Debug 布局中包含了如下一些视图。

- Debug：用来显示程序执行过程中的调用栈。在 Debug 标签页的工具栏上有一些功能按钮，提供了继续、暂停、终止、单步执行、逐过程执行和返回等功能。

- Variables 和 Breakpoints 标签页：在 Variables 标签页中可以显示当前代码作用域内的所有变量值，Breakpoints 标签页中列出了程序中所有的断点。
- Editor：Debug 布局中的编辑器与 Java 布局中的编辑器一样，只不过在 Debug 布局中当前执行的代码会高亮显示。
- Outline：该视图可以显示当前项目的结构图。
- Console/Tasks/Properties：这 3 个视图位于 Debug 布局的左下角，其中 Console（命令行）视图是最常用的一个，在程序调试过程中许多重要的信息都显示在 Console 视图中。

2.3.2 Logcat

Logcat 是 Android SDK 中的一个通用日志工具，在程序的运行过程中可以通过 Logcat 打印状态信息和错误信息等。Logcat 另外一个重要的用途是在程序启动和初始化的过程中向开发者报告进展状况。

当应用程序在模拟器中加载并启动时，Eclipse 会自动切换到 Debug 布局，关于程序运行状态的各种信息就会出现在右下方的 Logcat 视图中。为了更加方便地浏览 Logcat 视图中的内容，可以单击 Logcat 视图右上角的最大化按钮。Logcat 视图中的信息按照消息产生的顺序出现，最开始是关于模拟器启动的消息，接着是 Android 操作系统启动的消息，然后是各种应用程序启动消息，最后才是与加载程序启动相关的消息。

在 Logcat 视图的工具栏中可以看到标记为 V、D、I、W 和 E 的几个按钮，作用是对消息进行过滤。

- V(Verbose)显示所有类型的消息。
- D(Debug)显示 Debug、Information、Warning 和 Error 消息。
- I(Information)只显示 Information、Warning 和 Error 消息。
- W(Warning)只显示 Warning 和 Error 消息。
- E(Error)只显示 Error 消息。

Logcat 视图中包含了如下列：

- Time：用于显示消息产生的时间。
- Priority(这一列并没有在标题栏中显式地标出)消息的级别(取值为 D、I、W 或者 E，分别代表 Debug、Information、Warning 和 Error)。
- pid：产生消息的进程 ID。
- tag：消息产生来源的简短描述。
- Message：消息的详细内容。

★注意：在程序开发过程中，如果需要多人协作进行错误的调试，那么就要对 Logcat 日志进行共享。导出 Logcat 日志的方法非常简单，首先在 Logcat 视图中选中想要导出的日志内容，然后单击 Logcat 视图右上角的向下箭头，在弹出菜单的最下方有一个名为 Exports Selection as Text 的菜单项，执行这个菜单项，就可以将选中的日志保存成一个文本文件。

在 Android LogCat 中显示和添加过滤器的方法如下：

（1）执行菜单命令 Window→Show View→Other→Android→LogCat，选择"确定"即可。

（2）在下面的 LogCat 窗口中单击+号图标创建过滤器。如要显示 System.out 输出的信息，可以在 Filter Name 处任意写一个名字，在 By Log Tag 处输入 System.out，其他默认即可。

2.4 Android 系统架构

Android 系统架构分为 4 层（如图 2-3 所示），从上到下分别是应用程序层、应用程序框架层、系统运行库层以及 Linux 内核层。

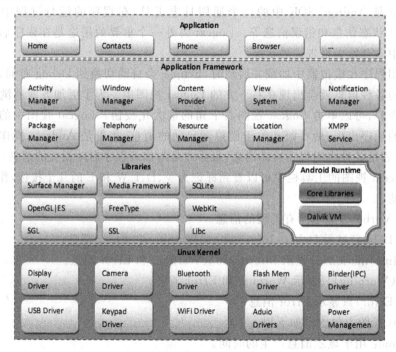

图 2-3　Android 系统架构图

1. 应用程序层

Android 平台不仅是操作系统，也包含了许多应用程序，例如 SMS 短信客户端程序、电话拨号程序、图片浏览器、Web 浏览器等应用程序。这些应用程序都是用 Java 语言编写的，并且这些应用程序都可以被其他应用程序所替换，这一点不同于其他手机操作系统固化在系统内部的系统软件。

2. 应用程序框架层

应用程序框架层是 Android 开发的基础，很多核心应用程序也是通过这一层来实现其核心功能的，该层简化了组件的重用，开发人员可以直接使用其提供的组件来进行快速的应用程序开发,也可以通过继承来实现个性化的拓展。
- Activity Manager(活动管理器):管理各个应用程序生命周期以及通常的导航回退功能。
- Window Manager(窗口管理器):管理所有的窗口程序。
- Content Provider(内容提供器):使得不同应用程序之间存取或者分享数据。
- View System(视图系统):构建应用程序的基本组件。
- Notification Manager(通告管理器):使得应用程序可以在状态栏中显示自定义的提示信息。
- Package Manager(包管理器):Android 系统内的程序管理。
- Telephony Manager(电话管理器):管理所有的移动设备功能。
- Resource Manager(资源管理器):提供应用程序使用的各种非代码资源,例如本地化字符串、图片、布局文件、颜色文件等。
- Location Manager(位置管理器):提供位置服务。
- XMPP Service(XMPP 服务):提供 Google Talk 服务。

3. 系统运行库层

系统运行库层可以分成两部分,分别是系统库和 Android 运行时环境。系统库是应用程序框架的支撑,是连接应用程序框架层与 Linux 内核层的重要纽带,主要分为如下几种类型。

- Surface Manager:执行多个应用程序时,负责管理显示与存取操作间的互动,另外也负责 2D 绘图与 3D 绘图进行显示合成。
- Media Framework:多媒体库,基于 PacketVideo OpenCore;支持多种常用的音频、视频格式录制和回放,编码格式包括 MPEG4、MP3、H.264、AAC、ARM。
- SQLite:小型的关系型数据库引擎。
- OpenGL|ES:根据 OpenGL ES 1.0 API 标准实现的 3D 绘图函数库。
- FreeType:提供点阵字与向量字的描绘与显示。
- WebKit:一套网页浏览器的软件引擎。
- SGL:底层的 2D 图形渲染引擎。
- SSL:在 Android 上的通信过程中实现握手。
- Libc:从 BSD 继承来的标准 C 系统函数库,专门为基于 Embedded Linux 的设备定制。

Android 应用程序采用 Java 语言编写,程序在 Android 运行时环境中执行,其运行时分为核心库和 Dalvik 虚拟机两部分。

- 核心库:提供了 Java 语言 API 中的大多数功能,同时也包含了 Android 的一些核心 API,例如 android.os、android.net、android.media 等。
- Dalvik 虚拟机:每个 Android 应用程序都有一个专有的进程,并且不是多个程序运行在一个虚拟机中,而是每个 Android 程序都有一个 Dalvik 虚拟机的实例,并在该实例中执行。Dalvik 虚拟机是一种基于寄存器的 Java 虚拟机,而不是传统的基于栈的虚拟机,它进行了内存资源使用的优化,并且具有支持多个虚拟机的特点。Android 程序在虚拟机中执行的并非编译后的字节码,而是通过转换工具 dx 将 Java 字节码转成 dex 格式的中间码。

4. Linux 内核层

Android 基于 Linux 2.6 内核,其核心系统服务如安全性、内存管理、进程管理、网络协议以及驱动模型都依赖于 Linux 内核。

2.5 创建第一个 Android 项目

Android 软件开发工具包(Software Development Kit,SDK)可以轻松地创建一个包含了默认项目目录和文件的工程,有使用 Eclipse 集成开发环境创建和使用命令行创建两种方式。

2.5.1 使用 Eclipse 创建项目

使用装有 ADT 插件的 Eclipse 创建一个新工程的步骤如下。

步骤 1:在 Eclipse 中,选择 File→New→Project,选择建立 Android Project,然后单击 Next 按钮。

步骤 2:在 Application Name 文本框中输入项目名称(比如 MyFirstApp),然后单击 Next 按钮 。

步骤 3:选择一个构建目标。被选中的版本将作为要编译用户开发的应用的版本。

★注意:建议尽可能选择最新版本。虽然可以创建支持较旧版本的应用,但是选择最新版本可以更加轻松地优化应用。使用最新的 Android 设备有更佳的用户体验。

步骤 4:单击 Next 按钮,设置应用程序的其他细节(如图 2-4 所示)。

步骤 5:单击 Finish 按钮,完成项目创建,这个项目中包含了一些默认的文件(不同版本下创建的目录结构和自动生成的代码有所不同)。

2.5.2 使用命令行创建项目

如果没有使用安装了 ADT 插件的 Eclipse 开发工具,也可以在命令行中使用 SDK 工具提供的命令创建工程。步骤所下。

步骤 1:打开命令行工具(在"运行"中输入 cmd 即可)。

步骤 2:进入 Android SDK 工具所在的目录。

步骤 3:执行 android list targets。

命令行中会列出你使用 SDK 下载的 Android 平台,找到适合应用的平台,给目标 ID 做标记。

步骤 4:执行以下命令:

```
android create project --target<target-id> --name MyFirstApp
```

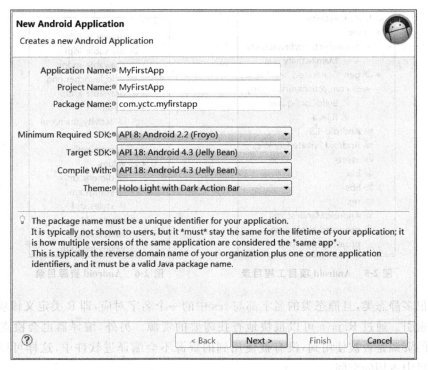

图 2-4　创建新项目对话框

```
--path<path-to-workspace>/MyFirstApp --activity MyFirstActivity
--package com.yctu.myfirstapp
```

用目标列表中的一个 ID 值(参考步骤 3)代替＜target-id＞,并更换＜path-to-workspace＞的位置作为 Android 项目的保存路径。

★注意：将 tools/ 目录加入环境变量中的 path 变量中,会提高工作效率。

2.6　Android Project 项目结构

创建 Android 项目时,系统自动生成 MainActivity.java 文件(继承于 Activity 类),项目工程目录如图 2-5 所示,资源目录如图 2-6 所示。

因为几乎所有的活动都是与用户交互的,所以 Activity 类关联了创建窗口,并重写了两个方法。

(1) onCreate()方法：用来初始化 Activity,调用 setContentView(View)方法绑定界面资源。

(2) onCreateOptionsMenu()方法：默认创建一个菜单。

项目工程目录包含以下目录：

(1) src 目录：用来存放项目的源代码(.java)。

(2) gen 目录：存放 R.java 文件,该文件在建立项目时自动生成,属于只读模式。其

图 2-5 Android 项目工程目录

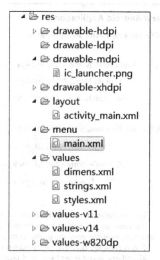
图 2-6 Android 资源目录

中包含很多静态类，且静态类的名字都与 res 中的一个名字对应，即 R 类定义该项目所有资源的索引。通过 R.java 可以很快地查找需要的资源。另外，编译器也会检查 R.java 列表中的资源是否被使用到，没有被使用到的资源不会编译进软件中，这样可以减少应用在手机中占用的空间。

（3）Android 4.3 目录（现有 SDK 版本）和 Android Private Libraries 目录：Android 4.3 目录中包含了一个 android.jar 包（Java 归档文件），其中包含构建应用程序所需的所有的 Android SDK 库（如 Views、Controls）和 API。通过 android.jar 将应用程序绑定到 Android SDK 和 Android Emulator，这允许使用所有 Android 的库和包，且使应用程序在适当的环境中调试。Android Private Libraries 目录中包含了第三方 JAR 包，这是最新版本的 ADT 所特有的，它将第三方的 JAR 包规整到这个目录下。

（4）Assets 目录：用于存放音频文件、视频文件或用户不经常修改的文件。

（5）bin 目录：包含编译生成 apk 的应用程序。

（6）libs 目录：包含引用的第三方的库文件。

（7）res 目录：包含了各种资源文件，并且将它们编译进应用程序。向此目录添加资源时，会被 R.java 自动记录。

（8）AndroidMainfest.xml 文件：当前项目的配置文件，包含编码格式、应用的图标、程序的版本号以及指定该程序用到的服务等。

（9）project.properties 文件：记录项目工程的环境信息。

下面介绍 Android 项目 src 目录中的程序源码。

MainActivity.java 代码如下：

```
1   package com.yctc.myfirstactivity;
2   import android.os.Bundle;
3   import android.app.Activity;
4   import android.view.Menu;
```

```
5   public class MainActivity extends Activity {
6       @Override
7       protected void onCreate(Bundle savedInstanceState){
8           super.onCreate(savedInstanceState);
9           setContentView(R.layout.activity_main);
10      }
11      @Override
12      public boolean onCreateOptionsMenu(Menu menu){
13          getMenuInflater().inflate(R.menu.main, menu);
14          return true;
15      }
16  }
```

对以上代码的说明如下。

第 1 行：本类所在的包路径。

第 2~4 行：引入相关类。导入类库的 3 种方式如下：

- 手动导入。例如声明一个 Button，输入 import android.widget.Button。
- 声明类型时，由 Eclipse 自动导入包。例如声明一个 Button，写 Button 类型的时候不写全类名，利用快捷键 Alt＋/自动完成。
- 使用导包快捷键。利用快捷键 Shift＋Ctrl＋O 快速完成导入。

第 5 行：创建一个类，并继承 Activity 类。

第 6 行：@Override 表示下面的 onCreate()函数（方法），是重写了基类 Activity 中的 onCreate()方法；如果没有这个标识，编译代码时会认为这是开发者自定义的函数。

第 7 行：重写 Activity 类的生命周期中的 onCreate()方法。

第 8 行：调用父类的 onCreate()函数。

第 9 行：利用当前的 Activity 类中的 setContentView()来显示布局。

第 12 行：重写 Activity 类的 onCreateOptionsMenu()建立 MENU 功能键菜单方法（有的版本不会自动生成）。

第 13 行：装载菜单项布局文件。

XML 布局文件代码如下：

```
1   <RelativeLayout xmlns:android="http://schemas.android.com/apk/res/android"
2       xmlns:tools="http://schemas.android.com/tools"
3       android:layout_width="match_parent"
4       android:layout_height="match_parent"
5       android:paddingBottom="@dimen/activity_vertical_margin"
6       android:paddingLeft="@dimen/activity_horizontal_margin"
7       android:paddingRight="@dimen/activity_horizontal_margin"
8       android:paddingTop="@dimen/activity_vertical_margin"
9       tools:context=".MainActivity">
10      <TextView
11          android:id="@+id/myText"
12          android:layout_width="wrap_content"
```

```
13            android:layout_height="wrap_content"
14            android:text="@string/hello_world"/>
15 </RelativeLayout>
```

对以上代码的说明如下：

第 1 行：描述 XML 的版本以及编码格式。

第 2 行：定义命名空间。

第 3、4 行：设置布局宽、高为匹配父容器。

第 5～8 行：引用尺寸资源，设置内容与边框下、左、右、上的距离。

第 9 行：这一句不会被打包进 apk。只是 ADT 的布局编辑器在当前的布局文件里面设置对应的渲染上下文，说明当前的布局所在的渲染上下文是 activity name 对应的那个活动，如果这个活动在 manifest 文件中设置了主题，那么 ADT 的布局编辑器会根据这个主题来渲染当前的布局（所见即所得的效果）。

第 10～14 行：设置 TextView 组件的文本内容，/> 表示 TextView 设置结束。

第 15 行：整个相对布局设置结束。

★注意：给组件添加 ID 属性的定义格式为 android:id="@+id/name"，这里的 name 是自定义的，不是索引变量。@+表示新声明，@表示引用。为组件添加了新声明的 ID 之后，系统会自动在 gen 目录下创建相应的 R 资源类变量，就可以在源代码中通过 R 资源引用组件。

在 AndroidManifest.xml 文件中的 application 节点设置 theme（主题），属性 theme 应用到整个应用程序中。代码如下：

```
<application
    android:allowBackup="true"
    android:icon="@drawable/ic_launcher"
    android:label="@string/app_name"
    android:theme="@android:style/Theme.Black">
```

使用 Java 代码或者在 AndroidManifest.xml 中对活动的主题进行设置，主题仅应用到当前活动中。在 AndroidMainifest.xml 中设置活动主题的代码如下：

```
<activity
    android:name="com.example.projectspinner.MainActivity"
    android:label="@string/app_name"
    android:theme="@android:style/Theme.Black">
```

在当前活动的 onCreate() 方法中设置主题的 Java 代码如下：

```
@Override
protected void onCreate(Bundle savedInstanceState){
    super.onCreate(savedInstanceState);
    setTheme(android.R.style.Theme_Black);
    setContentView(R.layout.activity_main);
}
```

2.7 Android 资源使用

Android 中的资源是可以在代码中使用的外部文件,这些文件作为应用程序的一部分被编译到应用程序当中。各种资源都被保存到 Android 项目的 res 目录下对应的子目录中,可以在 Java 文件中使用,也可以在其他 XML 资源中使用。

R.java 文件在根包中定义了一个顶级类 public static final class R 以及若干内部类,R.java 将内部静态类创建为一个命名空间,以保持字符串资源 ID。在 XML 文件中可以通过@[<包名>.]<文件夹>/<文件名>的语法格式使用字符串资源。R.java 文件代码如下:

```
public final class R {
    public static final class attr {
    }
    public static final class color {
        public static final int verdigris=0x7f040000;
        public static final int white=0x7f040001;
    }
    public static final class dimen {
        public static final int activity_horizontal_margin=0x7f050000;
        public static final int activity_vertical_margin=0x7f050001;
    }
    public static final class drawable {
        public static final int bg_game=0x7f020000;
        public static final int bg_options=0x7f020001;
        ……
    }
    public static final class id {
        public static final int action_settings=0x7f09000e;
        public static final int editName=0x7f090009;
        ……
    }
    public static final class layout {
        public static final int main_menu=0x7f030000;
        public static final int option=0x7f030001;
    }
    public static final class menu {
        public static final int main=0x7f080000;
        public static final int options=0x7f080001;
    }
    public static final class string {
        public static final int action_settings=0x7f060000;
        public static final int app_name=0x7f060001;
```

```
        ……
    }
    public static final class style {
        public static final int AppBaseTheme=0x7f070000;
        public static final int AppTheme=0x7f070001;
    }
}
```

Android 还支持子 XML 元素,例如＜b＞、＜i＞、＜string＞节点下其他简单的 HTML 文本格式,在文本视图绘制时可以使用这种复合 HTML 字符串来设置文本样式。

★注意:在 Android 中,资源文件的文件名不能是大写字母,必须是以小写字母开头,由小写字母 a~z、0~9 或下划线"_"组成。

2.7.1 字符串资源

Android 可以将字符串声明在位于/res/values 子目录下的配置文件中,文件名可以任意指定(默认为 strings.xml)。代码如下:

```
<?xml version="1.0" encoding="utf-8"?>
<resources>
    <string name="app_name">第一个 Android 应用</string>
    <string name="action_settings">Settings</string>
    <string name="hello_world">Hello world!</string>
</resources>
```

在应用程序中通过 String str=getResources().getString(R.string.hello_world)的方法引用字符串。

2.7.2 数组资源

指定一个字符串数组作为/res/values 子目录下所有文件中的资源,使用一个名为 string-array 的 XML 节点,此节点是 resources 的子节点。代码如下:

```
<?xml version="1.0" encoding="utf-8"?>
<resources>
    <string-array name="test_array">
        <item>one</item>
        <item>two</item>
        <item>three</item>
        <item>four</item>
    </string-array>
    …
</resources>
```

在应用程序中通过 String strings[] = this.getResources().getStringArray(R.array.test_array)的方法引用字符串数组。

2.7.3 颜色资源

在 Android 中，颜色值通过 RGB(红、绿、蓝)三原色和一个透明度(Alpha)值表示，必须以 ♯ 号开头。其中，Alpha 值可以省略。通常情况下，颜色值的使用有 ♯RGB、♯ARGB、♯RRGGBB 和 ♯AARRGGBB 4 种方式，采用十六进制数值。在表示透明度时，0 表示完全透明，f 表示完全不透明。除此之外，Android 还在自己的资源文件中定义了一组基本颜色，这些 ID 也可通过 Android 的 android.R.color 进行访问。代码如下：

```
<?xml version="1.0" encoding="utf-8"?>
<resources>
    <color name="verdigris">#527F76</color>
    <color name="white">#FFFFFF</color>
</resources>
```

在应用程序中通过 int mainBlackGroundColor = this.getResources().getColor(R.color.red)的方法引用颜色。

2.7.4 尺寸资源

像素、英寸和磅值等都是可以在 XML 布局或 Java 代码中使用的尺寸，使用这些尺寸资源来本地化 Android UI 和设置样式无须更改源码。代码如下：

```
<resources>
    <dimen name="activity_horizontal_margin">16dp</dimen>
    <dimen name="activity_vertical_margin">16dp</dimen>
</resources>
```

在应用程序中通过 float dimen = this.getResources().getDimension(R.dimen.mysize_in_pixels)的方法引用尺寸。

2.7.5 Drawable 资源

Android 支持的图像格式包括 git、jpg 和 png，在 res/drawable 目录下的图像文件会生成唯一的 ID，以供应用程序引用。代码如下：

```
BitmapDrawable btn_bg=this.getResources().getDrawable(R.drawable.sample_image);
button.setBackgroundDrawanle(btn_bg);
```

Drawable 资源不仅可以直接使用图片作为资源，而且可以使用多种 XML 文件(可以被系统编译成 Drawable 子类的对象)作为资源。StateListDrawable 资源文件是定义在 XML 文件中的 Drawable 对象，也放在 res/drawable-xxx 目录中，能根据状态来呈现不同的图像。StateListDrawable 资源文件的根元素为<selector></selector>，在该元素中包括<item></item>元素,item 元素可以设置以下两个属性。
- android:color 或 android:drawable:用于指定颜色或者 Drawable 资源。
- android:state_xxx:用于指定一个特定的状态。常用的状态属性如表 2-2 所示。

表 2-2 StateListDrawable 资源常用的状态属性

属 性 名 称	描 述
android:state_active	是否处于激活状态
android:state_checked	是否处于选中状态
android:state_enabled	是否处于可用状态
android:state_first	是否处于开始状态
android:state_focused	是否处于获得焦点状态
android:state_last	是否处于结束状态
android:state_middle	是否处于中间状态
android:state_pressed	是否处于被按下状态
android:state_selected	是否处于被选择状态
android:state_window_focused	窗口是否处于获得焦点状态

例如,在 res/drawable-xxx 目录下创建一个 StateListDrawable 资源文件 edittext_focused.xml,可以根据编辑框是否获得焦点来改变编辑框内文字的颜色。代码如下:

```
<?xml version="1.0" encoding="utf-8"?>
<selector xmlns:android="http://schemas.android.com/apk/res/android">
    <item android:state_focused="true" android:color="#f50"/>
    <item android:state_focused="false" android:color="#080"/>
</selector>
```

在编辑框属性中可以通过 android:textColor="@drawable/edittext_focused" 的方法引用该资源。

再创建一个名称为 button_state.xml 的 StateListDrawable 资源文件,可以根据按钮的可用状态来使用不同的图片作为背景。代码如下:

```
<?xml version="1.0" encoding="utf-8"?>
<selector xmlns:android="http://schemas.android.com/apk/res/android">
    <item android:drawable="@drawable/red" android:state_enabled="true"/>
    <item android:drawable="@drawable/grey" android:state_enabled="false"/>
</selector>
```

在按钮的属性中可以通过 android:background="@drawable/button_state" 的方法引用该资源。

2.7.6 样式和主题资源

样式资源主要用于对组件的显示样式进行设置,Android 中的样式定义在 res 目录下的 values、values-v11 和 values-v14 中。例如,定义一个设置文本大小和颜色的样式,代码如下:

```xml
<style name="title">
    <item name="android:textSize">22sp</item>
    <item name="android:textColor">#0a0</item>
</style>
```

样式还支持继承的功能,代码如下:

```xml
<style name="basic">
    <item name="android:textSize">22sp</item>
    <item name="android:textColor">#0a0</item>
</style>
<style name="title" parent="basic">
    <item name="android:padding">5dp</item>
    <item name="android:gravity">center</item>
</style>
```

在按钮的属性中可以通过 style="@style/title" 的方法引用该样式。

★**注意**：当一个样式继承另一个样式后,如果在子样式中出现与父样式相同的属性,将使用在子样式中定义的属性值。

主题资源通常用于改变窗口外观,与样式资源的定义方法类似。不同的是,主题资源不能用于单个的视图组件,而是对所有活动起作用。例如,定义一个改变所有窗口背景的主题,代码如下:

```xml
<style name="bg">
    <item name="android:background">@drawable/background</item>
</style>
```

在 Android 中,使用主题资源有以下两种方法。

一是在 AndroidManifest.xml 文件中使用主题资源。代码如下:

android:theme="@style/bg"

二是在 Java 文件中使用主题资源。代码如下:

```java
public void onCreate(Bundle savedInstanceState){
    super.onCreate(savedInstanceState);
    setTheme(R.style.bg);
    setContentView(R.layout.activity_main);
}
```

2.7.7 布局资源

在 Android 中,屏幕的视图通常以资源的形式从 XML 文件中加载,这些 XML 文件称为布局资源。布局资源是 Android 的 UI 编程中最重要的一种资源。在/res/layout 下新建一个 main.xml 布局资源文件时,会自动在 R.layout 中生成一个 R.layout.main 的 ID 来标示这个布局资源。代码如下:

```xml
<LinearLayout xmlns:android="http://schemas.android.com/apk/res/android"
    xmlns:tools="http://schemas.android.com/tools"
    android:layout_width="fill_parent"
    android:layout_height="fill_parent">
    <TextView
        android:id="@+id/tv"
        android:layout_width="wrap_content"
        android:layout_height="wrap_content"
        android:text="@string/hello_world"/>
</LinearLayout>
```

在代码中可以使用布局文件。代码如下:

```java
@Override
protected void onCreate(Bundle savedInstanceState){
    super.onCreate(savedInstanceState);
    setContentView(R.layout.activity_main);
    TextView tv= (TextView)findViewById(R.id.tv);
}
```

在列表、网格等显示组件中显示自定义布局,可以先定义布局格式。代码如下:

```xml
<LinearLayout xmlns:android="http://schemas.android.com/apk/res/android"
    android:layout_width="fill_parent"
    android:layout_height="fill_parent"
    android:orientation="horizontal">
    <!--头像 -->
    <ImageView
        android:id="@+id/player"
        android:layout_width="30dp"
        android:layout_height="30dp"/>
    <!--名称 -->
    <TextView
        android:id="@+id/username"
        android:layout_width="130dp"
        android:layout_height="30dp"
        android:paddingLeft="10dp"/>
    <!--密码 -->
    <TextView
        android:id="@+id/password"
        android:layout_width="150dp"
        android:layout_height="30dp"/>
</LinearLayout>
```

以上布局可以应用于简单适配器中。代码如下:

```java
adapter=new SimpleAdapter(this, data, R.layout.list_item, new String[]
```

```
        { "player", "username", "password" },
        new int[] {R.id.player, R.id.username, R.id.password });
```

2.7.8 原始资源

除了支持任意 XML 文件之外，Android 还支持使用原始文件，这些原始文件位于 res/raw 下，包括音频、视频或文本文件等需要本地化或需要通过资源 ID 引用的原始文件资源。

与 res/xml 下的 XML 文件不同，这些文件按照原样转移到应用程序包中，但是每个文件在 R.java 中都会生成一个标识符。例如，解析 res/raw/下的 test.txt 文本文件的代码如下：

```
String getStringFromRawFile(Activity activity){
    InputStream is=activity.getResources().openRawResource(R.raw.test);
    String myText=convertStreamToString(is);
    is.close();
    return myText;
}
String convertStreamToString(InputStream is){
    ByteArrayOutputStream baos=new ByteArrayOutputStream();
    int num=is.read();
    while(num !=-1){
        baos.write(num);
        num=is.read();
    }
    return baos.toString();
}
```

2.7.9 原始资产

Android 提供了一个 assets 目录，可以将要包含在包中的文件放在这里，这个目录与 src 目录具有相同的级别，assets 目录中的文件不会在 R.java 中生成资源 ID，必须制定文件路径才能读取。文件路径是以 asset 开头的绝对路径，可以使用 AssetManager 类访问这些文件。代码如下：

```
String getStringFromAssetFile(Activity activity){
    AssetManager am=activity.getAssets();
    InputStream is=am.open("test.txt");
    String s=convertStreamToString(is);
    is.close();
    reutn s;
}
```

2.7.10 其他 XML 文件

Android 还允许将任意 XML 文件用作资源,提供了一种快速方式来根据所生成的资源 ID 引用这些文件。例如存储在 res/xml 子目录下的 XML 文件。代码如下:

```
<root>
    <sub>
        Hello world from an sub element
    </sub>
</root>
```

与处理其他 XML 资源文件一样,Android 将编译此 XML 文件,然后将其放入应用程序包中。如果需要解析这些文件,需要使用一个 XmlPullParser 实例。代码如下:

```
private String getEventsFromXmlFile(Activity activity){
    StringBuffer sb=new StringBuffer();
    XmlResourceParser xpp=activity.getResources().getXml(R.xml.test);
    int eventType=xpp.next().getEventType();
    while(eventType !=XmlParser.END_DOCUMENT){
        if(eventType==XmlPullParser.START_DOCUMENT){
            sb.append("Start document");
        } else if(eventType==XmlPullParser.START_TAG){
            sb.append("Start tag"+xpp.getName());
        } else if(eventType==XmlPullParser.ENG_TAG){
            sb.append("end tag"+xpp.getName());
        } else if(eventType==XmlPullParser.TEXT){
            sb.append("Text:"+xpp.getText());
        }
        eventType=xpp.next().getEventType();
    }
    sb.append("End document");
    return sb.toString();
}
```

2.8 屏幕方向改变的应对策略

可以为一个活动指定一个特定的方向,指定之后即使转动屏幕方向,显示方向也不会跟着改变。具体方法如下:

(1) 指定为竖屏,在 AndroidManifest.xml 文件中设置指定的活动的屏幕方向属性如下:

android:screenOrientation="portrait"

或者在 onCreate()方法中指定,代码如下:

```
setRequestedOrientation(ActivityInfo.SCREEN_ORIENTATION_PORTRAIT);
```

(2) 指定为横屏,在 AndroidManifest.xml 文件中设置指定的活动的屏幕方向属性如下:

```
android:screenOrientation="landscape"
```

或者在 onCreate()方法中指定,代码如下:

```
setRequestedOrientation(ActivityInfo.SCREEN_ORIENTATION_LANDSCAPE);
```

★注意:如果在＜activity＞中配置 android:screenOrientation 属性,会使 android:configChanges="orientation"的设置失效。

2.9 Android 中常用的计量单位

Android 支持下列单位:
- px(像素):屏幕上的点。
- in(英寸):长度单位。
- mm(毫米):长度单位。
- pt(磅):1/72 英寸。
- dp(与密度无关的像素):基于屏幕密度的抽象单位。在每英寸 160 点的显示器上 1dp=1px。
- dip:与 dp 相同,多用于 Android/oPhone 应用中。
- sp(与刻度无关的像素):与 dp 类似,但可以根据用户的字体大小首选项进行缩放。

为了使用户界面能够在现在和将来的显示器类型上正常显示,建议使用 sp 作为文字大小的单位,将 dip 作为其他元素的单位。dp 与密度无关,sp 除与密度无关外,还与 scale 无关,使用 dp 和 sp,系统会根据屏幕密度的变化自动进行转换。

2.10 Android 中的国际化

Android 对 i18n(即 internationalization,国际化)和 L10n(即 Localization,本地化)提供了非常好的支持。Android 没有专门的 API 来提供国际化,而是通过对不同资源的命名来达到国际化。只要在 res/目录下重新定义 values-国家编号,例如 values-en-rUS 英语(美国)、values-en-rGB 英语(英国)、values-zh-rCN 中文简体、values-zh-rTW 中文繁体、values-fr-rFR 法文、values-de-rDE 德文等。

★提示:配置选项包括语言代号和地区代号。表示中文和中国的配置选项是 zh-rCN,表示英文和美国的配置选项是 en-rUS,zh 和 en 表示中文和英文,CN 和 US 表示中国和美国,前面的 r 是必需的。

运行程序之前,更改手机语言(执行 settings(设置)→language & keyboards(语言与键盘))。

2.11 消息提示与对话框

有些情况下需要向用户弹出提示消息,例如显示错误信息、收到短消息等。Android 提供两种弹出消息的方式:消息提示框 Toast 和对话框 Alert。

2.11.1 用 Toast 类显示消息

Toast 类用于在屏幕中显示一些快速提示信息,该消息提示框没有控制按钮,并且不会获得焦点,经过一段时间自动消失。Toast 类的常用方法如表 2-3 所示。

表 2-3 Toast 类的常用方法

方法	描述
setDuration	设置消息提示持续的时间
setGravity(int gravity,int xOffset,int yOffset)	设置消息提示框显示的位置
setMargin(float horizontalMargin,float verticalMargin)	设置消息提示框的页边距
setText(CharSequence s)	设置显示的文本内容
setView(View view)	设置在消息提示框显示的视图
Show()	显示消息提示框

创建一个 Toast 对象,通常有两种方法:
(1) 调用该类的 makeText()方法创建。代码如下:

```
Toast.makeText(MainActivity.this, "消息文本", Toast.LENGTH_SHORT).show();
```

(2) 使用构造方法进行创建。代码如下所示:

```
Toast toast=new Toast(MainActivity.this);            //建立 Toast 对象
toast.setDuration(Toast.LENGTH_LONG);                //设置显示时间
toast.setGravity(Gravity.CENTER, 0, 0);              //设置对齐方式
LinearLayout toast_layout=new LinearLayout(MainActivity.this);
                                                     //创建线性布局
toast_layout.setOrientation(1);   //设置线性布局方向(0 为水平,1 为垂直)
ImageView toast_icon=new ImageView(MainActivity.this);  //创建 ImageView 对象
toast_icon.setImageResource(R.drawable.ic_launcher);
                                                     //设置 ImageView 显示的图像
toast_layout.addView(toast_icon);                    //添加 ImageView 对象到线性布局
TextView toast_text=new TextView(MainActivity.this); //创建 TextView 对象
toast_text.setText("消息文本!");                     //设置 TextView 对象文本
toast_layout.addView(toast_text);                    //添加 TextView 对象到线性布局
toast.setView(toast_layout);                         //设置消息框中显示的视图
toast.show();                                        //显示消息框
```

2.11.2 用 AlertDialog 类实现对话框

AlertDialog 类不仅能生成带按钮的对话框,还可以生成带列表的对话框和自定义视图的对话框。AlertDialog 类常用的方法如表 2-4 所示。

表 2-4 AlertDialog 类的常用方法

方 法	描 述
setTitle(CharSequence title)	设置对话框标题
setIcon(Drawable icon)	设置对话框图标(Drawable 对象)
setIcon(int resId)	设置对话框图标(资源 ID)
setMessage(CharSequence message)	设置对话框显示的内容
setButton()	为对话框添加按钮

使用 AlertDialog 类一般只能生成带多个按钮的提示对话框,如果要生成带列表的对话框,需要使用 AlertDialog.Builder 类,该类提供的常用方法如表 2-5 所示。

表 2-5 AlertDialog.Builder 类的常用方法

方 法	描 述
setTitle(CharSequence title)	设置对话框标题
setIcon(Drawable icon)	设置对话框图标(Drawable 对象)
setIcon(int resId)	设置对话框图标(资源 ID)
setMessage(CharSequence message)	设置对话框显示的内容
setNegativeButton()	为对话框添加取消按钮
setPositiveButton()	为对话框添加确定按钮
setNeutralButton()	为对话框添加中立按钮
setItems()	为对话框添加列表项
setSingleChoiceItems()	为对话框添加单选列表项
setMultiChoiceItems()	为对话框添加多选列表项
setView(View view)	设置对话框自定义视图

举例:AlertDialog 类应用

步骤 1:修改 res/layout 目录下的 XML 布局文件,添加 4 个按钮组件用于显示对话框。

步骤 2:在主活动文件中分别为 4 个按钮添加事件处理。

(1) 显示多个按钮的对话框代码如下:

```
Btn_multibtn.setOnClickListener(new OnClickListener(){
    @Override
    public void onClick(View v){
        AlertDialog alert=new AlertDialog.Builder(MainActivity.this).create();
        alert.setTitle("提示信息");
        alert.setIcon(R.drawable.ic_launcher);
        alert.setMessage("显示多个按钮的对话框!");
```

```java
        alert.setButton(DialogInterface.BUTTON_NEGATIVE,"取消",
                new DialogInterface.OnClickListener(){
            public void onClick(DialogInterface dialog, int which){
            }
        });
        alert.setButton(DialogInterface.BUTTON_POSITIVE,"确定",
                new DialogInterface.OnClickListener(){
            public void onClick(DialogInterface dialog, int which){
            }
        });
        alert.setButton(DialogInterface.BUTTON_NEUTRAL,"忽略",
                new DialogInterface.OnClickListener(){
            public void onClick(DialogInterface dialog, int which){
            }
        });
        alert.show();
    }
});
```

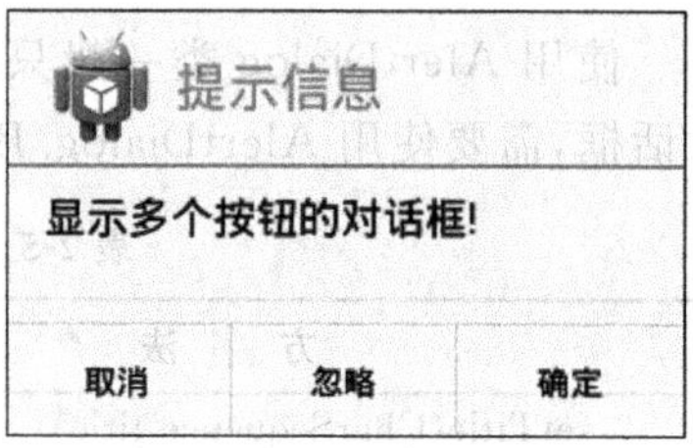

运行效果如图 2-7 所示。

（2）显示带列表项的对话框代码如下：

图 2-7 多按钮对话框

```java
btn_list.setOnClickListener(new View.OnClickListener(){
    @Override
    public void onClick(View v){
        final String items[]=new String[]{"返回基地","攻击敌方","选择武器","装备升级"};
        Builder builder=new AlertDialog.Builder(MainActivity.this);
        builder.setTitle("行动选择");
        builder.setIcon(R.drawable.dialog_ico);
        builder.setItems(items, new DialogInterface.OnClickListener(){
            public void onClick(DialogInterface dialog, int which){
                Toast.makeText(MainActivity.this, "您的选择是:"+items[which],
                        Toast.LENGTH_SHORT).show();
            }
        });
        builder.create().show();
    }
});
```

运行效果如图 2-8 所示。

（3）显示带单选列表的对话框代码如下：

```java
btn_select_single.setOnClickListener(new View.OnClickListener(){
    @Override
    public void onClick(View v){
```

```
        final String items[]=new String[]{"急救药品","反坦导弹","机枪子弹","交
        通工具"};
        Builder builder=new AlertDialog.Builder(MainActivity.this);
        builder.setTitle("装备选择");
        builder.setIcon(R.drawable.dialog_ico);
        builder. setSingleChoiceItems ( items, 0, new DialogInterface.
        OnClickListener(){
            public void onClick(DialogInterface dialog, int which){
                Toast.makeText(MainActivity.this, "您的选择是:"+items[which] ,
                Toast.LENGTH_SHORT).show();
            }
        });
        builder.setPositiveButton("确定", null);
        builder.create().show();
    }
});
```

运行效果如图 2-9 所示。

图 2-8 列表项对话框

图 2-9 单选列表对话框

(4) 显示带多选列表的对话框代码如下：

```
btn_select_multi.setOnClickListener(new View.OnClickListener(){
    @Override
    public void onClick(View v){
        final String items[]=new String[]{"坦克世界","空当接龙","开心乐园","超
        级战神"};
        final boolean itemsChecked[]=new boolean[]{false, true, false, true};
        Builder builder=new AlertDialog.Builder(MainActivity.this);
        builder.setTitle("游戏选择");
        builder.setIcon(R.drawable.dialog_ico);
        builder.setMultiChoiceItems(items, itemsChecked,
            new DialogInterface.OnMultiChoiceClickListener(){
                @Override
```

```
            public void onClick (DialogInterface dialog, int which, boolean
        isChecked){
            itemsChecked[which]=isChecked;          //改变列表项的选择状态
        }
    });
    builder.setPositiveButton("确定", new DialogInterface.OnClickListener(){
        @Override
        public void onClick(DialogInterface dialog, int which){
            String result="";                        //保存选择项
            for(int i=0; i<itemsChecked.length; i++){  //遍历数组
                if(itemsChecked[i]){                   //如果选项被选中
                    result+=items[i]+"、";             //将选项内容添加到 result 中
                }
            }
            if(!"".equals(result.trim())){             //如果 result 不为空
                result=result.substring(0, result.length()-1);
                                                       //去掉最后面的"、"符号
                Toast.makeText(MainActivity.this, "您选择了：["+result+"]",
                    Toast.LENGTH_LONG).show();
            }
        }
    });
    builder.create().show();
    }
});
```

运行效果如图 2-10 所示。

图 2-10 多选列表对话框

2.11.3 基础实例：自定义视图对话框

本例实现的对话框显示内容为自定义视图。具体实现步骤如下。

步骤1：准备图片资源（如图2-11所示），复制图片资源到res/drawable-mdpi目录。

图2-11 自定义视图对话框图片资源

步骤2：在res/drawable-mdpi目录中建立按钮背景切换的资源文件button_ok.xml、button_cancel.xml。其中button_ok.xml文件的代码如下：

```xml
<?xml version="1.0" encoding="utf-8"?>
<selector xmlns:android="http://schemas.android.com/apk/res/android">
    <item android:drawable="@drawable/quitokover" android:state_pressed="true"/>
    <item android:drawable="@drawable/quitok" android:state_focused="false"/>
</selector>
```

步骤3：在res/layout目录中新建dialog.xml布局文件。代码如下：

```xml
<?xml version="1.0" encoding="utf-8"?>
<RelativeLayout xmlns:android="http://schemas.android.com/apk/res/android"
    xmlns:tools="http://schemas.android.com/tools"
    android:layout_width="fill_parent"
    android:layout_height="wrap_content"
    android:background="@drawable/quitdialog_bg">
    <Button
        android:id="@+id/btn_ok"
        android:layout_width="50dp"
        android:layout_height="30dp"
        android:background="@drawable/button_ok"
        android:layout_marginTop="90dp"
        android:layout_marginLeft="65dp"/>
    <Button
        android:id="@+id/btn_cancel"
        android:layout_width="50dp"
        android:layout_height="30dp"
        android:background="@drawable/button_cancel"
        android:layout_toRightOf="@+id/btn_ok"
        android:layout_marginLeft="30dp"
        android:layout_alignTop="@+id/btn_ok"/>
</RelativeLayout>
```

步骤4：在res/layout目录下的主布局文件中添加按钮组件，在主活动文件中获取按钮组件并添加事件处理。代码如下：

```
btn_view.setOnClickListener(new View.OnClickListener(){
    @Override
    public void onClick(View v){
        View view=getLayoutInflater().inflate(R.layout.dialog, null);
                                                //获取视图
        Button btn_ok=(Button)view.findViewById(R.id.btn_ok);
                                                //获取视图中的按钮
        btn_ok.setOnClickListener(new View.OnClickListener(){
                                                //添加事件侦听
            @Override
            public void onClick(View v){
                finish();
            }
        });
        new AlertDialog.Builder(MainActivity.this)
            .setView(view).create().show();
    }
});
```

运行效果如图 2-12 所示。

图 2-12 自定义视图对话框效果

2.12 本章小结

本章主要讲解 Android 平台开发环境搭建、程序调试的方法以及项目创建、项目结构、关键资源、计量单位、国际化、消息提示与对话框等基础知识。需要注意的是，Android SDK 的升级较快，不同 SDK 版本下创建的项目文件结构会有所调整。

2.13 思考与练习

（1）新建一个 Android 程序，运行后在屏幕中显示"学习 Android 游戏开发！"，并部署到真机上。

（2）开发一个 Android 程序，用自己设计的 logo 图片，在进入程序、退出程序时在控制台有提示信息输出。程序支持国际化（中文、英语（美国））。

第 3 章

Android 游戏开发之视图界面

学习目标：
- 掌握 5 种布局管理器的使用。
- 掌握游戏开发的常用组件。
- 熟悉使用 XML 和 Java 代码混合控制 UI 界面。
- 掌握 Android 的生命周期。
- 掌握 Activity 页面切换及传递数据的方法。
- 掌握 Android 的事件处理机制。

本章导读：

在移动应用开发中，不论是基于哪个平台，都会用到系统控件（组件）。在游戏开发中一般会自定义（代码实现）符合游戏题材的组件。游戏除了丰富的玩法之外，最注重的就是界面(UI)。如果这些也都利用系统组件做开发，游戏就会趋于单调，用户体验差。但是一些基础控件也经常会用到。本章重点讲解 Android 游戏开发中常用的界面布局、基础组件、程序单元和事件处理机制。

3.1 界面布局

Android 中提供了 5 种布局：LinearLayout（线性布局）、FrameLayout（单帧布局）、AbsoluteLayout（绝对布局）、TableLayout（表格布局）、RelativeLayout（相对布局），其中最常用的的是 LinearLayout、TableLayout 和 RelativeLayout。这些布局管理器可以互相嵌套使用。

3.1.1 线性布局

线性布局是按照水平或垂直的顺序将子元素（可以是控件或布局）依次按照顺序排列，即每一个元素都位于前面一个元素之后。通过属性 android:orientation 可以设置布局管理器内组件的排列方式，通过属性 android:gravity 可以设置布局管理器内组件的对齐方式，通过 android:background 属性可以设置背景图片或背景颜色。

新建项目，修改 res/layout 目录下的 XML 布局文件。代码如下：

```xml
<LinearLayout xmlns:android="http://schemas.android.com/apk/res/android"
    xmlns:tools="http://schemas.android.com/tools"
    android:layout_width="fill_parent"
    android:layout_height="fill_parent"
    android:orientation="vertical"
    tools:context=".MainActivity">
<Button
    android:layout_width="fill_parent"
    android:layout_height="wrap_content"
    android:text="按钮 1"/>
<Button
    android:layout_width="fill_parent"
    android:layout_height="wrap_content"
    android:text="按钮 2"/>
<Button
    android:layout_width="fill_parent"
    android:layout_height="wrap_content"
    android:text="按钮 3"/>
</LinearLayout>
```

将 android:orientation 属性值设置为 vertical(垂直),运行效果如图 3-1 所示。如果将 android:orientation 属性值设置为 horizontal(水平),并将各按钮的 android:layout_width 属性值和 android:layout_height 属性值互换,运行效果如图 3-2 所示。

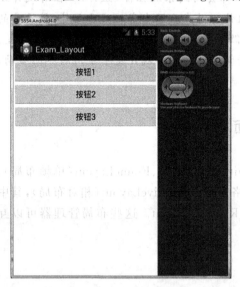
图 3-1 垂直线性布局

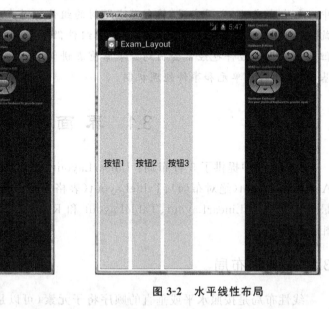
图 3-2 水平线性布局

★注意:对文本、颜色等值的设置,最好采用资源引用的方式。

3.1.2 表格布局

表格布局适用于多行多列的布局格式,每个 TableLayout 是由多个 TableRow 组成,一个 TableRow 就表示 TableLayout 中的每一行，这一行可以由多个子元素组成。TableLayout 和 TableRow 都是 LinearLayout 的子类。TableRow 实际是一个横向的线性布局，且所有子元素宽度和高度一致。

★**注意**：在 TableLayout 中，单元格可以为空，但是不能跨列，不能有相邻的单元格为空。

在 TableLayout 布局中，一列的宽度由该列中最宽的那个单元格指定，而该表格的宽度由父容器指定。TableLayout 中的特有属性如下：

- android:collapseColumns：设置隐藏列序号（从 0 开始，列序号间用逗号隔开）。
- android:shrinkColumns：设置允许收缩列序号（从 0 开始，列序号间用逗号隔开）。
- android:stretchColumns ：设置允许拉伸的列序号（从 0 开始，列序号间用逗号隔开）。

新建项目，修改 res/layout 目录下的 XML 布局文件。代码如下：

```xml
<TableLayout xmlns:android="http://schemas.android.com/apk/res/android"
    xmlns:tools="http://schemas.android.com/tools"
    android:layout_width="fill_parent"
    android:layout_height="fill_parent"
    android:gravity="center_vertical"
    android:stretchColumns="0, 3"
    tools:context=".MainActivity">
    <TableRow
        android:layout_width="wrap_content"
        android:layout_height="wrap_content">
        <TextView/>
        <TextView
            android:layout_width="wrap_content"
            android:layout_height="wrap_content"
            android:text="用户名:"
            android:textSize="24sp"/>
        <EditText
            android:id="@+id/userName"
            android:layout_width="wrap_content"
            android:layout_height="wrap_content"
            android:minWidth="200sp"
            android:textSize="24sp"/>
        <TextView/>
    </TableRow>
    <TableRow
        android:layout_width="wrap_content"
        android:layout_height="wrap_content">
        <TextView/>
```

```xml
<TextView
    android:layout_width="wrap_content"
    android:layout_height="wrap_content"
    android:text="密  码:"
    android:textSize="24sp"/>
<EditText
    android:id="@+id/passWord"
    android:layout_width="wrap_content"
    android:layout_height="wrap_content"
    android:inputType="textPassword"
    android:textSize="24sp"/>
<TextView/>
</TableRow>
<TableRow
    android:layout_width="wrap_content"
    android:layout_height="wrap_content">
<TextView/>
<Button
    android:id="@+id/btn_Login"
    android:layout_width="wrap_content"
    android:layout_height="wrap_content"
    android:text="登  录"/>
<Button
    android:id="@+id/btn_Quit"
    android:layout_width="wrap_content"
    android:layout_height="wrap_content"
    android:text="退  出"/>
<TextView/>
</TableRow>
</TableLayout>
```

运行结果如图 3-3 所示。

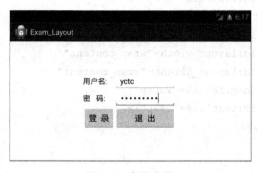

图 3-3 表格布局

3.1.3 相对布局

相对布局是按照子元素之间的位置关系完成布局的，作为 Android 的 5 种布局中最灵活也是最常用的一种布局方式，适合比较复杂的界面设计。相对布局支持常用的 XML 属性，例如，android:gravity 属性用于设置布局管理器中子组件的对齐方式，android:ignoreGravity 属性用于指定不受 gravity 属性影响的组件。相对布局在内部类 RelativeLayout.LayoutParams 中提供了常用的位置设置属性。

新建项目，修改 res/layout 目录下的 XML 布局文件。代码如下：

```xml
<LinearLayout xmlns:android="http://schemas.android.com/apk/res/android"
    xmlns:tools="http://schemas.android.com/tools"
    android:layout_width="fill_parent"
    android:layout_height="fill_parent"
    android:orientation="vertical"
    tools:context=".MainActivity">
    <ImageView
        android:layout_width="match_parent"
        android:layout_height="wrap_content"
        android:layout_weight="1"
        android:contentDescription="@string/app_name"
        android:scaleType="centerCrop"
        android:src="@drawable/toppic"/>
    <RelativeLayout
        android:layout_width="match_parent"
        android:layout_height="wrap_content"
        android:layout_weight="2"
        android:background="@drawable/bottompic">
        <!--中间位置图片 -->
        <ImageView
            android:id="@+id/imageButton_enter"
            android:layout_width="wrap_content"
            android:layout_height="wrap_content"
            android:layout_centerInParent="true"
            android:src="@drawable/enter"/>
        <!--上方位置图片 -->
        <ImageView
            android:id="@+id/imageButton_setting"
            android:layout_width="wrap_content"
            android:layout_height="wrap_content"
            android:layout_above="@id/imageButton_enter"
            android:layout_alignLeft="@id/imageButton_enter"
            android:src="@drawable/setting"/>
        <!--下方位置图片 -->
```

```xml
<ImageView
    android:id="@+id/imageButton_exit"
    android:layout_width="wrap_content"
    android:layout_height="wrap_content"
    android:layout_alignLeft="@id/imageButton_enter"
    android:layout_below="@id/imageButton_enter"
    android:src="@drawable/exit"/>
<!--左方位置图片 -->
<ImageView
    android:id="@+id/imageButton_help"
    android:layout_width="wrap_content"
    android:layout_height="wrap_content"
    android:layout_alignTop="@id/imageButton_enter"
    android:layout_toLeftOf="@id/imageButton_enter"
    android:src="@drawable/help"/>
<!--右方位置图片 -->
<ImageView
    android:id="@+id/imageButton_board"
    android:layout_width="wrap_content"
    android:layout_height="wrap_content"
    android:layout_alignTop="@id/imageButton_enter"
    android:layout_toRightOf="@id/imageButton_enter"
    android:src="@drawable/board"/>
</RelativeLayout>
</LinearLayout>
```

★**注意**：在引用其他子元素之前，引用的 ID 必须已经存在，否则将出现异常。

修改主程序类文件，添加如下代码：

```java
public class MainActivity extends Activity {
    @Override
    protected void onCreate(Bundle savedInstanceState){
        super.onCreate(savedInstanceState);
        requestWindowFeature(Window.FEATURE_NO_TITLE);           //去除标题栏
        getWindow().setFlags(WindowManager.LayoutParams.FLAG_KEEP_SCREEN_ON,
            WindowManager.LayoutParams.FLAG_KEEP_SCREEN_ON);     //保持亮度
        getWindow().setFlags(WindowManager.LayoutParams.FLAG_FULLSCREEN,
            WindowManager.LayoutParams.FLAG_FULLSCREEN);         //设置全屏
        setContentView(R.layout.main);
    }
}
```

运行结果如图 3-4 所示。

3.1.4 帧布局

将所有的子元素放在整个界面的左上角，每加入一个组件，都将创建一个空白的区

图 3-4　相对布局

域,后面的子元素直接覆盖前面的子元素。

3.1.5　绝对布局

绝对布局中通过设置 android:layout_x 和 android:layout_y 属性,将所有子元素的坐标位置固定下来,layout_x 表示横坐标,layout_y 表示纵坐标。屏幕左上角为坐标 (0,0),横向往右为正方向,纵向往下为正方向。实际应用中,这种布局用得比较少,因为 Android 终端机型较多,屏幕大小、分辨率等都不一样,如果用绝对布局,可能会导致在有的终端上显示不全等问题。

3.2　游戏开发常用组件

3.2.1　按钮类组件

Android 提供的按钮类组件主要包括普通按钮(Button)、图片按钮(ImageButton)、单选按钮(RadioButton)和复选框(CheckBox)。

1. 普通按钮

普通按钮通常用于触发一些特定事件,需要为按钮添加单击事件监听器。在 XML 布局文件中添加普通按钮组件的代码如下:

```
<Button
    android:id="@+id/btn_ok"
    android:layout_width="fill_parent"
    android:layout_height="wrap_content"
    android:text="@string/btn_ok"/>
<Button
    android:id="@+id/btn_cancel"
    android:layout_width="fill_parent"
    android:layout_height="wrap_content"
    android:text="@string/btn_cancel"/>
```

字符串资源文件 string.xml 代码如下：

```xml
<?xml version="1.0" encoding="utf-8"?>
<resources>
    <string name="btn_ok">确定</string>
    <string name="btn_cancel">取消</string>
</resources>
```

在 Android 中，为按钮添加单击事件监听器的方法主要有 3 种。

方法一：在当前类使用点击监听器接口。代码如下：

```java
public class MainActivity extends Activity implements OnClickListener {
    private Button btn_ok, btn_cancel;           //声明两个按钮对象
    private TextView tv;                          //声明文本视图对象
    @Override
    public void onCreate(Bundle savedInstanceState){
        super.onCreate(savedInstanceState);
        setContentView(R.layout.main);
        btn_ok= (Button)findViewById(R.id.btn_ok);        //获取 btn_ok 按钮
        btn_cancel= (Button)findViewById(R.id.btn_cancel); //获取 btn_cancel 按钮
        btn_ok.setOnClickListener(this);     //将 btn_ok 按钮绑定在单击监听器上
        btn_cancel.setOnClickListener(this);
                                             //将 btn_cancel 按钮绑定在单击监听器上
    }
    //使用单击监听器必须重写其抽象函数
    public void onClick(View v){
        if(v==btn_ok){
            Toast toast=Toast.makeText(MainActivity.this, "您单击了[确定]按钮",
            Toast.LENGTH_SHORT);
            toast.show();
        }else if(v==btn_cancel){
            ......                                        //要执行的代码
        }
    }
}
```

先用当前类使用单击监听器接口（onClickListener），重写单击监听器的抽象函数（onClick）；然后对需要监听的按钮绑定监听操作。因为定义了两个按钮，所以在 onClick 函数中对传入的视图进行按钮匹配判断，让不同的按钮响应不同的处理事件。

方法二：使用内部类实现单击监听器进行监听。修改后的代码如下：

```java
public class ButtonProject extends Activity {
    private Button btn_ok, btn_cancel;
    private TextView tv;
    @Override
    public void onCreate(Bundle savedInstanceState){
```

```
super.onCreate(savedInstanceState);
setContentView(R.layout.main);
btn_ok=(Button)findViewById(R.id.btn_ok);
btn_cancel=(Button)findViewById(R.id.btn_cancel);
btn_ok.setOnClickListener(new OnClickListener(){
    @Override
    Public void onClick(View arg0){
        ……                                          //要执行的代码
    }
});
btn_cancel.setOnClickListener(new OnClickListener(){
    @Override
    Public void onClick(View arg0){
        ……                                          //要执行的代码
    }
});
    }
}
```

利用内部类的形式也需要重写单击监听器的抽象函数,然后在 onClick 中处理事件,这里不用判断视图了,因为一个按钮对应一个监听器。

方法三：在活动中编写包含 View 类型参数的方法

首先在布局文件中为按钮添加 android:onClick="myClick"属性,然后在活动中编写一个 myClick()方法,关键代码如下：

```
Public void myClick(View view){
    ……                                              //要执行的代码
}
```

2. 图片按钮

图片按钮与普通按钮的功能和使用方法基本相同,只是图片按钮没有 text 属性,并且还可以为其指定 android:src 属性,用于设置要显示的图片。

如果不设置 android:background 属性,按钮的图片将显示在一个灰色的按钮上,这时的图片按钮将会随着用户的动作而改变；如果设置 android:background 属性,将不会随着用户的动作而改变。如果要让其随着用户的动作而改变,就需要使用 StateListDrawable 资源来对其进行设置。

3. 单选按钮

单选按钮是一种单个圆形双状态的按钮。单选按钮组(RadioGroup)是单选按钮的组合框,可以容纳多个单选按钮。在没有单选按钮组的情况下,单选按钮可以全部都选中；当多个单选按钮被单选按钮组包含的情况下,单选按钮只可以选择一个。并用 setOnCheckedChangeListener 来对单选按钮进行监听。

字符串资源文件代码如下：

```xml
<?xml version="1.0" encoding="utf-8"?>
<resources>
    <string name="app_name">ProjectButton</string>
    <string name="action_settings">Settings</string>
    <string name="lab_quest">请选择性别:</string>
    <string name="sex_man">男</string>
    <string name="sex_woman">女</string>
    <string name="btn_submit">提交</string>
</resources>
```

布局文件代码如下：

```xml
<LinearLayout xmlns:android="http://schemas.android.com/apk/res/android"
    xmlns:tools="http://schemas.android.com/tools"
    android:layout_width="fill_parent"
    android:layout_height="fill_parent"
    android:orientation="vertical"
    tools:context=".MainActivity">
    <TextView
        android:layout_width="fill_parent"
        android:layout_height="wrap_content"
        android:text="@string/lab_quest"/>
    <RadioGroup
        android:id="@+id/sex_radioGroup"
        android:layout_width="wrap_content"
        android:layout_height="wrap_content"
        android:orientation="horizontal">
        <RadioButton
            android:id="@+id/sex_man"
            android:layout_width="wrap_content"
            android:layout_height="wrap_content"
            android:text="@string/sex_man"/>
        <RadioButton
            android:id="@+id/sex_woman"
            android:layout_width="wrap_content"
            android:layout_height="wrap_content"
            android:text="@string/sex_woman"/>
    </RadioGroup>
    <Button
        android:id="@+id/btn_submit"
        android:layout_width="wrap_content"
        android:layout_height="wrap_content"
        android:text="@string/btn_submit"/>
```

```
    </LinearLayout>
```

主程序文件代码如下：

```
public class MainActivity extends Activity {
    private RadioGroup sex_radiogroup;
    @Override
    protected void onCreate(Bundle savedInstanceState) {
        super.onCreate(savedInstanceState);
        setContentView(R.layout.activity_main);
        sex_radiogroup= (RadioGroup)findViewById(R.id.sex_radioGroup);
        sex_radiogroup.setOnCheckedChangeListener(new OnCheckedChangeListener(){
            @Override
            public void onCheckedChanged(RadioGroup arg0, int arg1){
                if(arg1==R.id.sex_man){
                    Toast toast= Toast.makeText(MainActivity.this, "您选中的是
                    男", Toast.LENGTH_SHORT);
                    toast.show();
                } else {
                    Toast toast= Toast.makeText(MainActivity.this, "您选中的是
                    女", Toast.LENGTH_SHORT);
                    toast.show();
                }
            }
        });
    }
```

以上在改变单选按钮组时对值的获取，也可以采用如下代码：

```
public void onCheckedChanged(RadioGroup arg0, int arg1){
    RadioButton radio= (RadioButton)findViewById(arg1);
    Toast toast=Toast.makeText(MainActivity.this, "您选中的是"+ radio.getText
    (), Toast.LENGTH_SHORT);
    toast.show();
}
```

如果要在单击其他按钮时获取选中的值，首先要遍历当前单选按钮组，用 isCheck()
方法判断该按钮是否被选中。实现代码如下：

```
btn_submit= (Button)findViewById(R.id.btn_submit);
btn_submit.setOnClickListener(new OnClickListener(){
    Override
    public void onClick(View arg0){
        for(int i=0; i<sex_radiogroup.getChildCount(); i++){
            RadioButton radio= (RadioButton)sex_radiogroup.getChildAt(i);
            if(radio.isChecked()){
                Toast toast= Toast.makeText(MainActivity.this, "您选中的是"+
```

```
                radio.getText(), Toast.LENGTH_SHORT);
                toast.show();
            }
        }
    }
});
```

4. 复选框

复选框是一种双状态按钮的特殊类型,可以选中或者不选中。可以先在布局文件中定义复选框,然后对每一个复选框进行事件监听(setOnCheckedChangeListener),通过isChecked 来判断选项是否被选中。

字符串资源文件代码如下:

```xml
<?xml version="1.0" encoding="utf-8"?>
<resources>
    <string name="app_name">复选框测试</string>
    <string name="hello">你喜欢的运动是:</string>
    <string name="football">足球</string>
    <string name="basketball">篮球</string>
    <string name="volleyball">排球</string>
</resources>
```

布局文件代码如下:

```xml
<LinearLayout xmlns:android="http://schemas.android.com/apk/res/android"
    android:orientation="vertical"
    android:layout_width="fill_parent"
    android:layout_height="fill_parent">
    <TextView
        android:layout_width="fill_parent"
        android:layout_height="wrap_content"
        android:text="@string/hello"
        android:textSize="20sp"
        android:textStyle="bold"
        android:textColor="#FFFFFF"/>
    <CheckBox
        android:id="@+id/chk_football"
        android:layout_width="wrap_content"
        android:layout_height="wrap_content"
        android:text="@string/football"
        android:textSize="16sp"/>
    <CheckBox
        android:id="@+id/chk_basketball"
        android:layout_width="wrap_content"
```

```xml
        android:layout_height="wrap_content"
        android:text="@string/basketball"
        android:textSize="16sp"/>
    <CheckBox
        android:id="@+id/chk_volleyball"
        android:layout_width="wrap_content"
        android:layout_height="wrap_content"
        android:text="@string/volleyball"
        android:textSize="16sp"/>
</LinearLayout>
```

主程序文件代码如下：

```java
public class MainActivity extends Activity {
    private CheckBox chk_football, chk_basketball, chk_volleyball;
    @Override
    protected void onCreate(Bundle savedInstanceState){
        super.onCreate(savedInstanceState);
        setContentView(R.layout.activity_main);
        chk_football=(CheckBox)findViewById(R.id.chk_football);
        chk_basketball=(CheckBox)findViewById(R.id.chk_basketball);
        chk_volleyball=(CheckBox)findViewById(R.id.chk_volleyball);
        chk_football.setOnCheckedChangeListener(checkBox_listener);
        chk_basketball.setOnCheckedChangeListener(checkBox_listener);
        chk_volleyball.setOnCheckedChangeListener(checkBox_listener);
    }
    private OnCheckedChangeListener checkBox_listener=
            new OnCheckedChangeListener(){
        @Override
        public void onCheckedChanged(CompoundButton arg0, boolean arg1){
            if(arg1){
                Toast toast=Toast.makeText(MainActivity.this, "选中了:"
                    +arg0.getText().toString(), Toast.LENGTH_SHORT);
                toast.show();
            }
        }
    };
}
```

3.2.2 文本类组件

在 Android 中有文本框（TextView）、编辑框（EditText）和自动完成文本框（AutoCompleteTextView）3 种文本类控件，用户通过这些组件可以把数据传给 Android 应用，然后得到用户想要的数据。

1. 文本框

文本框用于在屏幕上显示文本，相当于 Java 中的标签（JLabel）。Android 中的文本框可以显示单行文本、多行文本以及带图像的文本。文本框常用的 XML 属性如表 3-1 所示。

表 3-1　文本框组件常用的 XML 属性

XML 属性	描　　述
@android:autoLink	指定是否将指定格式的文本转换为可单击的超链接形式，其属性值有 none、web、email、phone、map 和 all
@android:drawableBottom	在文本框内文本的底端绘制指定图像，该图像放在 res/drawable 目录下，通过"@drawable/文件名（不包括文件扩展名）"设置；同类属性还有 drawableTop、drawableLeft 和 drawableRight
@android:gravity	设置文本框内文本的对齐方式，其属性值可以同时指定，各属性值之间用竖线隔开（例如 right|top）
@android:hint	指定当文本框内容为空时，默认显示的提示文本
@android:inputType	指定文本框显示内容的文本类型，其可选值有 textPassword、textEmailAddress、phone 和 date 等，可同时指定多个，各属性值之间用竖线隔开
@android:singleLine	指定文本框为单行模式，其属性为布尔值，当为 true 时，表示该文本框不会换行，超出的部分将被省略，同时在结尾处添加"…"
@android:text	指定文本框中显示的文本内容，可以直接在该属性中指定，但建议通过引用 values/string.xml 文件中定义文本常量的方式指定
@android:textColor	设置文本框内文本的颜色，其属性值可以是 #rgb、#argb、#rrggbb、#aarrggbb 格式，也可以通过资源文件引用的方式指定
@android:textSize	设置文本框内文本的大小
@android:width	指定文本的宽度，以像素为单位
@android:height	指定文本的高度，以像素为单位

2. 编辑框

编辑框用于在屏幕上显示文本输入框，可以输入单行文本和多行文本，还可以输入指定格式的文本（如密码、电话号码、E-mail 地址等）。编辑框是文本框的子类，表 3-1 中列出的 XML 属性同样适用于编辑框组件。

★注意：在编辑框组件中，android:inputType 属性可以帮助输入法显示合适的类型。

举例：会员注册界面

步骤 1：完成 res/layout 目录下的 XML 布局文件。代码如下：

```
<?xml version="1.0" encoding="utf-8"?>
<TableLayout xmlns:android="http://schemas.android.com/apk/res/android"
```

```xml
android:id="@+id/tableLayout1"
android:layout_width="fill_parent"
android:layout_height="fill_parent"
android:background="@drawable/bg_pic">
<!--第一行 会员昵称 -->
<TableRow android:id="@+id/tableRow1"
    android:layout_width="wrap_content"
    android:layout_height="wrap_content">
<TextView
    android:layout_width="wrap_content"
    android:layout_height="wrap_content"
    android:inputType="textEmailAddress"
    android:text="会员昵称:"
    android:height="50px"/>
<EditText android:id="@+id/nickname"
    android:hint="请输入会员昵称"
    android:layout_width="300px"
    android:layout_height="wrap_content"
    android:singleLine="true"/>
</TableRow>
<!--第二行 输入密码 -->
<TableRow android:id="@+id/tableRow2"
    android:layout_width="wrap_content"
    android:layout_height="wrap_content">
<TextView
    android:layout_width="wrap_content"
    android:layout_height="wrap_content"
    android:inputType="textEmailAddress"
    android:text="输入密码:"
    android:height="50px"/>
<EditText android:id="@+id/pwd"
    android:layout_width="300px"
    android:inputType="textPassword"
    android:layout_height="wrap_content"/>
</TableRow>
<!--第三行 确认密码 -->
<TableRow android:id="@+id/tableRow3"
    android:layout_width="wrap_content"
    android:layout_height="wrap_content">
<TextView
    android:layout_width="wrap_content"
    android:layout_height="wrap_content"
    android:inputType="textEmailAddress"
    android:text="确认密码:"
```

```xml
            android:height="50px"/>
        <EditText android:id="@+id/repwd"
            android:layout_width="300px"
            android:layout_height="wrap_content"
            android:inputType="textPassword"/>
    </TableRow>
    <!--第四行 电子邮件 -->
    <TableRow android:id="@+id/tableRow4"
        android:layout_width="wrap_content"
        android:layout_height="wrap_content">
        <TextView
            android:layout_width="wrap_content"
            android:layout_height="wrap_content"
            android:inputType="textEmailAddress"
            android:text="E-mail:"
            android:height="50px"/>
        <EditText android:id="@+id/email"
            android:layout_width="300px"
            android:layout_height="wrap_content"
            android:inputType="textEmailAddress"/>
    </TableRow>
    <!--添加水平线性布局 -->
    <LinearLayout
        android:orientation="horizontal"
        android:layout_width="wrap_content"
        android:layout_height="wrap_content">
        <Button android:text="注册"
            android:id="@+id/btn_register"
            android:layout_width="wrap_content"
            android:layout_height="wrap_content"/>
        <Button android:text="重置"
            android:id="@+id/btn_reset"
            android:layout_width="wrap_content"
            android:layout_height="wrap_content"/>
    </LinearLayout>
</TableLayout>
```

步骤 2：在主活动的 onCreate() 方法中，为"注册"按钮添加单击事件监听，在日志面板（LogCat）中显示输入的内容。代码如下：

```java
Button btn_register= (Button)findViewById(R.id.btn_register);
btn_register.setOnClickListener(new OnClickListener(){
    @Override
    public void onClick(View v){
```

```java
        EditText nicknameET=(EditText)findViewById(R.id.nickname);
                                            //获取会员昵称编辑框组件
        String nickname=nicknameET.getText().toString();  //获取输入的会员昵称
        EditText pwdET=(EditText)findViewById(R.id.pwd);  //获取密码编辑框组件
        String pwd=pwdET.getText().toString();            //获取输入的密码
        EditText emailET=(EditText)findViewById(R.id.email);
                                            //获取E-mail编辑框组件
        String email=emailET.getText().toString();        //获取输入的E-mail地址
        Log.i("编辑框的应用","会员昵称:"+nickname);
        Log.i("编辑框的应用","密码:"+pwd);
        Log.i("编辑框的应用","E-mail地址:"+email);
    }
});
```

以上例子中,编辑框为默认外观样式,也可以改成图片背景样式。具体操作步骤如下。

步骤1：准备两张图片(png格式),一张是EditText获得焦点后的边框背景,另一张是没有获得焦点时的背景,然后在drawable里添加一个XML文件(文件名自定为selector_edittext_bg.xml)。代码如下：

```xml
<?xml version="1.0" encoding="utf-8"?>
<selector xmlns:android="http://schemas.android.com/apk/res/android">
    <item android:drawable="@drawable/edit_focus" android:state_focused="true"/>
    <item android:drawable="@drawable/edit_normal"/>
</selector>
```

步骤2：在values文件夹下新建一个style.xml文件(有的开发环境中,此文件为自动生成,此时应根据情况添加样式代码)。代码如下：

```xml
<?xml version="1.0" encoding="utf-8"?>
<resources>
    <style name="my_edittext_style" parent="@android:style/Widget.EditText">
        <item name="android:background">@drawable/selector_edittext_bg</item>
    </style>
</resources>
```

步骤3：在EditText组件上使用新建的样式。代码如下：

```xml
<EditText
    android:id="@+id/firstNum"
    android:layout_width="fill_parent"
    android:layout_height="wrap_content"
    android:inputType="number"
    style="@style/my_edittext_style"
    android:padding="15dp"/>
```

另外,也可以定义成形状背景。具体操作步骤如下。

步骤1:在 drawable 里添加一个 XML 文件(文件名自定 rounded_edittext.xml)。代码如下:

```xml
<?xml version="1.0" encoding="utf-8"?>
<shape xmlns:android="http://schemas.android.com/apk/res/android">
    <!--边缘线的宽度和颜色 -->
    <stroke
        android:width="1px"
        android:color="#969696"/>
    <!--中间渐变,0°从左往右,正值为逆时针,270°为从上到下 -->
    <gradient
        android:angle="270"
        android:centerColor="#e9e9e9"
        android:endColor="#d8d8d8"
        android:startColor="#ffffff"/>
    <!--设置4个角的角度 -->
    <corners
        android:bottomLeftRadius="5dp"
        android:bottomRightRadius="5dp"
        android:topLeftRadius="5dp"
        android:topRightRadius="5dp"/>
    <!--设置 padding -->
    <padding
        android:bottom="10dp"
        android:left="10dp"
        android:right="10dp"
        android:top="10dp"/>
</shape>
```

步骤2:在 EditText 组件上添加属性引用。代码如下所示:

```xml
<EditText
    android:id="@+id/firstNum"
    android:layout_width="fill_parent"
    android:layout_height="wrap_content"
    android:inputType="number"
    android:background="@drawable/rounded_edittext"/>
```

3. 自动完成文本框

自动完成文本框用于实现允许用户输入一定字符后显示一个下拉菜单,供用户从中选择并自动填写该文本框。

自动完成文本框组件继承自编辑框,支持编辑框提供的属性。同时,该组件还支持如表 3-2 所示的 XML 属性。

表 3-2　自动完成文本框的部分 XML 属性

XML 属性	描　　述
@android:completionHint	为弹出的下拉菜单指定提示标题
@android:completionThreshold	指定用户至少输入几个字符才会显示提示
@android:dropDownHeight	指定下拉菜单的高度
@android:dropdownHorizontalOffset	指定下拉菜单与文本之间的水平偏移；下拉菜单默认与文本框左对齐
@android:dropDownVerticalOffset	指定下拉菜单与文本之间的垂直偏移；下拉菜单默认紧跟文本框
@android:dropDownWidth	指定下拉菜单的宽度
@android:popupBackground	设置下拉菜单背景

举例：带自动提示功能的搜索框

步骤 1：完成 res/layout 目录下的 XML 布局文件。代码如下：

```
<?xml version="1.0" encoding="utf-8"?>
<LinearLayout xmlns:android="http://schemas.android.com/apk/res/android"
    android:layout_width="fill_parent"
    android:layout_height="fill_parent"
    android:orientation="horizontal">
    <AutoCompleteTextView
        android:id="@+id/search_auto"
        android:layout_width="wrap_content"
        android:layout_height="wrap_content"
        android:layout_weight="7"
        android:completionHint="输入搜索内容"
        android:completionThreshold="2"
        android:text="">
    </AutoCompleteTextView>
    <Button
        android:id="@+id/btn_search"
        android:layout_width="wrap_content"
        android:layout_height="wrap_content"
        android:layout_marginLeft="10dp"
        android:layout_weight="1"
        android:text="搜索"/>
</LinearLayout>
```

步骤 2：在主活动文件中定义字符串常量和 AutoCompleteTextView 组件变量，用于保存下拉菜单中显示的列表项。代码如下：

```
private static final String COUNTRIS[]={"light", "lightwave", "jilinlightwave",
                                        "ligh", "lightdoc" };
```

```
private AutoCompleteTextView search_auto;
```

步骤 3：在主活动的 onCreate()方法中获取组件，创建保存下拉菜单显示项的 ArrayAdapter 适配器，最后将该适配器与自动完成文本框相关联。代码如下：

```
search_auto=(AutoCompleteTextView)findViewById(R.id.search_auto);
ArrayAdapter<String>adapter=new ArrayAdapter<String>(this,android.R.layout.simple_dropdown_item_1line,COUNTRIS);
search_auto.setAdapter(adapter);
```

步骤 4：获取"搜索"按钮组件，添加单击事件监听器，通过消息框显示完成文本框中输入的内容。代码如下：

```
Button btn_search=(Button)findViewById(R.id.btn_search);      //获取搜索按钮组件
btn_search.setOnClickListener(new OnClickListener(){          //为搜索按钮添加单击事件
    @Override
    public void onClick(View v){
        Toast.makeText(MainActivity.this,search_auto.getText().toString(),
        Toast.LENGTH_SHORT).show();
    }
});
```

3.2.3　进度条类组件

在 Android 开发中，常用的进度条主要有 3 种：普通进度条（ProgressBar）、拖动条（SeekBar）和星级评分条（RatingBar）。

1. 普通进度条

普通进度条组件支持的常用 XML 属性如表 3-3 如示。

表 3-3　普通进度条组件常用的 XML 属性

XML 属性	描　述
android:max	设置进度条的最大值
android:progress	指定进度条已完成的进度数
android:progressDrawable	设置进度条轨道的绘制形式
android:progressBarStyle	默认进度条样式

普通进度条组件提供的常用方法如表 3-4 所示：

表 3-4　普通进度条组件常用的方法

XML 属性	描　述
setProgress(int progress)	设置进度值
incrementProgressBy(int diff)	指定增加的进度
getMax()	返回进度条的范围的上限

续表

XML 属性	描述
getProgress()	返回进度
getSecondaryProgress()	返回次要进度
isIndeterminate()	指示进度条是否在不确定模式下
setIndeterminate(boolean indeterminate)	设置进度条是否在不确定模式下
setVisibility(int v)	设置该进度条是否可见

举例：水平进度条和圆形进度条

步骤1：在 res/layout 目录的 XML 布局文件中添加普通进度条组件。代码如下：

```xml
<!--水平进度条 -->
<ProgressBar
    android:id="@+id/proHorizontal"
    style="@android:style/Widget.ProgressBar.Horizontal"
    android:layout_width="fill_parent"
    android:layout_height="wrap_content"
    android:max="100"
    android:progress="10"/>
<!--圆形进度条 -->
<ProgressBar
    android:id="@+id/proCircle"
    style="@android:attr/progressBarStyleLarge"
    android:layout_width="wrap_content"
    android:layout_height="wrap_content"
    android:max="100"/>
```

在上面的代码中，通过 style 属性设置普通进度条的风格，常用的 style 属性如表 3-5 所示。

表 3-5 普通进度条的 style 属性的可选值

XML 属性	描述
@android:style/Widget.ProgressBar.Horizontal	粗水平长条进度条
@android:style/Widget.ProgressBar.Large	大跳跃、旋转画面的进度条
@android:style/Widget.ProgressBar.Small	小跳跃、旋转画面的进度条
@android:attr/progressBarStyleLarge	大圆形进度条
@android:attr/progressBarStyleSmall	小圆形进度条
@android:attr/progressBarStyleHorizontal	细水平长条进度条

步骤2：在主活动中定义所需变量。代码如下：

```java
private ProgressBar proHorizontal, proCircle;    //定义进度条变量
private int mProStatus=0;                         //完成进度
private Handler mHandler;                         //处理消息的 Handler 类对象
```

步骤 3：在主活动的 onCreate()方法中，通过匿名内部类实例化 Handler 类对象，重写其 handleMessage()方法，实现当进度没完成时更新进度。代码如下：

```
proHorizontal=(ProgressBar)findViewById(R.id.proHorizontal);    //获取水平进度条
proCircle=(ProgressBar)findViewById(R.id.proCircle);            //获取圆形进度条
mHandler=new Handler(){
    @Override
    public void handleMessage(Message msg){
        super.handleMessage(msg);
        if(msg.what==0x100){                                    //判断传值
            proHorizontal.setProgress(mProStatus);              //更新进度
        } else {
            Toast.makeText(MainActivity.this, "时间完成！", Toast.LENGTH_SHORT).show();
            proHorizontal.setVisibility(View.GONE);             //隐藏进度条，不占用空间
            proCircle.setVisibility(View.GONE);
        }
    }
};
```

步骤 4：开启一个线程，用于模拟耗时操作，向 Handler 对象发送消息。代码如下：

```
new Thread(new Runnable(){
    @Override
    public void run(){
        while(true){
            mProStatus+=Math.random() * 10;                     //改变进度
            Message msg=new Message();                          //实例化消息对象
            if(mProStatus<100){
                msg.what=0x100;                                 //整型变量赋值
                mHandler.sendMessage(msg);                      //发送消息
            } else {
                msg.what=0x110;
                mHandler.sendMessage(msg);
                break;
            }
            try {
                Thread.sleep(200);                              //线程休眠 200ms
            } catch(InterruptedException e){
                e.printStackTrace();
            }
        }
    }
}).start();
```

2. 拖动条

拖动条的属性和普通进度条很相似,普通进度条是所有进度条的父类,和普通进度条不同的是拖动条有监听器(OnseekBarChangeListener),通常用于实现对某种数值的调节。另外,拖动条组件可以使用 android:thumb 属性(Drawable 对象)改变拖动滑块的外观。

举例：应用拖动条改变值

步骤 1：在 res/layout 目录下的 XML 布局文件中添加拖动条组件。代码如下：

```
<TextView
    android:id="@+id/lab_SeekBar"
    android:layout_width="wrap_content"
    android:layout_height="wrap_content"/>
<SeekBar
    android:id="@+id/num_SeekBar"
    android:layout_width="match_parent"
    android:layout_height="wrap_content"
    android:max="100"/>
```

步骤 2：在主活动中定义所需变量。代码如下：

```
private SeekBar num_SeekBar;                              //拖动条变量
```

步骤 3：在主活动的 onCreate()方法中,为拖动条添加 setOnSeekBarChangeListener 事件监听器,并且重写 onStopTrackingTouch()、onStartTrackingTouch()和 onProgressChanged()方法。代码如下：

```
final TextView lab_SeekBar= (TextView)findViewById(R.id.lab_SeekBar);
                                                          //获取显示文本框
num_SeekBar= (SeekBar)findViewById(R.id.num_SeekBar);     //获取拖动条
num_SeekBar.setOnSeekBarChangeListener(new OnSeekBarChangeListener(){
    @Override
    public void onStopTrackingTouch(SeekBar seekBar){
        Toast.makeText(MainActivity.this,"结束滑动",Toast.LENGTH_SHORT).show();
    }
    @Override
    public void onStartTrackingTouch(SeekBar seekBar){
        Toast.makeText(MainActivity.this,"开始滑动",Toast.LENGTH_SHORT).show();
    }
    @Override
    public void onProgressChanged(SeekBar seekBar, int progress, boolean fromUser){
        lab_SeekBar.setText("当前值:"+progress);          //修改文本视图的值
    }
});
```

3. 星级评分条

星级评分条与拖动条类似，允许用户通过拖动来改变进度（通过星星图标表示进度）。通常使用星级评分条表示对某种事物的支持度或对某种服务的满意度等。星级评分条组件支持的 XML 属性如表 3-6 所示。

表 3-6 星级评分条支持的 XML 属性

XML 属性	描述
android:isIndicator	指定该星级评分条是否允许用户改变（true 允许，false 不允许）
android:numStars	指定该星级评分条中星的数量
android:rating	指定该星级评分条默认的星级
android:stepSize	指定该星级评分条需要改变多少个星级（默认为 0.5 个）

星级评分条还提供 3 个常用的方法：
- getRating()方法：获取等级（选中几颗星）。
- getStepSize()方法：获取每次最少要改变多少个星级。
- getProgress()方法：获取进度（getRating()方法的返回值乘以 getStepSize()方法的返回值）。

3.2.4 选项卡组件

Android 中的选项卡组件主要由 TabHost（选项容器）、TabWidget（选项切换卡）和 FrameLayout（帧布局）3 个组件组成，有两种用法，一种是继承系统自带的 TabHost，即继承 TabActivity 类；另一种是自定义 TabHost。

举例：自定义 TabHost 应用

步骤 1：在 res/layout 目录下的 XML 布局文件中添加组件。代码如下：

```
<?xml version="1.0" encoding="utf-8"?>
<TabHost xmlns:android="http://schemas.android.com/apk/res/android"
    android:id="@+id/my_TabHost"
    android:layout_width="fill_parent"
    android:layout_height="fill_parent">
    <LinearLayout
        android:layout_width="fill_parent"
        android:layout_height="fill_parent"
        android:orientation="vertical">
        <!--使用系统 ID -->
        <TabWidget
            android:id="@android:id/tabs"
            android:layout_width="fill_parent"
            android:layout_height="wrap_content">
        </TabWidget>
        <!--使用系统 ID -->
```

```xml
        <FrameLayout
            android:id="@android:id/tabcontent"
            android:layout_width="fill_parent"
            android:layout_height="fill_parent">
        </FrameLayout>
    </LinearLayout>
</TabHost>
```

步骤 2：编写各标签页显示的内容所对应的 XML 布局文件，例如名称为 tab1.xml 的文件。代码如下：

```xml
<?xml version="1.0" encoding="utf-8"?>
<LinearLayout xmlns:android="http://schemas.android.com/apk/res/android"
    android:id="@+id/player_login"
    android:layout_width="wrap_content"
    android:layout_height="wrap_content"
    android:orientation="vertical">
    <TextView
        android:layout_width="fill_parent"
        android:layout_height="wrap_content"
        android:text="游戏玩家登录    [2014-05-01 8:20]"/>
    <TextView
        android:layout_width="fill_parent"
        android:layout_height="wrap_content"
        android:text="游戏玩家登录    [2014-05-06 7:10]"/>
</LinearLayout>
```

步骤 3：在主活动中定义所需变量，在主活动的 onCreate() 方法中添加如下代码：

```java
private TabHost my_TabHost;
my_TabHost=(TabHost)findViewById(R.id.my_TabHost);
my_TabHost.setup();
LayoutInflater inflater=LayoutInflater.from(this);
inflater.inflate(R.layout.tab1, my_TabHost.getTabContentView());
inflater.inflate(R.layout.tab2, my_TabHost.getTabContentView());
inflater.inflate(R.layout.tab3, my_TabHost.getTabContentView());
my_TabHost .addTab(my_TabHost
         .newTabSpec("tab01")
         .setIndicator(my_TabHost.newTabSpec("tab01").setIndicator("游戏玩
          家登录")
         .setContent(R.id.player_login));
my_TabHost .addTab(my_TabHost.newTabSpec("tab02").setIndicator("游戏玩家信息")
         .setContent(R.id.player_info));
my_TabHost .addTab(my_TabHost.newTabSpec("tab03").setIndicator("游戏玩家退出")
         .setContent(R.id.play_exit));
```

★注意：自定义 TabHost 除了 TabHost 的 id 可以自定义外，其他的必须使用系统的 id。

如果要在标题中显示图片，则可以用多态方法 setIndicator(CharSequence label, Drawable icon)。代码如下：

```
.setIndicator("", getResources().getDrawable(R.drawable.ic_launcher)
```

★注意：在 Android 的 SDK 4.0 及以上版本中，只有文字标题会显示，图标是不显示的。如果将文字标题设置为空字符串，则此时图标可显示。因此，只能是自定义标签卡布局，创建一个包含 ImageView 和 TextView 组件的界面布局文件 tab_indicator.xml (layout/tab_indicator.xml)，然后用 setIndicator(View view)方法来设置 TabSpec 的界面布局。

3.2.5 列表类组件

在 Android 开发中，常用的列表类组件有列表选择框(Spinner)、列表视图(ListView)和网格视图(GridView)3 种。

1. 列表选择框

如果要显示的列表项是已知的，可以将其保存在数组资源文件中(res/values 目录中)，这里命名为 players.xml。代码如下：

```
<?xml version="1.0" encoding="utf-8"?>
<resources>
    <string-array name="player_name">
        <item>玩家 1</item>
        <item>玩家 2</item>
        <item>玩家 3</item>
        <item>玩家 4</item>
        <item>玩家 5</item>
    </string-array>
</resources>
```

在布局文件中添加 Spinner 组件定义。代码如下：

```
<Spinner
    android:id="@+id/my_Spinner"
    android:layout_width="wrap_content"
    android:layout_height="wrap_content"
    android:entries="@array/player_name"
    android:prompt="@string/info"/>
```

其中，android:entries 用于指定列表项，如果不在布局文件中指定该属性，可以在 Java 代码中通过为其指定适配器的方式设置；android:prompt 属性用于指定列表选择框标题。

★注意：在 Android 4.2 版本中，在默认的主题（Theme. Holo）下，不显示 android：prompt 属性的效果，采用 Theme. Black 主题则可以看到。

在屏幕上显示列表选择框组件后，可以用 getSelectedItem 方法获取选中值。代码如下：

```
Spinner my_Spinner= (Spinner)findViewById(R.id.my_Spinner);
my_Spinner.getSelectedItem();
```

如果要在用户选择不同的列表项后执行相应处理，则需添加 setOnItemSelectedListener 事件监听器重写未实现的两个方法。代码如下：

```
my_Spinner.setOnItemSelectedListener(new OnItemSelectedListener(){
    @Override
    public void onItemSelected(AdapterView<?> arg0, View arg1, int pos, long id){
        String result=arg0.getItemAtPosition(pos).toString();
        System.out.println("选中:"+result);
    }
    @Override
    public void onNothingSelected(AdapterView<?>arg0){
        //TODO Auto-generated method stub
    }
});
```

在使用列表选择框时，如果不在布局文件中指定显示项，可以采用为其指定适配器的方式。通常可以分以下两个步骤实现。

步骤 1：创建一个适配器对象，通常使用 ArrayAdapter 类。有以下两种方法。

（1）通过数组资源文件创建。代码如下：

```
ArrayAdapter<CharSequence>adapter=ArrayAdapter.createFromResource(
    this, R.array.player_name, android.R.layout.simple_dropdown_item_1line);
```

（2）通过在 Java 文件中使用字符串数组创建，代码如下：

```
String[] players=new String[] { "玩家 1", "玩家 2", "玩家 3", "玩家 4", "玩家 5" };
    ArrayAdapter<String>adapter=new ArrayAdapter<String>(this,
    android.R.layout.simple_spinner_item, players);
```

步骤 2：设置下拉时的选项样式，将适配器与选择列表框相关联。代码如下：

```
adapter.setDropDownViewResource(android.R.layout.simple_spinner_dropdown_item);
my_Spinner.setAdapter(adapter);
```

2. 列表视图

列表视图是一个常用的组件，其数据内容以列表形式直观地展示出来。比如做一个游戏的排行榜,对话列表等都可以使用列表来实现。列表视图的优点是列表中的数据可以自适应屏幕大小。列表视图组件常用的 XML 属性如表 3-7 所示。

表 3-7 列表视图支持的 XML 属性

XML 属性	描 述
android:divider	设置分隔条(可以用颜色,也可以用 Drawable 资源)
android:dividerHeight	设置分隔条高度
android:entries	通过数组资源指定列表项
android:headerDividersEnabled	设置是否在 header View 之后绘制分隔条。使用该属性时,需要通过组件提供的 addHeaderView()方法
android:footerDividersEnabled	设置是否在 footer View 之前绘制分隔条。使用该属性时,需要通过组件提供的 addFooterView()方法

除了可以直接用 android:entries 属性为列表视图组件指定数组资源作为列表项外,还可以使用两种方法添加列表,一种是直接使用列表视图组件创建,另一种是让 Activity 继承 ListActivity 实现。在列表中定义的数据都通过适配器来映射到列表视图上,列表视图中常用的适配器有两种:

- ArrayAdapter：最简单的适配器,只能显示一行文字。
- SimpleAdapter：具有很好的扩展性的适配器,可以显示自定义内容。

举例：使用列表视图创建列表

步骤 1：在 res/layout 目录下的 XML 布局文件中添加列表视图组件。代码如下:

```
<ListView
        android:id="@+id/my_listView"
        android:layout_width="fill_parent"
        android:layout_height="fill_parent"
        android:divider="@drawable/greendivider">
</ListView>
```

步骤 2：在 onCreate()方法中添加如下代码:

```
final ListView my_ListView=(ListView)findViewById(R.id.my_listView);
    my_ListView.addHeaderView(line());
    String[] goods=new String[] { "列表项 1", "列表项 2", "列表项 3", "列表项 4"};
    ArrayAdapter<String> adapter= new ArrayAdapter<String> (this, android.
        R.layout.simple_list_item_checked, goods);
    my_ListView.setAdapter(adapter);
    my_ListView.addFooterView(line());
//图片线条函数
private View line(){
    ImageView image=new ImageView(this);
    image.getResources().getDrawable(R.drawable.line1);
    return image;
}
```

为了在单击列表视图中各列表项时获取选择的值,需要添加 OnItemClickListener 事件监听器,具体方法与列表选择框组件相同。

举例:继承 ListActivity 实现列表

实现类中无须调用 setContentView() 方法来显示页面。代码如下:

```
public class MainActivity extends ListActivity {
    @Override
    protected void onCreate(Bundle savedInstanceState){
        super.onCreate(savedInstanceState);
        String[] goods=new String[] { "列表项1","列表项2","列表项3","列表项4" };
        ArrayAdapter<String> adapter=new ArrayAdapter<String>(this, android.R.layout.simple_list_item_checked, goods);
        setListAdapter(adapter);
    }
}
```

如要在单击列表视图中各列表项时获取选择的值,则应重写 onListItemClick() 方法。代码如下:

```
@Override
protected void onListItemClick(ListView l, View v, int position, long id){
    super.onListItemClick(l, v, position, id);
    String result=l.getItemAtPosition(position).toString();
    Toast.makeText(this, result, Toast.LENGTH_SHORT).show();
}
```

3. 网格视图

网格视图按照行、列分布的方式显示多个组件,常用的 XML 属性如表 3-8 所示。

表 3-8 网格视图支持的 XML 属性

XML 属性	描 述
android:columnWidth	设置列的宽度
android:horizontalSpacing	设置各元素之间的水平间距
android:numColumns	设计列的数目
android:stretchMode	设置拉伸模式,其中属性值可以是 none(不拉伸)、spacingWidth(仅拉伸元素之间的间距)、columnWidth(仅拉伸表格元素本身)或 spacingWidthUniform(表格元素本身、元素之间的间距一起拉伸)

举例:网格视图的使用

步骤 1:在布局文件中添加网格视图组件。代码如下:

```
<GridView
```

```
        android:id="@+id/pic_GridView"
        android:layout_width="fill_parent"
        android:layout_height="wrap_content"
        android:numColumns="4"
        android:stretchMode="columnWidth">
</GridView>
```

步骤 2：编写用于布局网格内容的 XML 文件 items.xml。代码如下：

```
<?xml version="1.0" encoding="utf-8"?>
<LinearLayout
    xmlns:android="http://schemas.android.com/apk/res/android"
    android:orientation="vertical"
    android:layout_width="match_parent"
    android:layout_height="match_parent">
    <ImageView
        android:id="@+id/image"
        android:paddingLeft="10px"
        android:scaleType="fitCenter"
        android:layout_height="wrap_content"
        android:layout_width="wrap_content"/>
    <TextView
        android:layout_width="wrap_content"
        android:layout_height="wrap_content"
        android:padding="5px"
        android:layout_gravity="center"
        android:id="@+id/title"/>
</LinearLayout>
```

步骤 3：在主活动类的 onCreate() 方法中，使用 SimpleAdapter 简单适配器与网格视图组件相关联。代码如下：

```
pic_GridView= (GridView)findViewById(R.id.pic_GridView);
//定义并初始化保存图片 id 的数组
int imageID[]={ R.drawable.img01, R.drawable.img02, R.drawable.img03,
                R.drawable.img04, R.drawable.img05, R.drawable.img06,
                R.drawable.img07, R.drawable.img08, R.drawable.img09,
                R.drawable.img10, R.drawable.img11, R.drawable.img12 };
//定义并初始化保存说明文字的数组
String title[]={ "装备1", "装备2", "装备3", "装备4", "装备5", "装备6", "装备7",
                "装备8", "装备9", "装备10", "装备11", "装备12", };
//创建一个 List 集合
List<Map<String, Object>>listItem=new ArrayList<Map<String, Object>>();
//通过 for 循环将图片 id 和列表项文字放到 Map 中，并添加到 list 集合中
for(int i=0; i<imageID.length; i++){
    Map<String, Object>map=new HashMap<String, Object>();
```

```
        map.put("image", imageID[i]);
        map.put("title", title[i]);
        listItem.add(map);
}
//配置适配器
SimpleAdapter adapter= new SimpleAdapter(this, listItem, R.layout.items, new
String[] { "title", "image" }, new int[] {R.id.title, R.id.image });
pic_GridView.setAdapter(adapter);                    //将适配器与网格视图关联
```

3.2.6 日期、时间类组件

在 Android 中提供了一些与日期和时间相关的组件,常用的有日期选择器(DatePicker)、时间选择器(TimePicker)和计时器(Chronometer)。

举例:日期/时间选择器使用

步骤 1:在 res/layout 目录下的 XML 布局文件中添加日期选择器和时间选择器组件。代码如下:

```
<DatePicker
    android:id="@+id/my_DatePicker"
    android:layout_width="match_parent"
    android:layout_height="309dp"
    android:scrollbars="vertical"/>
<TimePicker
    android:id="@+id/my_TimePicker"
    android:layout_width="match_parent"
    android:layout_height="match_parent"/>
```

步骤 2:设置相关变量。代码如下:

```
private DatePicker my_DatePicker;
private TimePicker my_TimePicker;
private int year;
private int month;
private int day;
private int hour;
private int minute;
```

步骤 3:在主活动的 onCreate()方法中获取组件,并设置时间组件为 24 小时制式显示。代码如下:

```
my_DatePicker= (DatePicker)findViewById(R.id.my_DatePicker);
my_TimePicker= (TimePicker)findViewById(R.id.my_TimePicker);
my_TimePicker.setIs24HourView(true);
```

步骤 4:创建日历对象并获取年、月、日、小时和分钟数。代码如下:

```
Calendar calendar=Calendar.getInstance();
year=calendar.get(Calendar.YEAR);              //获取当前年份
month=calendar.get(Calendar.MONTH);            //获取当前月份
day=calendar.get(Calendar.DAY_OF_MONTH);       //获取当前日
hour=calendar.get(Calendar.HOUR_OF_DAY);       //获取当前小时数
minute=calendar.get(Calendar.MINUTE);          //获取当前分钟数
```

步骤 5：编写 show()方法，用于通过消息框显示选择的日期和时间。代码如下：

```
private void show(int year, int month, int day, int hour, int minute){
    String str=year+"年"+(month+1)+"月"+day+"日"+hour+":"+minute;
                                            //获取拾取器设置的日期和时间
    Toast.makeText(this, str, Toast.LENGTH_SHORT).show();}
                                            //显示消息提示框
```

步骤 6：初始化时间选择组件。代码如下：

```
my_TimePicker.setCurrentHour(hour);
my_TimePicker.setCurrentMinute(minute);
```

步骤 7：初始化日期选择组件，并在初始化时设置 OnDateChangeListener 事件监听器。代码如下：

```
my_DatePicker.init(year, month, day, new OnDateChangedListener(){
    @Override
    public void onDateChanged(DatePicker arg0, int year, int month, int day){
        MainActivity.this.year=year;        //改变 year 属性的值
        MainActivity.this.month=month;      //改变 month 属性的值
        MainActivity.this.day=day;          //改变 day 属性的值
        show(year, month, day, hour, minute);  //通过消息框显示日期时间
    }
});
```

步骤 8：为时间选择器添加 setOnTimeChangedListener 事件监听器。代码如下：

```
my_TimePicker.setOnTimeChangedListener(new OnTimeChangedListener(){
    @Override
    public void onTimeChanged(TimePicker view, int hourOfDay, int minute){
        MainActivity.this.hour=hourOfDay;   //改变 hour 属性的值
        MainActivity.this.minute=minute;    //改变 minute 属性的值
        show(year, month, day, hourOfDay, minute);  //显示选择的日期和时间
    }
});
```

★**注意**：通过 DatePicker 对象获取的月份是 0～11，所以需要将获取的结果加 1，才能代表真正的月份。

计时器组件继承自文本视图，可显示从某个时间开始共过去多长时间的文本。常用

的方法如下:
- setBase():设置计时器的起始时间。
- setFormat():设置显示时间的格式(默认为 MM:SS 格式,可将参数设置为"已用时间:%s")。
- start():开启计时器。
- stop():停止计时器。

举例:计时器应用

步骤 1:在 res/layout 目录下的 XML 布局文件中添加计时器组件。代码如下:

```
<Chronometer
    android:id="@+id/my_chronometer"
    android:layout_width="wrap_content"
    android:layout_height="wrap_content"
    android:text="@string/txtTime"/>
```

步骤 2:在主活动的 onCreate()方法中获取组件,并设置时间显示格式及添加监听器。代码如下:

```
final Chronometer ch= (Chronometer)findViewById(R.id.my_chronometer);
    ch.setBase(SystemClock.elapsedRealtime());     //设置起始时间
    ch.setFormat("已用时间:%s");                    //设置显示时间的格式
    ch.start();                                    //开启计时器
    ch.setOnChronometerTickListener(new OnChronometerTickListener(){
        @Override
        public void onChronometerTick(Chronometer chronometer){
            if(SystemClock.elapsedRealtime()-ch.getBase()>=20000){
                ch.stop();                         //停止计时器
            }
        }
    });
```

3.3 基本程序单元——活动

  Android 有 Activity(活动)、BroadcastReceiver(广播接收器)、Content Provider(内容提供器)和 Service(服务)这 4 种组件,它们均是独立的,可以互相调用,协调工作。活动(Activity)是 Android 程序中最基本的模块,提供了和用户交互的可视化界面。每个活动都被给予一个默认的窗口以进行绘制,这个窗口一般是满屏的,但也可以是一个小的、位于其他窗口之上的浮动窗口。活动是作为一个对象存在的,与 Android 中其他对象类似,也支持很多 XML 属性。活动常用的 XML 属性如表 3-9 所示。

表 3-9　活动常用的 XML 属性

XML 属性	描述
android:label	设置显示的名称,一般在 Launcher(启动器)里面显示
android:icon	指定显示的图标,在 Launcher 里面显示
android:screenOrientation	指定当前活动是以横屏或竖屏显示
android:allowTaskReparenting	是否允许活动更换从属的任务(例如,从短信息任务切换到浏览器任务)
android:alwaysRetainTaskState	设置系统是否会自动清理任务中除了根活动以外的活动
android:clearTaskOnLaunch	当根活动为 true,且用户离开任务并返回时,任务会清除直到根活动
android:configChanges	当配置列表发生修改时,是否调用 onConfigurationChanged()方法
android:excludeFromRecents	是否可被显示在最近打开的活动列表中
android:exported	是否允许活动被其他程序调用
android:launchMode	设置活动的启动方式(standard、singleTop、singleTask 和 singleInstance,其中前两个为一组,后两个为一组)
android:finishOnTaskLaunch	当用户重新启动这个任务时是否关闭已打开的活动
android:noHistory	当用户切换到其他屏幕时,是否需要移除这个活动
android:process	一个活动运行时所在的进程名,与应用程序的包名一致
android:windowSoftInputMode	定义软键盘弹出的模式

活动主要有 4 种状态:
- 运行状态(Running):该活动位于前台,此时处于可见并可和用户交互的激活状态。
- 暂停状态(Paused):其他活动位于前台,该活动依然可见,只是不能获得焦点。
- 停止状态(Stopped):该活动不可见,失去焦点。
- 销毁状态(Killed):该活动结束,或活动所在的进程被结束。

★注意:在以上 4 种状态中,运行状态和暂停状态是可见的,而停止状态和销毁状态是不可见的。

3.3.1　Android 生命周期

Android 程序创建时,会自动在 Java 源文件中重写活动类中的 onCreate()方法,该方法是创建活动时必须调用的一个方法。其他操纵生命周期的方法如表 3-10 所示。

表 3-10　Android 生命周期方法

方法	描述	是否可被杀死	下一个操作
onCreate()	初始化,创建视图,绑定列表的数据等等	否	onStart()
onRestart()	在活动被停止后调用	否	onStart()

续表

方法	描述	是否可被杀死	下一个操作
onStart()	在活动被用户可见之前调用	否	onResume()或者onStop()
onResume()	在活动和用户交互之前调用(快速,持久化)	否	onPause()
onPause()	在系统要激活另一个活动时调用(快速)	是	onResume()或者onStop()
onStop()	在活动不再可见时调用	是	onRestart()或者nDestroy()
onDestroy()	在活动被销毁时调用	是	

以上这些方法的先后执行顺序构成了一个完整的生命周期,如图3-5所示。

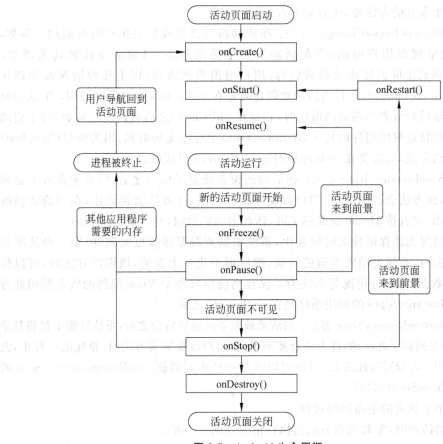

图3-5 Android生命周期

该完整生命周期又可以分成以下3个嵌套生命周期循环。

(1) 前台生命周期:自onResume()调用起,至相应的onPause调用为止。在此期间,活动位于前台最上面并与用户进行交互,活动会经常在暂停和恢复之间进行状态转换。例如,当设备转入休眠状态或者有新的活动启动时,将调用onPause()方法;而当活动获得结果或者接收到新的Intent(意图)时,会调用onResume()方法。

(2) 可视生命周期:自onStart()调用开始,直到相应的onStop()调用结束。在此期间,用户可以在屏幕上看到活动,尽管它也许并不是位于前台或者也不与用户进行交互。在这两个方法之间可以保留用来向用户显示这个活动所需的资源。例如,当用户看不到显示的内容时,可以在 onStart()中注册一个 BroadcastReceiver 广播接收器来监控可能影响 UI 的变化,而在 onStop()中注销。onStart()和 onStop()方法可以随着应用程序是否被用户可见而被多次调用。

(3) 完整生命周期:自第一次调用 onCreate()开始,直至调用 onDestroy()为止。活动在 onCreate()中设置所有"全局"状态以完成初始化,而在 onDestroy()中释放所有系统资源。例如,如果活动有一个线程在后台运行,从网络上下载数据,它会在 onCreate()中创建线程,而在 onDestroy()中销毁线程。

除了几个常见的方法外,还有如下方法。

(1) onWindowFocusChanged 方法:在活动窗口获得或失去焦点时被调用。例如,创建时首次呈现在用户面前,当前活动被其他活动覆盖,当前活动转到其他活动,按 Home 键回到主屏而使活动退居后台,用户退出当前活动,以上几种情况都会调用 onWindowFocusChanged,并且当活动被创建时是在 onResume()之后被调用,当活动被覆盖或者退居后台或者当前活动退出时,它是在 onPause()之后被调用。例如,程序启动时想要获取视特定视图组件的尺寸,在 onCreate()中可能无法取到,因为窗口(Window)对象还没创建完成,就需要在 onWindowFocusChanged 里获取。

(2) onSaveInstanceState 方法:在活动被覆盖或退居后台之后,因系统资源不足而将其杀死时,此方法会被调用;在用户改变屏幕方向时,此方法会被调用;在当前活动跳转到其他活动,或者按 Home 键回到主屏,活动退居后台时,此方法会被调用。

第一种情况无法保证什么时候发生,系统根据资源紧张程度去调度;第二种情况是屏幕翻转方向时,系统先销毁当前的活动,然后再重建一个新的,调用此方法时,可以保存一些临时数据;第三种情况是系统为了保存当前窗口各个 View 组件的状态调用此方法。onSaveInstanceState 的调用顺序是在 onPause()之前。

(3) onRestoreInstanceState 方法:在活动被覆盖或退居后台之后,系统资源不足将其杀死,然后用户又回到了此活动,此方法会被调用;在用户改变屏幕方向时,重建的过程中,此方法会被调用。可以重写此方法,以便可以恢复一些临时数据。onRestoreInstanceState 的调用顺序是在 onStart()之后。

举例:单个活动的生命周期过程

(1) 启动活动时,生命周期方法的执行情况如图 3-6 所示。

tag	Message
LifeCycleActivity	onCreate called.
LifeCycleActivity	onStart called.
LifeCycleActivity	onResume called.

图 3-6 启动活动的生命周期

在系统调用了 onCreate()和 onStart()之后,调用了 onResume(),自此,活动进入了运行状态。

(2) 跳转到其他活动(或按下 Home 键回到主屏)时,生命周期方法的执行情况如

图 3-7 所示。

```
tag                 Message
LifeCycleActivity   onSaveInstanceState called. put param: 1
LifeCycleActivity   onPause called.
LifeCycleActivity   onStop called.
```

图 3-7 跳转到其他活动的生命周期

可以看到,此时 onSaveInstanceState 方法在 onPause()之前被调用了,并且注意,退居后台时,在 onPause()后 onStop()被调用。

(3) 从后台回到前台时,生命周期方法的执行情况如图 3-8 所示。

```
tag                 Message
LifeCycleActivity   onRestart called.
LifeCycleActivity   onStart called.
LifeCycleActivity   onResume called.
```

图 3-8 从后台到前台时活动的生命周期

当从后台回到前台时,系统先调用 onRestart()方法,然后调用 onStart()方法,最后调用 onResume 方法,活动又进入了运行状态。

(4) 修改 TargetActivity 在 AndroidManifest.xml 中的配置。

将 android:theme 属性设置为@android:style/Theme.Dialog,跳转行为就变为了 TargetActivity 覆盖到 LifeCycleActivity 之上,此时生命周期方法的执行情况如图 3-9 所示。

```
tag                 Message
LifeCycleActivity   onSaveInstanceState called. put param: 1
LifeCycleActivity   onPause called.
```

图 3-9 不完全覆盖时活动的生命周期

注意:还有一种情况就是,单击返回按钮,或按下锁屏键,执行的效果也是如此。可以看到,此时 LifeCycleActivity 的 OnPause()方法被调用,并没有调用 onStop()方法,因为此时的 LifeCycleActivity 没有退居后台,只是被覆盖或被锁屏;onSaveInstanceState 会在 onPause()之前被调用。

(5) 按回退键,使 LifeCycleActivity 从被覆盖回到前面,或者按解锁键解锁屏幕,此时生命周期方法的执行情况如图 3-10 所示。

```
tag                 Message
LifeCycleActivity   onResume called.
```

图 3-10 按回退键或解锁屏幕时活动的生命周期

此时只有 onResume()方法被调用,直接再次进入运行状态。

(6) 退出,此时生命周期方法的执行情况如图 3-11 所示。

最后 onDestroy()方法被调用,标志着 LifeCycleActivity 的终结。

在所有的过程中,并没有 onRestoreInstanceState 的出现,因为该方法只有在杀死不

```
tag              Message
LifeCycleActivity   onPause called.
LifeCycleActivity   onStop called.
LifeCycleActivity   onDestroy called.
```

图 3-11　退出时活动的生命周期

在前台的活动之后用户回到此活动或者用户改变屏幕方向的这两个重建过程中被调用。

生命周期只有 10s 左右，如果在 onReceive()操作超过 10s，就会报告 ANR（Application No Response,程序无响应）的错误信息,如果需要完成一项比较耗时的工作，应该通过发送 Intent 给 Service，由 Service 来完成。这里不能使用子线程来解决，因为广播接收器的生命周期很短,子线程可能还没有结束,广播接收器就先结束了。广播接收器一旦结束,此时广播接收器所在的进程很容易在系统需要内存时被优先杀死,因为它属于空进程（没有任何活动组件的进程）。如果它的宿主进程被杀死,那么正在工作的子线程也会被杀死,所以采用子线程来解决是不可靠的。

举例：两个活动的生命周期过程

（1）从一个活动通过 Intent 切换到另一个活动,此时生命周期方法的执行情况如图 3-12 所示。

```
INFO/System.out(339): MainActivity------->onPause()
INFO/System.out(339): Another------->onCreate()
INFO/System.out(339): Another------->onStart()
INFO/System.out(339): Another------->onResume()
INFO/System.out(339): MainActivity------->onStop()
```

图 3-12　切换活动时的生命周期

（2）按回退键返回,此时生命周期方法的执行情况如图 3-13 所示。

```
INFO/System.out(339): Another------->onPause()
INFO/System.out(339): MainActivity------->onRestart()
INFO/System.out(339): MainActivity------->onStart()
INFO/System.out(339): MainActivity------->onResume()
INFO/System.out(339): Another------->onStop()
INFO/System.out(339): Another------->onDestroy()
```

图 3-13　按回退键时活动的生命周期

（3）第二个活动使用了 finish 方法返回,此时生命周期方法的执行情况如图 3-14 所示。

```
INFO/System.out(366): Another------->onPause()
INFO/System.out(366): MainActivity------->onRestart()
INFO/System.out(366): MainActivity------->onStart()
INFO/System.out(366): MainActivity------->onResume()
INFO/System.out(366): Another------->onStop()
INFO/System.out(366): Another------->onDestroy()
```

图 3-14　使用 finish 方法时活动的生命周期

★**注意**：在当前活动调出对话框（Dialog),活动不会执行生命周期中的任何方法。

3.3.2 用 Intent 切换页面

Intent 起着一个媒体中介的作用，专门提供组件互相调用的相关信息，实现调用者与被调用者之间的解耦。

举例：闪屏切换界面

步骤 1：复制所需图片资源到 res/drawable-mdpi 目录下，修改 res/layout 目录下闪屏界面的 XML 布局文件，添加背景。代码如下：

```
<LinearLayout xmlns:android="http://schemas.android.com/apk/res/android"
    xmlns:tools="http://schemas.android.com/tools"
    android:layout_width="fill_parent"
    android:layout_height="fill_parent"
    android:orientation="vertical"
    android:background="@drawable/splash_bg">
</LinearLayout>
```

步骤 2：修改主活动页面名称为 SplashActivity，并在 AndroidManifest.xml 文件中修改相关设置。

步骤 3：创建第二个活动页面，这里命名为 MenuActivity。

步骤 4：在 SplashActivity 页面的 onCreate()方法中添加如下代码：

```
protected void onCreate(Bundle savedInstanceState){
    super.onCreate(savedInstanceState);
    requestWindowFeature(Window.FEATURE_NO_TITLE);    //隐藏标题栏
    getWindow(). setFlags ( WindowManager. LayoutParams. FLAG _ FULLSCREEN,
WindowManager.LayoutParams.FLAG_FULLSCREEN);         //设置全屏
    setContentView(R.layout.activity_main);
    new Handler().postDelayed(new Runnable(){         //使用 Handler 对象的延时方法
        @Override
        public void run(){
            Intent intent=new Intent(SplashActivity.this, MenuActivity.class);
            startActivity(intent);                    //执行页面切换
            finish();                                 //关闭当前页面
        }
    }, 2000);                                         //延时 2s
}
```

为了使闪屏切换的效果更加自然，可以应用添加不透明动画的方法。具体步骤如下。

步骤 1：修改 res/layout 目录下闪屏界面的 XML 布局文件，添加图像组件。代码如下：

```
<LinearLayout xmlns:android="http://schemas.android.com/apk/res/android"
    xmlns:tools="http://schemas.android.com/tools"
```

```xml
android:layout_width="fill_parent"
android:layout_height="fill_parent"
android:gravity="center"
android:orientation="vertical">
<ImageView
    android:id="@+id/splash"
    android:layout_width="fill_parent"
    android:layout_height="fill_parent"
    android:src="@drawable/splash_bg"/>
</LinearLayout>
```

步骤 2：修改 SplashActivity 页面 onCreate() 方法中的代码：

```java
protected void onCreate(Bundle savedInstanceState){
    super.onCreate(savedInstanceState);
    requestWindowFeature(Window.FEATURE_NO_TITLE);   //隐藏标题栏
    getWindow().setFlags(WindowManager.LayoutParams.FLAG_FULLSCREEN,
        WindowManager.LayoutParams.FLAG_FULLSCREEN);  //设置全屏
    setContentView(R.layout.activity_main);
    ImageView splashImage=(ImageView)this.findViewById(R.id.splash);
                                                     //获取图像组件
    AlphaAnimation animation=new AlphaAnimation(0.1f, 1.0f);
                                                     //创建不透明度动画
    animation.setDuration(2000);                     //设置过渡时间为 2s
    splashImage.startAnimation(animation);           //播放动画
    animation.setAnimationListener(new AnimationListener(){
                                                     //设置动画的事件侦听
        @Override
        public void onAnimationStart(Animation animation){
            //TODO: Auto-generated method stub
        }
        @Override
        public void onAnimationRepeat(Animation animation){
            //TODO: Auto-generated method stub
        }
        @Override
        public void onAnimationEnd(Animation animation){
            Intent intent=new Intent(SplashActivity.this, MenuActivity.class);
            startActivity(intent);                   //执行页面切换
            finish();                                //关闭当前页面
        }
    });
}
```

★提示：关于 AlphaAnimation 的详细用法见 4.7 节。

3.3.3 用 Intent 实现活动间简单参数传递

Intent 是 Android 程序中传输数据的核心对象,负责对应用中一次操作的动作、动作涉及的数据和附加数据进行描述。因此,将要传递的数据保存在 Intent 中,就可以将其传递到另一个活动中。

举例:传递运算结果

步骤 1:在 res/values 目录下的 strings.xml 文件中添加相应用的字符串资源。代码如下:

```
<string name="str_add">加</string>
<string name="btn_cal">计算结果</string>
<string name="title_activity_second">SecondActivity</string>
```

步骤 2:在 activity_main.xml 布局文件中添加两个编辑框组件、一个文本视图组件和一个按钮。代码如下:

```
<?xml version="1.0" encoding="utf-8"?>
<LinearLayout xmlns:android="http://schemas.android.com/apk/res/android"
    android:orientation="vertical"
    android:layout_width="fill_parent"
    android:layout_height="fill_parent">
<EditText
    android:id="@+id/firstNum"
    android:layout_width="fill_parent"
    android:layout_height="wrap_content"
    android:inputType="number"/>
<TextView
    android:layout_width="fill_parent"
    android:layout_height="wrap_content"
    android:text="@string/str_add"
    android:textSize="20sp"/>
<EditText
    android:id="@+id/secondNum"
    android:layout_width="fill_parent"
    android:layout_height="wrap_content"
    android:inputType="number"/>
<Button
    android:id="@+id/btnCalc"
    android:layout_width="fill_parent"
    android:layout_height="wrap_content"
    android:text="@string/btn_cal"/>
</LinearLayout>
```

步骤 3:在 activity_second.xml 布局文件中,添加一个编辑框组件来显示从

FirstActivity 中传过来的值。代码如下：

```xml
<?xml version="1.0" encoding="utf-8"?>
<LinearLayout
    xmlns:android="http://schemas.android.com/apk/res/android"
    android:layout_width="fill_parent"
    android:layout_height="fill_parent">
    <EditText
        android:id="@+id/result"
        android:layout_width="fill_parent"
        android:layout_height="wrap_content"
        android:inputType="number"/>
</LinearLayout>
```

步骤 4：在主程序文件 MainActivity.java 中添加如下代码：

```java
private EditText firstNum, secondNum;
private Button btnCalc;
    @Override
    public void onCreate(Bundle savedInstanceState){
        super.onCreate(savedInstanceState);
        setContentView(R.layout.activity_main);
        firstNum=(EditText)findViewById(R.id.firstNum);
        secondNum=(EditText)findViewById(R.id.secondNum);
        btnCalc=(Button)findViewById(R.id.btnCalc);
        btnCalc.setOnClickListener(new View.OnClickListener(){
        @Override
        public void onClick(View v){
            String num1=firstNum.getText().toString();
            String num2=secondNum.getText().toString();
            Intent intent=new Intent();
            intent.putExtra("one", num1);
            intent.putExtra("two", num2);
            intent.setClass(MainActivity.this, SecondActivity.class);
            startActivity(intent);
        }
    });
}
```

步骤 5：在 SecondActivity.java 中添加如下代码：

```java
private EditText result;
public void onCreate(Bundle savedInstanceState){
    super.onCreate(savedInstanceState);
    setContentView(R.layout.activity_second);
    result=(EditText)findViewById(R.id.result);
```

```
        Intent intent=getIntent();
        String num1=intent.getStringExtra("one");
        String num2=intent.getStringExtra("two");
        int ret=Integer.parseInt(num1)+Integer.parseInt(num2);
        result.setText(ret+"");
    }
```

★**注意**：同时在 AndroidManifest.xml 配置文件中检查是否已对 SecondActivity 进行注册。

3.3.4 Bundle 类在活动传值中的使用

可以将要保存的数据存放在 Bundle 对象中（类似于 Map，用于存放名值对形式的值）。在 Android 中提供了各种常用类型的 putXxx()/getXxx()方法，例如 putString()/getString()和 putInt()/getInt()。putXxx()用于往 Bundle 对象中放入数据，Bundle 的内部使用了 HashMap<String,Object>类型的变量来存放 putXxx()方法放入的值；getXxx()方法用于从 Bundle 对象里获取数据。

举例：用户注册界面

步骤 1：将 res/layout 目录下的主页面 XML 布局文件设置为垂直线性布局，添加需要的组件。代码如下：

```xml
<TextView
        android:id="@+id/textView1"
        android:layout_width="wrap_content"
        android:layout_height="wrap_content"
        android:text="用户名:"/>
    <EditText
        android:id="@+id/user"
         android:minWidth="200px"
        android:layout_width="wrap_content"
        android:layout_height="wrap_content"/>
    <TextView
        android:id="@+id/textView2"
        android:layout_width="wrap_content"
        android:layout_height="wrap_content"
        android:text="密码:"/>
    <EditText
        android:id="@+id/pwd"
         android:minWidth="200px"
        android:inputType="textPassword"
        android:layout_width="wrap_content"
        android:layout_height="wrap_content"/>
    <TextView
        android:id="@+id/textView3"
```

```xml
        android:layout_width="wrap_content"
        android:layout_height="wrap_content"
        android:text="确认密码:"/>
    <EditText
        android:id="@+id/repwd"
        android:minWidth="200px"
        android:inputType="textPassword"
        android:layout_width="wrap_content"
        android:layout_height="wrap_content"/>
    <TextView
        android:id="@+id/textView3"
        android:layout_width="wrap_content"
        android:layout_height="wrap_content"
        android:text="E-mail 地址:"/>
    <EditText
        android:id="@+id/email"
        android:minWidth="400px"
        android:layout_width="wrap_content"
        android:layout_height="wrap_content"/>
    <Button
        android:id="@+id/submit"
        android:layout_width="wrap_content"
        android:layout_height="wrap_content"
        android:text="提交"/>
</LinearLayout>
```

步骤 2：设置要跳转的目标页面为垂直线性布局，添加需要的组件。代码如下：

```xml
<TextView
    android:id="@+id/user"
    android:layout_width="wrap_content"
    android:layout_height="wrap_content"
    android:padding="10px"
    android:text="用户名:"/>
<TextView
    android:id="@+id/pwd"
    android:layout_width="wrap_content"
    android:layout_height="wrap_content"
    android:padding="10px"
    android:text="密码:"/>
<TextView
    android:id="@+id/email"
    android:padding="10px"
    android:layout_width="wrap_content"
    android:layout_height="wrap_content"
```

```
            android:text="E-mail:"/>
```

步骤3：在主活动的onCreate方法中获取组件，实现数据收集与传递。代码如下：

```
Button submit=(Button)findViewById(R.id.submit);
submit.setOnClickListener(new View.OnClickListener(){
@Override
public void onClick(View v){
    String user=((EditText)findViewById(R.id.user)).getText().toString();
    String pwd=((EditText)findViewById(R.id.pwd)).getText().toString();
    String repwd=((EditText)findViewById(R.id.repwd)).getText().toString();
    String email=((EditText)findViewById(R.id.email)).getText().toString();
    if(!"".equals(user)&& !"".equals(pwd)&& !"".equals(email)){
      if(!pwd.equals(repwd)){
        Toast.makeText(MainActivity.this, "两次密码不同!", Toast.LENGTH_LONG).
        show();
        ((EditText)findViewById(R.id.pwd)).setText("");
        ((EditText)findViewById(R.id.repwd)).setText("");
        ((EditText)findViewById(R.id.pwd)).requestFocus();
      }else{
        Intent intent=new Intent(MainActivity.this, RegisterActivity.class);
        Bundle bundle=new Bundle();
        bundle.putCharSequence("user", user);
        bundle.putCharSequence("pwd", pwd);
        bundle.putCharSequence("email", email);
        intent.putExtras(bundle);
        startActivity(intent);
      }
    }else{
      Toast.makeText(MainActivity.this, "信息不完整!", Toast.LENGTH_LONG).
      show();
    }
  }
});
```

步骤4：在跳转的目标活动文件的onCreate方法中接收并显示数据。代码如下：

```
Intent intent=getIntent();
Bundle bundle=intent.getExtras();
TextView user=(TextView)findViewById(R.id.user);
user.setText("用户名:"+bundle.getString("user"));
TextView pwd=(TextView)findViewById(R.id.pwd);
pwd.setText("密码:"+bundle.getString("pwd"));
TextView email=(TextView)findViewById(R.id.email);
email.setText("E-mail:"+bundle.getString("email"));
```

以上完成了一个简单的利用 Bundle 对象在两个活动之间传值的例子。但在 Android 开发过程中,有时需要在一个活动中调用另一个活动,当用户在第二个活动中选择完成后,程序自动返回到第一个活动中,第一个活动必须能够获取并显示用户在第二个活动中选择的结果;或者是在第一个活动中将一些数据传递到第二个活动,由于某些原因,又要返回到第一个活动中,并显示传递的数据。这时,也可以通过 Intent 和 Bundle 来实现,与在两个活动之间交换数据不同的是,此处需要使用 startActivityForResult 方法来启动另一个活动。

举例:页面之间的数据回调

修改用户注册界面实例,具体步骤如下。

步骤 1:在主活动 MainActivity.java 中定义常量,用于设置 requestCode 请求码(根据业务自行设定)。代码如下:

```
private final int CODE=0x066;
```

步骤 2:将原来用 startActivity 方法启动新活动的代码修改为用 startActivityForResult 方法实现。修改后的代码如下:

```
startActivityForResult(intent, CODE);
```

步骤 3:在跳转页面的布局文件中添加"返回"按钮。代码如下:

```
<Button
    android:id="@+id/btn_return"
    android:layout_width="wrap_content"
    android:layout_height="wrap_content"
    android:text="返回"/>
```

步骤 4:在 RegisterActivity.java 中获取新添加的"返回"按钮并为其添加单击事件监听器,重写 onClick 方法。关键代码如下:

```
Button btn_return= (Button)findViewById(R.id.btn_return);
btn_return.setOnClickListener(new OnClickListener(){
    @Override
    public void onClick(View v){
        setResult(0x666, intent);
        finish();
    }
});
```

步骤 5:再次打开 MainActivity.java,重写 onActivityResult 方法,判断请求码和结果码是否与预先设置的相同,如果相同,则清空"密码"编辑框和"确认密码"编辑框内容。关键代码如下:

```
@Override
protected void onActivityResult(int requestCode, int resultCode, Intent data){
    super.onActivityResult(requestCode, resultCode, data);
```

```
if(requestCode==CODE && resultCode==0x666){
    ((EditText)findViewById(R.id.pwd)).setText("");
    ((EditText)findViewById(R.id.repwd)).setText("");
    ((EditText)findViewById(R.id.pwd)).setFocusable(true);
}
}
```

在 onActivityResult(int requestCode, int resultCode, Intent data)方法中，参数 data 中保存传回的数据。

3.3.5 用 Intent 实现活动间传递对象参数

活动之间传递数据时，往往会将一些值封装成对象，然后将整个对象传递过去。传递对象的时候有两种情况：一种是实现 Parcelable 接口，一种是实现 Serializable 接口。实现 Serializable 接口是 JavaSE 本身就支持的，Parcelable 接口是 Android 特有的功能，还可以用在进程间通信(IPC)，除了基本类型外，只有实现了 Parcelable 接口的类才能被放入 Parcel 中。

在使用内存的时候，Parcelable 比 Serializable 效率高，Serializable 在序列化的时候会产生大量的临时变量。但 Parcelable 在外界有变化的情况下不能很好地保证数据的持续性，所以不能在要将数据存储在磁盘上的情况下使用。

举例：应用 Serializable 接口传递对象

步骤 1：完成 res/layout 目录下主页面的 XML 布局。代码如下：

```
<?xml version="1.0" encoding="utf-8"?>
<LinearLayout xmlns:android="http://schemas.android.com/apk/res/android"
    android:layout_width="fill_parent"
    android:layout_height="fill_parent"
    android:orientation="vertical">
    <TextView
        android:layout_width="fill_parent"
        android:layout_height="wrap_content"
        android:layout_gravity="center_horizontal"
        android:padding="20px"
        android:text="计算您的标准体重"/>
    <LinearLayout
        android:id="@+id/linearLayout1"
        android:layout_width="match_parent"
        android:layout_height="wrap_content"
        android:gravity="center_vertical">
        <TextView
            android:id="@+id/textView1"
            android:layout_width="wrap_content"
            android:layout_height="wrap_content"
```

```xml
            android:text="性别："/>
        <RadioGroup
            android:id="@+id/sex"
            android:layout_width="wrap_content"
            android:layout_height="wrap_content"
            android:orientation="horizontal">
            <RadioButton
                android:id="@+id/radio0"
                android:layout_width="wrap_content"
                android:layout_height="wrap_content"
                android:checked="true"
                android:text="男"/>
            <RadioButton
                android:id="@+id/radio1"
                android:layout_width="wrap_content"
                android:layout_height="wrap_content"
                android:text="女"/>
        </RadioGroup>
    </LinearLayout>
    <LinearLayout
        android:id="@+id/linearLayout1"
        android:layout_width="match_parent"
        android:layout_height="wrap_content"
        android:gravity="center_vertical">
        <TextView
            android:id="@+id/textView1"
            android:layout_width="wrap_content"
            android:layout_height="wrap_content"
            android:text="身高："/>
        <EditText
            android:id="@+id/stature"
            android:layout_width="wrap_content"
            android:layout_height="wrap_content"
            android:minWidth="100px">
        </EditText>
        <TextView
            android:id="@+id/textView2"
            android:layout_width="wrap_content"
            android:layout_height="wrap_content"
            android:text="cm"/>
    </LinearLayout>
    <Button
        android:id="@+id/button1"
        android:layout_width="wrap_content"
```

```
        android:layout_height="wrap_content"
        android:text="确定"/>
</LinearLayout>
```

步骤 2：编写一个实现 java.ip.Serializable 接口的 Java 类，创建变量并实现其 setter 和 getter 方法。代码如下：

```
public class Info implements Serializable {
private static final long serialVersionUID=1L;
private String sex="";                        //性别
private int stature=0;                        //身高
   public String getSex(){
       return sex;
   }
   public void setSex(String sex){
       this.sex=sex;
   }
   public int getStature(){
       return stature;
   }
   public void setStature(int stature){
       this.stature=stature;
   }
}
```

步骤 3：在主活动 MainActivity.java 的 onCreate()方法中添加如下代码：

```
Button button=(Button)findViewById(R.id.button1);
button.setOnClickListener(new OnClickListener(){
    @Override
    public void onClick(View v){
        Info info=new Info();                 //实例化一个保存输入基本信息的对象
        if("".equals(((EditText)findViewById(R.id.stature)).getText().
toString())){
        Toast.makeText(MainActivity.this,"请输入您的身高!",Toast.LENGTH_
SHORT).show();
            return;
        }
        int stature=Integer.parseInt(((EditText)findViewById(R.id.stature))
            .getText().toString());
        RadioGroup sex=(RadioGroup)findViewById(R.id.sex);
                                              //获取设置性别的单选按钮组
                                              //获取单选按钮组的值
        for(int i=0;i<sex.getChildCount();i++){
            RadioButton r=(RadioButton)sex.getChildAt(i);
                                              //根据索引值获取单选按钮
```

```
            if(r.isChecked()){                          //判断单选按钮是否被选中
                info.setSex(r.getText().toString());    //获取被选中的单选按钮的值
                break;                                  //跳出 for 循环
            }
        }
        info.setStature(stature);                       //设置身高
        Bundle bundle=new Bundle();                     //实例化一个 Bundle 对象
        bundle.putSerializable("info", info);           //将信息保存到 Bundle 对象中
        Intent intent=new Intent(MainActivity.this, ResultActivity.class);
        intent.putExtras(bundle);                       //将 bundle 保存到 Intent 对象中
        startActivity(intent);                          //启动 intent 对应的活动

        }
    });
 }
}
```

步骤 4：创建一个活动用于显示传递的数据。XML 布局文件代码如下：

```
<?xml version="1.0" encoding="utf-8"?>
<LinearLayout xmlns:android="http://schemas.android.com/apk/res/android"
    android:layout_width="match_parent"
    android:layout_height="match_parent"
    android:orientation="vertical">
    <TextView
        android:id="@+id/sex"
        android:layout_width="wrap_content"
        android:layout_height="wrap_content"
        android:padding="10px"
        android:text="性别"/>
    <TextView
        android:id="@+id/stature"
        android:layout_width="wrap_content"
        android:layout_height="wrap_content"
        android:padding="10px"
        android:text="身高"/>
    <TextView
        android:id="@+id/weight"
        android:padding="10px"
        android:layout_width="wrap_content"
        android:layout_height="wrap_content"
        android:text="标准体重"/>
</LinearLayout>
```

在程序文件中添加如下代码：

```
TextView sex=(TextView)findViewById(R.id.sex);           //获取显示性别的文本框
TextView stature=(TextView)findViewById(R.id.stature);   //获取显示身高的文本框
TextView weight=(TextView)findViewById(R.id.weight);     //获取显示标准体重的文本框
Intent intent=getIntent();                               //获取 Intent 对象
Bundle bundle=intent.getExtras();                        //获取传递的数据包
Info info=(Info)bundle.getSerializable("info");          //获取一个可序列化的 info 对象
sex.setText("您是一位"+info.getSex()+"士");               //获取性别并显示到相应文本框
stature.setText("您的身高是"+info.getStature()+"厘米");
                                                         //获取身高并显示到相应文本框
weight.setText("您的标准体重是"+getWeight(info.getSex(), info.getStature())+"
公斤");
//获取体重的函数
private String getWeight(String sex, float stature){
    String weight="";                                    //保存体重
    NumberFormat format=new DecimalFormat();
    if(sex.equals("男")){                                //计算男士标准体重
        weight=format.format((stature-80) * 0.7);
    }else{                                               //计算女士标准体重
        weight=format.format((stature-70) * 0.6);
    }
    return weight;
}
```

3.4 Android 事件处理

图形界面的应用程序都是通过事件来实现人机交互的。在 Android 手机和平板电脑上,主要包括键盘事件和触摸事件两大类。键盘事件包括按下、弹起等,触摸事件包括按下、滑动等。Android 组件提供了事件处理的相关方法。

3.4.1 处理键盘事件

一个标准的 Android 设备,包含了多个能够触发事件的物理按钮,各个可用的物理按钮能够触发的事件说明如表 3-11 所示。

表 3-11 Android 设备可用的物理按键

物理按键	KeyEvent	说明
电源键	KEYCODE_POWER	启动或唤醒设备,切换界面到锁定的屏幕
回退键	KEYCODE_BACK	返回到前一个界面
菜单键	KEYCODE_MENU	显示当前应用的可用菜单
搜索键	KEYCODE_SEARCH	返回到 HOME 界面

续表

物理按键	KeyEvent	说　　明
HOME 键	KEYCODE_HOME	在当前应用中启动搜索
音量键	KEYCODE_VOLUME_UP KEYCODE_VOLUME-DOWN	控制当前上下文音量（手机铃声、通话音量等）
方向键	KEYCODE_DPAD_CENTER KEYCODE_DPAD_UP KEYCODE_DPAD_DOWN KEYCODE_DPAD_LEFT KEYCODE_DPAD_RIGHT	有些设备中包含的方向键用于移动光标等

Android 中控件在处理物理按键事件时提供的回调方法有 onKeyUp()、onKeyDown、onKeyLongPress()。例如，重写 onKeyDown()方法来拦截用户按回退键的事件，代码如下：

```
public class ForbiddenBackActivity extends Activity {
    @Override
    protected void onCreate(Bundle savedInstanceState){
        super.onCreate(savedInstanceState);
        setContentView(R.layout.main);              //设置页面布局
    }
    @Override
    public boolean onKeyDown(int keyCode, KeyEvent event){
        if(keyCode==KeyEvent.KEYCODE_BACK){
            return true;                            //屏蔽回退键
        }
        return super.onKeyDown(keyCode, event);
    }
}
```

3.4.2　处理触摸事件

目前主流的手机和平板电脑都取消了外置键盘，提供了更大的屏幕。这些设备都需要通过触摸来操作，对于触摸屏上的按钮，可以使用 OnClickListener 和 OnLongClickListener 两个监听器分别处理用户短时间单击和长时间单击操作。

应用举例：编写 ButtonTouchEventActivity 类（继承自 Activity 类），为其增加 OnClickListener 事件监听器。

具体代码如下：

```
public class ButtonTouchEventActivity extends Activity {
    public void onCreate(Bundle savedInstanceState){
        super.onCreate(savedInstanceState);
        setContentView(R.layout.main);              //设置页面布局
```

```java
Button button= (Button)findViewById(R.id.button);   //获得按钮控件
button.setOnClickListener(new OnClickListener(){
    public void onClick(View v){            //处理用户短时间单击按钮事件
        Toast.makeText(ButtonTouchEventActivity.this, getText(R.string.short_click), Toast.LENGTH_SHORT).show();
    }
});
```

View 类是其他 Android 控件的父类。在该类中定义了 setOnTouchListener()方法用来为控件设置触摸事件监听器。

应用举例：编写 ScreenTouchEventActivity 类（继承自 Activity 类），实现 OnTouchListener 接口，增加触摸事件监听器，重写 onTouch()方法。

具体代码如下：

```java
public class ScreenTouchEventActivity extends Activity implements OnTouchListener {
    @Override
    protected void onCreate(Bundle savedInstanceState){
        super.onCreate(savedInstanceState);              //调用父类方法
        LinearLayout layout=new LinearLayout(this);      //定义线性布局
        layout.setOnTouchListener(this);                 //设置触摸事件监听器
        layout.setBackgroundResource(R.drawable.background);
                                                         //设置背景图片
        setContentView(layout);                          //使用布局
    }
    @Override
    public boolean onTouch(View v, MotionEvent event){
        Toast.makeText(this, "发生触摸事件", Toast.LENGTH_LONG).show();
        return true;
    }
}
```

3.5 综合实例一：游戏菜单及选项设置界面

3.5.1 功能描述

本例实现一款少儿益智类小游戏"开心跳级"的主菜单和选项设置界面，实现自定义图片背景，菜单按钮有图片交换效果。

3.5.2 关键技术

本例的关键是实现触摸按钮时，能根据触摸状态切换不同的资源图片。这里有两种方法：一种是用 event.getAction()方法获取用户按键状态，用 setBackgroundResource()

方法实现图片资源的设置。代码如下：

```
if(event.getAction()==MotionEvent.ACTION_DOWN)
    button[0].setBackgroundResource(R.drawable.button_bg_down);
else if(event.getAction()==MotionEvent.ACTION_UP)
    button[0].setBackgroundResource(R.drawable.button_bg);
```

另一种是直接建立图片切换资源文件。代码如下：

```
<item
    android:state_pressed="true"
    android:drawable="@drawable/button_bg_down">
</item>
```

3.5.3 实现过程

步骤1：新建项目 GameMenu，在 res/values 目录下的 strings.xml 文件中定义需要的字符串资源。代码如下：

```
<string name="app_name">开心跳级</string>
<string name="gameButton">开始游戏</string>
<string name="rank">排行榜</string>
<string name="options">选项设置</string>
<string name="moreApp">更 多 &#8230;</string>
<string name="exit">退 出</string>
```

步骤2：在 res/values 目录下的 colors.xml 文件中定义需要的颜色资源。代码如下：

```
<?xml version="1.0" encoding="utf-8"?>
<resources>
    <color name="verdigris">#527F76</color>
    <color name="white">#FFFFFF</color>
</resources>
```

步骤3：复制图片素材到 res 目录下的 drawable-mdpi 文件夹；修改 res/layout 目录下的 XML 布局文件，添加背景图片。代码如下：

```
<?xml version="1.0" encoding="utf-8"?>
<LinearLayout xmlns:android="http://schemas.android.com/apk/res/android"
    android:layout_width="fill_parent"
    android:layout_height="fill_parent"
    android:orientation="vertical"
    android:background="@drawable/bg_game">
</LinearLayout>
```

步骤4：在主活动文件的 onCreate 方法中添加隐藏标题栏、设置全屏及保持屏幕亮度属性。代码如下：

```java
protected void onCreate(Bundle savedInstanceState){
    super.onCreate(savedInstanceState);
    requestWindowFeature(Window.FEATURE_NO_TITLE);      //隐藏标题栏
    getWindow().setFlags(WindowManager.LayoutParams.FLAG_FULLSCREEN,
        WindowManager.LayoutParams.FLAG_FULLSCREEN);    //设置全屏
    getWindow().setFlags(WindowManager.LayoutParams.FLAG_KEEP_SCREEN_ON,
        WindowManager.LayoutParams.FLAG_KEEP_SCREEN_ON);  //保持屏幕亮度
    setContentView(R.layout.activity_main);
}
```

除了可以在代码中设置隐去应用的标题栏和全屏设置，还可以在项目的AndroidManifest.xml文件中对活动的属性进行配置来实现。

隐藏应用标题栏：

android:theme="@android:style/Theme.Black.NoTitleBar">

全屏设置：

android:theme="@android:style/Theme.Black.NoTitleBar.Fullscreen"

步骤5：在布局文件main_menu.xml中添加logo图片及菜单按钮。代码如下：

```xml
<?xml version="1.0" encoding="utf-8"?>
<LinearLayout xmlns:android="http://schemas.android.com/apk/res/android"
    android:layout_width="fill_parent"
    android:layout_height="fill_parent"
    android:background="@drawable/bg_game"
    android:orientation="vertical">
<!--菜单logo图标-->
    <ImageView
        android:layout_width="fill_parent"
        android:layout_height="50pt"
        android:layout_gravity="center"
        android:layout_marginBottom="15pt"
        android:layout_marginTop="8pt"
        android:src="@drawable/game_logo"/>
<!--"开始游戏"按钮-->
    <Button
        android:id="@+id/gameButton"
        android:layout_width="70pt"
        android:layout_height="20pt"
        android:layout_gravity="center"
        android:layout_marginBottom="5pt"
        android:background="@drawable/button_bg"
        android:text="@string/gameButton"
```

```xml
        android:textColor="@color/verdigris"/>
<!--"排行榜"按钮-->
    <Button
        android:id="@+id/rankButton"
        android:layout_width="70pt"
        android:layout_height="20pt"
        android:layout_gravity="center"
        android:layout_marginBottom="5pt"
        android:background="@drawable/button_bg"
        android:text="@string/rank"
        android:textColor="@color/verdigris"/>
<!--"选项设置"按钮-->
    <Button
        android:id="@+id/optionButton"
        android:layout_width="70pt"
        android:layout_height="20pt"
        android:layout_gravity="center"
        android:layout_marginBottom="5pt"
        android:background="@drawable/button_bg"
        android:text="@string/options"
        android:textColor="@color/verdigris"/>
<!--"更多…"按钮-->
    <Button
        android:id="@+id/moreButton"
        android:layout_width="70pt"
        android:layout_height="20pt"
        android:layout_gravity="center"
        android:layout_marginBottom="5pt"
        android:background="@drawable/button_bg"
        android:text="@string/moreApp"
        android:textColor="@color/verdigris"/>
<!--"退出"按钮-->
    <Button
        android:id="@+id/exitButton"
        android:layout_width="70pt"
        android:layout_height="20pt"
        android:layout_gravity="center"
        android:background="@drawable/button_bg"
        android:text="@string/exit"
        android:textColor="@color/verdigris"/>
</LinearLayout>
```

图 3-15 游戏菜单界面

执行效果如图 3-15 所示。

步骤 6：在主活动程序文件中实现 OnTouchListener 接口，加入 onTouch(View v,

MotionEvent event)方法,添加菜单图片切换功能。代码如下:

```java
public class MainActivity extends Activity implements OnTouchListener {
    private Button button[];                                      //声明按钮数据变量
    @Override
    protected void onCreate(Bundle savedInstanceState){
        super.onCreate(savedInstanceState);
        ......                          //此处省略隐藏标题栏、设置全屏和保持屏幕亮度的代码
        setContentView(R.layout.main_menu);
        button=new Button[5];                                     //实例化按钮数组
        button[0]=(Button)findViewById(R.id.gameButton);   //查找按钮组件
        button[1]=(Button)findViewById(R.id.rankButton);
        button[2]=(Button)findViewById(R.id.optionButton);
        button[3]=(Button)findViewById(R.id.moreButton);
        button[4]=(Button)findViewById(R.id.exitButton);
        for(int i=0; i<button.length; i++){
            button[i].setOnTouchListener(this);                   //设置按钮触屏监听
        }
    }
    @Override
    public boolean onTouch(View v, MotionEvent event){
        switch(v.getId()){                                        //获取组件 ID
        case R.id.gameButton:
            if(event.getAction()==MotionEvent.ACTION_DOWN)
                button[0].setBackgroundResource(R.drawable.button_bg_down);
            else if(event.getAction()==MotionEvent.ACTION_UP)
                button[0].setBackgroundResource(R.drawable.button_bg);
            break;
        case R.id.rankButton:
            if(event.getAction()==MotionEvent.ACTION_DOWN)
                button[1].setBackgroundResource(R.drawable.button_bg_down);
            else if(event.getAction()==MotionEvent.ACTION_UP)
                button[1].setBackgroundResource(R.drawable.button_bg);
            break;
        case R.id.optionButton:
            if(event.getAction()==MotionEvent.ACTION_DOWN)
                button[2].setBackgroundResource(R.drawable.button_bg_down);
            else if(event.getAction()==MotionEvent.ACTION_UP)
                button[2].setBackgroundResource(R.drawable.button_bg);
            break;
        case R.id.moreButton:
            if(event.getAction()==MotionEvent.ACTION_DOWN)
                button[3].setBackgroundResource(R.drawable.button_bg_down);
            else if(event.getAction()==MotionEvent.ACTION_UP)
```

```
            button[3].setBackgroundResource(R.drawable.button_bg);
            break;
        case R.id.exitButton:
            if(event.getAction()==MotionEvent.ACTION_DOWN)
            {
                button[4].setBackgroundResource(R.drawable.button_bg_down);
                finish();
            }
            break;
        }
        return false;
    }
}
```

执行效果如图 3-16 所示。

以上功能也可以采用另一种方法：在完成步骤 4 之后，只需配置一个选择器，由布局文件中的按钮组件调用即可。这种方法可以省略大量代码。在图片素材文件夹 drawable-mdpi 中新建 button.xml 文件并添加如下代码：

图 3-16　切换按钮图片效果

```xml
<?xml version="1.0" encoding="utf-8"?>
<selector xmlns:android="http://schemas.android.com/apk/res/android">
    <item android:drawable="@drawable/button_bg_down" android:state_pressed="true"/>
    <item android:drawable="@drawable/button_bg" android:state_focused="false"/>
</selector>
```

在布局文件 main_menu.xml 中可以由按钮组件直接调用。修改代码如下：

```xml
<Button
    …
    android:background="@drawable/button"
    …
/>
```

步骤 7：在 res/values 目录下的 strings.xml 文件中定义所需要的字符串资源。代码如下：

```xml
<string name="effect">播放震动效果</string>
<string name="sounds">开启背景音乐</string>
<string name="tips">显示提示信息</string>
<string name="power">敏捷度调节 (迟钝 -&gt; 敏捷)</string>
<string name="editINF">您在排行榜的昵称：</string>
<string name="initText">您的大名</string>
<string name="record">您的最高分数：</string>
```

```xml
<string name="ok">确定</string>
<string name="post">提交成绩</string>
<string name="tipButton">提示</string>
```

步骤 8：新建一个活动文件（OptionActivity.java），完成 XML 布局文件（FrameLayout）options.xml。代码如下：

```xml
<?xml version="1.0" encoding="utf-8"?>
<FrameLayout xmlns:android="http://schemas.android.com/apk/res/android"
    android:layout_width="fill_parent"
    android:layout_height="fill_parent"
    android:background="@drawable/bg_options"
    android:orientation="vertical">
<!--滚动视图 -->
    <ScrollView
        android:layout_width="fill_parent"
        android:layout_height="fill_parent"
        android:paddingTop="25pt">
        <LinearLayout
            android:layout_width="fill_parent"
            android:layout_height="wrap_content"
            android:orientation="vertical"
            android:paddingLeft="8pt"
            android:paddingRight="6pt">
            <CheckBox
                android:id="@+id/effect"
                android:layout_width="wrap_content"
                android:layout_height="wrap_content"
                android:shadowColor="#FFFFFF"
                android:text="@string/effect"
                android:textColor="#FFFFFF"/>
            <CheckBox
                android:id="@+id/sounds"
                android:layout_width="wrap_content"
                android:layout_height="wrap_content"
                android:text="@string/sounds"
                android:textColor="#FFFFFFFF"/>
            <CheckBox
                android:id="@+id/showTips"
                android:layout_width="wrap_content"
                android:layout_height="wrap_content"
                android:layout_marginBottom="4pt"
                android:text="@string/tips"
                android:textColor="#FFFFFFFF"/>
            <TextView
                android:layout_width="wrap_content"
```

```xml
            android:layout_height="wrap_content"
            android:text="@string/power"
            android:textColor="#FFFFFFFF"
            android:textSize="7pt"/>
        <SeekBar
            android:id="@+id/seekBar"
            android:layout_width="fill_parent"
            android:layout_height="wrap_content"
            android:layout_marginBottom="4pt"
            android:max="100"
            android:progress="0"/>
        <TextView
            android:layout_width="wrap_content"
            android:layout_height="wrap_content"
            android:text="@string/editINF"
            android:textColor="#FFFFFFFF"
            android:textSize="7pt"/>
        <EditText
            android:id="@+id/editName"
            android:layout_width="fill_parent"
            android:layout_height="wrap_content"
            android:layout_marginBottom="4pt"
            android:hint="@string/initText"
            android:textColor="@color/white"/>
        <TextView
            android:id="@+id/textViewRecord"
            android:layout_width="wrap_content"
            android:layout_height="wrap_content"
            android:layout_marginBottom="4pt"
            android:text="@string/record"
            android:textColor="#FFFFFFFF"
            android:textSize="7pt"/>
    <!--表格行布局-->
        <TableRow
            android:layout_width="fill_parent"
            android:layout_height="wrap_content">
            <Button
                android:id="@+id/okButton"
                android:layout_width="wrap_content"
                android:layout_height="wrap_content"
                android:layout_weight="1"
                android:text="@string/ok"
                android:textColor="@color/white"
                android:textSize="7pt"/>
            <Button
                android:id="@+id/postButton"
```

```xml
                    android:layout_width="wrap_content"
                    android:layout_height="wrap_content"
                    android:layout_weight="2"
                    android:text="@string/post"
                    android:textColor="@color/white"
                    android:textSize="7pt"/>
                <Button
                    android:id="@+id/tipButton"
                    android:layout_width="wrap_content"
                    android:layout_height="wrap_content"
                    android:layout_weight="1"
                    android:text="@string/tipButton"
                    android:textColor="@color/white"
                    android:textSize="7pt"/>
            </TableRow>
        </LinearLayout>
    </ScrollView>
</FrameLayout>
```

步骤9：在OptionActivity.java文件中获取相关组件，实现交互功能。代码如下：

```java
public class OptionActivity extends Activity implements OnClickListener {
    private CheckBox effectCheck, soundCheck, tipCheck;
    private SeekBar seekBar;
    private EditText editText;
    private TextView bestScore;
    private Button button;
    @Override
    public void onCreate(Bundle savedInstanceState){
        super.onCreate(savedInstanceState);
        ……                    //此处省略隐藏标题栏、设置全屏和保持屏幕亮度的代码
        setContentView(R.layout.options);
        effectCheck= (CheckBox)findViewById(R.id.effect);
        soundCheck= (CheckBox)findViewById(R.id.sounds);
        tipCheck= (CheckBox)findViewById(R.id.showTips);
        seekBar= (SeekBar)findViewById(R.id.seekBar);
        editText= (EditText)findViewById(R.id.editName);
        bestScore= (TextView)findViewById(R.id.textViewRecord);
        button= (Button)findViewById(R.id.okButton);
        button.setOnClickListener(this);
        bestScore.setText("Your  Best Score : "+0);
    }
    @Override
    public void onClick(View v){
        switch(v.getId()){
        case R.id.okButton:
```

```
            if(effectCheck.isChecked())
                System.out.println(effectCheck.getText()+"已被选中");
            if(soundCheck.isChecked())
                System.out.println(soundCheck.getText()+"已被选中");
            if(tipCheck.isChecked())
                System.out.println(tipCheck.getText()+"已被选中");
            float progress=seekBar.getProgress();
            System.out.println("设置的灵敏度为 :"+progress/seekBar.getMax()
            * 100+"%");
            String name=editText.getText().toString();
            System.out.println("用户名为 : "+name);
            break;
        default:
            break;
        }
    }
}
```

步骤 10：在 MainActivity.java 中添加页面跳转。代码如下：

```
button[2].setOnClickListener(new OnClickListener(){
    @Override
    public void onClick(View v){
        Intent intent=new Intent(MainActivity.this, OptionActivity.class);
        startActivity(intent);
    }
});
```

执行效果如图 3-17 所示。用同样方法完成帮助页面，执行效果如图 3-18 所示。

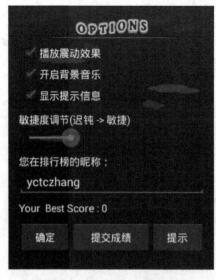

图 3-17　游戏选项设置界面

图 3-18　游戏帮助界面

单击"确定"按钮,显示输出信息如图 3-19 所示。

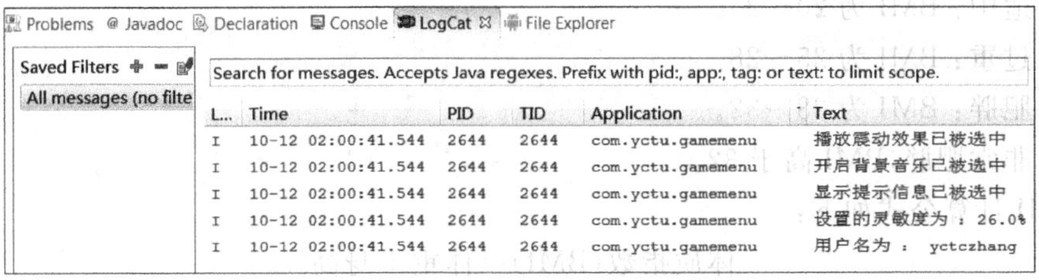

图 3-19 游戏选项设置输出信息

3.6 综合实例二: BMI 计算器

3.6.1 功能描述

本例实现一个成年人体质指数(BMI)计算器。根据用户输入的身高和体重数据,算出 BMI 值,并显示相关提示及建议信息。

3.6.2 关键技术

本例实现的关键是 BMI 的计算和格式化输出。这里采用的方法是用 DecimalFormat 类实例化之后的 format()方法实现数值的格式化。代码如下:

```
DecimalFormat df=new DecimalFormat("0.00");
double height=Double.parseDouble(txtHeight.getText().toString())/100;
double weight=Double.parseDouble(txtWeight.getText().toString());
double BMI=weight/(height * height);
labResult.setText("您的 BMI 健康指数为:"+df.format(BMI));
if(BMI>25){
    labSuggest.setText("您应该节食了!");
}else if(BMI<20){
    labSuggest.setText("您应该多吃点了!");
}else{
    labSuggest.setText("继续保持呀!");
}
```

3.6.3 准备知识

BMI 是 Body Mass Index 的缩写,中文是"体质指数"的意思,是以身高和体重计算出来的,原来的设计是一个用于公众健康研究的统计工具。成年人的 BMI 数值的体质结果如下所示:

- 过轻:BMI 低于 18.5。

- 正常：BMI 为 18.5～24.99。
- 适中：BMI 为 20～25。
- 过重：BMI 为 25～28。
- 肥胖：BMI 为 28～32。
- 非常肥胖，BMI 高于 32。

BMI 计算公式如下：

$$体质指数(BMI)=体重÷身高^2$$

其中，体重的单位是 kg，身高的单位是 m。

3.6.4 实现过程

步骤 1：新建项目 MyBMI，在 res/values 目录下的 strings.xml 文件中定义字符串资源。代码如下：

```
<string name="app_name">成年人 BMI 健康指数计算器</string>
<string name="height">身高 (cm)</string>
<string name="weight">体重 (kg)</string>
<string name="height_hint">请输入身高</string>
<string name="btn_cal">计算 BMI 值</string>
<string name="weight_hint">请输入体重</string>
<string name="menu_settings">设置</string>
```

步骤 2：完成 XML 布局文件 activity_main.xml。代码如下：

```
<LinearLayout xmlns:android="http://schemas.android.com/apk/res/android"
    xmlns:tools="http://schemas.android.com/tools"
    android:orientation="vertical"
    android:layout_width="fill_parent"
    android:layout_height="fill_parent"
    tools:context=".MainActivity">
    <TextView
        android:layout_width="fill_parent"
        android:layout_height="wrap_content"
        android:text="@string/height"/>
    <EditText
        android:id="@+id/height"
        android:layout_width="fill_parent"
        android:layout_height="wrap_content"
        android:inputType="number"
        android:hint="@string/height_hint"
        android:text=""/>
    <TextView
        android:layout_width="fill_parent"
        android:layout_height="wrap_content"
```

```xml
        android:text="@string/weight"/>
    <EditText
        android:id="@+id/weight"
        android:layout_width="fill_parent"
        android:layout_height="wrap_content"
        android:inputType="number"
        android:hint="@string/weight_hint"
        android:text=""/>
    <Button
        android:id="@+id/btn_cal"
        android:layout_width="fill_parent"
        android:layout_height="wrap_content"
        android:text="@string/btn_cal"/>
    <TextView
        android:id="@+id/result"
        android:layout_width="fill_parent"
        android:layout_height="wrap_content"
        android:text=""/>
    <TextView
        android:id="@+id/suggest"
        android:layout_width="fill_parent"
        android:layout_height="wrap_content"
        android:text=""/>
</LinearLayout>
```

运行效果如图 3-20 所示。

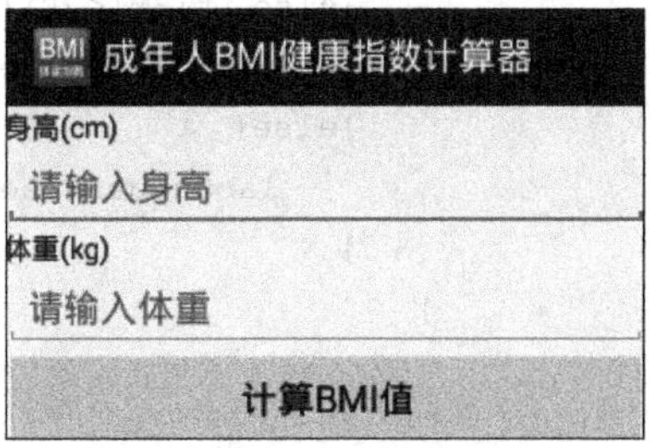

图 3-20 成年人 BMI 计算器布局

★**注意**：Android 1.5 版本之后添加了软件虚拟键盘的功能，在需要输入数据时，不必一定要打开键盘（如果有实体键盘）。直接用 android：inputType 参数来替换 android：password、android：singleLine、android：numeric、android：phoneNumber、android：capitalize、android：autoText、android：editable 这些属性。除此之外，Android 1.5 以后的版本还提供了更多的参数类型。

步骤 3：在 MainActivity.java 文件中获取相关组件并实现计算功能。代码如下：

```java
public class MainActivity extends Activity {
    private EditText txtHeight, txtWeight;
    private Button btn_cal;
    private TextView labResult, labSuggest;
    @Override
    protected void onCreate(Bundle savedInstanceState){
        super.onCreate(savedInstanceState);
        getWindow().setFlags(WindowManager.LayoutParams.FLAG_FULLSCREEN,
            WindowManager.LayoutParams.FLAG_FULLSCREEN);
        setContentView(R.layout.activity_main);
        txtHeight=(EditText)findViewById(R.id.height);
```

```java
        txtWeight=(EditText)findViewById(R.id.weight);
        labResult=(TextView)findViewById(R.id.result);
        labSuggest=(TextView)findViewById(R.id.suggest);
        btn_cal=(Button)findViewById(R.id.btn_cal);
        btn_cal.setOnClickListener(calBMI);
    }
    private OnClickListener calBMI=new OnClickListener(){
        @Override
        public void onClick(View v){
            DecimalFormat df=new DecimalFormat("0.00");
            double height = Double.parseDouble(txtHeight.getText().toString())/100;
            double weight=Double.parseDouble(txtWeight.getText().toString());
            double BMI=weight/(height * height);
            labResult.setText("您的 BMI 健康指数为:"+df.format(BMI));
            if(BMI>25){
                labSuggest.setText("您应该节食了!");
            }else if(BMI<20){
                labSuggest.setText("您应该多吃点了!");
            }else{
                labSuggest.setText("继续保持呀!");
            }
        }
    };
}
```

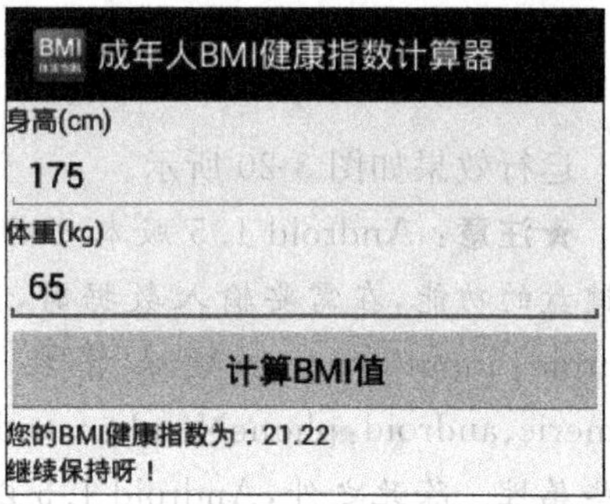

运行效果如图 3-21 所示。

步骤 4：在 strings.xml 字符串资源文件中添加字符串资源定义。代码如下：

```xml
<string name="f_result">您的 BMI 健康指数为:</string>
<string name="advice_heavy">您应该节食了!</string>
<string name="advice_light">您应该多吃点了!</string>
<string name="advice_good">继续保持呀!</string>
```

图 3-21 成年人 BMI 计算实例运行界面

步骤 5：修改 MainActivity.java 文件中显示计算结果。代码如下：

```java
labResult.setText(getText(R.string.f_result)+df.format(BMI));
    if(BMI>25){
        labSuggest.setText(R.string.advice_heavy);
    }else if(BMI<20){
        labSuggest.setText(R.string.advice_light);
    }else{
        labSuggest.setText(R.string.advice_good);
```

★**注意**：显示结果文本控件不能直接用 setText(R.string.f_result + df.format(BMI))，因为在参数有运算的情况下，R.string.f_result 就会默认为字符串在 R 文件中的地址。

3.6.5 实例扩展

1. 重构程序

Android 平台的开发者按照 MVC 模式，将显示界面所用的 XML 文件、显示资源所用的 XML 文件从程序代码中分隔开来。以上完成的项目，如果再增加几个功能按钮或菜单等内容，程序代码就会变得复杂，变得不容易阅读，也开始变得不容易维护。"将程序代码做变动以提高可读性或简化程序结构，而不影响输出结果"的过程，叫做"重构"。下面就来重构这个 BMI 应用程序。

把查找组件和为特定界面组件添加控制流程的两段程序代码分开整理成两个函数，将程序逻辑与界面组件的声明分离开来，以实现重构的目的。重构后的程序代码如下：

```
public class MainActivity extends Activity {
    ...                                             //此处省略变量定义代码
    @Override
    protected void onCreate(Bundle savedInstanceState){
        super.onCreate(savedInstanceState);
        setContentView(R.layout.activity_main);
        findviews();
        setListensers();
    }
    private void findviews(){
        ...                                         //此处省略部分组件获取代码
        btn_cal=(Button)findViewById(R.id.btn_cal);
    }
    private void setListensers(){
        btn_cal.setOnClickListener(calBMI);
    }
    private Button.OnClickListener calBMI=new Button.OnClickListener(){
        @Override
        public void onClick(View v){
            ...                                     //此处省略计算代码
        }
    };
}
```

以上代码中，将 calBMI 函数从原本声明成 OnClickListener 改为声明成 Button.OnClickListener。在 import Package 部分，由原来的 import android.view.View.

OnClickListener 变为 import android.widget.Button，在查阅程序时，整个 Button 界面组件和 OnClickListener 界面组件方法之间的关系变得更加清晰。

2. 添加菜单

为项目添加"关于"和"退出"两个功能菜单。在 Android SDK 21.0.1 及以上版本中，新建项目时，如果选择 Create activity 选项，则程序代码中自动覆写 onCreateOptionsMenu()和 onOptionsItemSelected()方法，并且直接调用在资源文件夹中生成的 menu 文件夹和 XML 布局文件。代码如下所示：

```java
@Override
public boolean onCreateOptionsMenu(Menu menu){
    getMenuInflater().inflate(R.menu.main, menu);
    return true;
}
@Override
public boolean onOptionsItemSelected(MenuItem item){
    int id=item.getItemId();
    if(id==R.id.action_settings){
        return true;
    }
    return super.onOptionsItemSelected(item);
}
```

在 res/menu 目录下的 main.xml 菜单布局文件代码如下：

```xml
<menu xmlns:android="http://schemas.android.com/apk/res/android"
    xmlns:tools="http://schemas.android.com/tools"
    tools:context="com.yctu.mybmi.MainActivity">
    <item
        android:id="@+id/action_settings"
        android:orderInCategory="100"
        android:showAsAction="never"
        android:title="@string/menu_settings"/>
</menu>
```

具体步骤如下。

步骤 1：在 strings.xml 字符串资源文件中添加相应字符串，其中字符"…"由占位符号"…"代替。代码如下：

```xml
<string name="menu_about">关于 &#8230;</string>
<string name="menu_quit">退出</string>
```

步骤 2：修改 XML 菜单资源文件。代码如下：

```xml
<menu xmlns:android="http://schemas.android.com/apk/res/android"
    xmlns:tools="http://schemas.android.com/tools"
```

```
            tools:context="com.yctu.mybmi.MainActivity">
    <item
        android:id="@+id/menu_about"
        android:icon="@android:drawable/ic_menu_help"
        android:orderInCategory="100"
        android:title="@string/menu_about"/>
    <item
        android:id="@+id/menu_quit"
        android:icon="@android:drawable/ic_menu_close_clear_cancel"
        android:orderInCategory="100"
        android:title="@string/menu_quit"/>
</menu>
```

以上代码中,android:icon="@android:drawable/ic_menu_help"是使用 Android 内置的图标素材,也可以用@drawable/方式引用自定义图标素材。Android 中部分内置图标如图 3-22 所示。

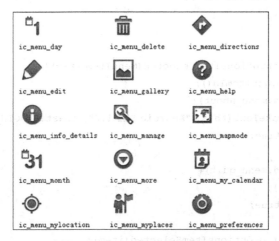

图 3-22 Android 部分内置图标

在 Android 4.0 系统中,创建菜单 Menu,通过 setIcon 方法给菜单添加图标是无效的,图标不能显示。原因是在 Android 4.0 及以上版本系统中,涉及菜单的源码类 MenuBuilder 做了改变,该类的部分源码如下:

```java
public class MenuBuilder implements Menu {
    ...
    private boolean mOptionalIconsVisible=false;
    ...
    void setOptionalIconsVisible(boolean visible){
        mOptionalIconsVisible=visible;
    }
    boolean getOptionalIconsVisible(){
        return mOptionalIconsVisible;
```

```
            }
            ...
        }
```

上面的代码中，mOptionalIconsVisible 成员初始值默认为 false，这就是为什么给菜单设置图标没有效果的原因。所以，只要在创建菜单时通过调用 setOptionalIconsVisible 方法设置 mOptionalIconsVisible 为 true 就可以了。但要想调用该方法，就需要创建 MenuBuilder 对象，而这里的问题是无法在开发的应用程序中创建 MenuBuilder 这个对象（MenuBuilder 为系统内部的框架类）。

因此要用反射机制，在创建菜单时反射调用 setOptionalIconsVisible 方法设置 mOptionalIconsVisible 为 true，就可以在菜单中显示添加的图标。代码如下：

```java
@Override
public boolean onCreateOptionsMenu(Menu menu){
    getMenuInflater().inflate(R.menu.main, menu);
    setIconEnable(menu, true);
    return true;
}
@Override
public boolean onOptionsItemSelected(MenuItem item){
    int id=item.getItemId();
    if(id==R.id.menu_about){
        Toast.makeText(this, "Android BMI 1.0", Toast.LENGTH_LONG).show();
        return true;
    }
    if(id==R.id.menu_quit){
        finish();
        return true;
    }
    return super.onOptionsItemSelected(item);
}
private void setIconEnable(Menu menu, boolean enable){
    try
    {
        Class<?> clazz = Class.forName("com.android.internal.view.menu.MenuBuilder");
        Method m=clazz.getDeclaredMethod("setOptionalIconsVisible", boolean.class);
        m.setAccessible(true);
        m.invoke(menu, enable);
    } catch(Exception e)
    {
        e.printStackTrace();
    }
}
```

MenuBuilder 实现 Menu 接口,创建菜单时,传进来的 menu 其实就是 MenuBuilder 对象(Java 语言的多态特征)。运行并按 Menu 功能键的效果如图 3-23 所示。

★**注意**:在 Android SDK 21.0.0 以前的版本中,并不会自动覆写 onCreateOptionsMenu 方法和生成布局资源文件。

图 3-23 图标菜单

3. 异常处理

为处理用户没有输入数据或程序中出现未知错误的情况,增强程序的适应性,用 try…catch 语句做错误处理。语法如下:

```
try {
    …                                    //功能代码
} catch(Exception e){
    //TODO: handle exception
}
```

步骤 1:在字符串资源文件中添加字符串资源定义。代码如下:

```
<string name="input_error">确认数据格式是否正确!</string>
```

步骤 2:修改计算按钮事件,结果信息采取调用字符串资源的方式。代码如下:

```
private Button.OnClickListener calBMI=new Button.OnClickListener(){
    @Override
    public void onClick(View v){
        DecimalFormat df=new DecimalFormat("0.00");
        try {
            double height = Double.parseDouble (txtHeight.getText ().toString
            ())/100;
            double weight=Double.parseDouble(txtWeight.getText().toString());
            double BMI=weight/(height * height);
            labResult.setText(getText(R.string.f_result)+df.format(BMI));
            if(BMI>25){
                labSuggest.setText(R.string.advice_heavy);
            } else if(BMI<20){
                labSuggest.setText(R.string.advice_light);
            } else {
                labSuggest.setText(R.string.advice_good);
            }
        } catch(Exception e){
            Toast.makeText (MainActivity.this, R.string.input_error, Toast.
            LENGTH_SHORT).show();
```

            }
        };

★注意：Toast 组件是在内部类中引用的，因此上下文用"外部类.this"方式。另外，也可以在程序中用判断语句来进行数据校检。

4. 使用体重数据

成年人身高的变化程度比起体重的变化程度小得多，如果在用户第一次输入身高和体重值后，程序能记住上次输入过的身高数据，就以可以减少用户重复输入的麻烦。

具体步骤如下：

步骤 1：定义所需字符串常量。代码如下：

```
private static final String PREF="BMI_PREF";
private static final String PREF_HEIGHT="BMI_HEIGHT";
```

步骤 2：覆写 onPause()方法，存储身高数据；编写 restorePrefs()方法用于获取存储在 SharedPreferences 中的身高数据，并在 onCreate()方法中调用 restorePrefs()。相关代码如下：

```
private void restorePrefs(){
    SharedPreferences settings= getSharedPreferences (PREF, Context.MODE_PRIVATE);
    String pref_height=settings.getString(PREF_HEIGHT, "");
    if(!"".equals(pref_height)){
        txtHeight.setText(pref_height);
        txtWeight.requestFocus();
    }
}
@Override
protected void onPause(){
    SharedPreferences settings = getSharedPreferences (PREF, Context.MODE_PRIVATE);
    settings.edit().putString(PREF_HEIGHT, txtHeight.getText().toString()).commit();
    super.onPause();
}
```

Java 语言中，用字符串类的 equals()方法来比较字符串可以有效避免空指针异常，用 requestFocus()方法设置焦点输入框。

★提示：关于 SharedPreferences 的详细用法见 6.1 节。

3.7 综合实例三：猜猜看

3.7.1 功能描述

本例实现一款简单的猜猜看小游戏。触屏单击 3 只玻璃瓶中的任意一只，显示玻璃瓶里面是否有卡通人物,并且将没有被单击的玻璃瓶设置为半透明效果,被单击的正常显示,同时根据结果显示提示文字。

3.7.2 关键技术

在实现本例时,关键是如何变换玻璃瓶图片。这里通过一个 for 循环和 Math 类的 random()方法来实现。代码如下:

```
public void gameReset(){
    for(int i=0; i<img.length; i++){
        int temp=img[i];                    //将数组元素 i 保存到临时变量中
        int index= (int)(Math.random() * 2);   //生成一个随机数
        img[i]=img[index];                  //将随机指定的数组元素内容赋值给数组元素 i
        img[index]=temp;                    //将临时变量的值赋给随机数组指定的数组元素
    }
}
```

3.7.3 实现过程

步骤 1:新建项目 ProjectGuessGame。准备游戏中所需要的图片资源,这里包括背景图片、Logo 图标、默认显示的玻璃瓶、有卡通人物的玻璃瓶和没有卡通人物的玻璃瓶 5 张图片(如图 3-24 所示)。并把图片资源复制到项目根目录下"res/drawable-mdpi"文件夹中。

图 3-24 "猜猜看"游戏的图片资源

步骤 2:在 res/values/string.xml 文件中定义字符串资源。代码如下:

```xml
<?xml version="1.0" encoding="utf-8"?>
<resources>
    <string name="app_name">猜猜看</string>
    <string name="game_info">猜猜哪个瓶子里有卡通人物?</string>
    <string name="game_again">再猜一次</string>
</resources>
```

步骤 3:修改 res/layout 目录下的 XML 布局文件。代码如下:

```xml
<LinearLayout xmlns:android="http://schemas.android.com/apk/res/android"
    xmlns:tools="http://schemas.android.com/tools"
    android:layout_width="match_parent"
    android:layout_height="match_parent"
```

```xml
    android:orientation="vertical">
    <!--第一行，放置文本框 -->
    <TextView
        android:id="@+id/game_info"
        android:layout_width="wrap_content"
        android:layout_height="wrap_content"
        android:layout_marginLeft="15dp"
        android:text="@string/game_info"/>
    <!--第二行，内嵌线性布局，包含3张图片 -->
    <LinearLayout
        android:layout_width="match_parent"
        android:layout_height="0dp"
        android:layout_weight="2"
        android:gravity="center">
        <ImageView
            android:id="@+id/img0"
            android:layout_width="wrap_content"
            android:layout_height="wrap_content"
            android:contentDescription="@string/app_name"
            android:src="@drawable/glass_default"/>
        <ImageView
            android:id="@+id/img1"
            android:layout_width="wrap_content"
            android:layout_height="wrap_content"
            android:contentDescription="@string/app_name"
            android:src="@drawable/glass_default"/>
        <ImageView
            android:id="@+id/img2"
            android:layout_width="wrap_content"
            android:layout_height="wrap_content"
            android:contentDescription="@string/app_name"
            android:src="@drawable/glass_default"/>
    </LinearLayout>
    <!--第三行，放置按钮 -->
    <Button
        android:id="@+id/game_again"
        android:layout_width="wrap_content"
        android:layout_height="wrap_content"
        android:layout_gravity="center"
        android:text="@string/game_again"/>
</LinearLayout>
```

整体采用线性布局方式。第一行添加了一个文本框组件，用于显示提示文字。第二行是一个水平线性布局，其中添加了3个ImageView组件，用于显示3只玻璃瓶的图片。

第三行是一个按钮组件,用于初始化游戏(再来一次)。

步骤4:在主活动文件中实现游戏规则。代码如下:

```java
private ImageView img0, img1, img2;
private TextView game_info;
private Button btn_again;
//定义保存图片数组
private int img[]=new int[]{ R.drawable.glass_story, R.drawable.glass_empty,
R.drawable.glass_empty };
@Override
protected void onCreate(Bundle savedInstanceState){
    super.onCreate(savedInstanceState);
    setContentView(R.layout.activity_main);
    game_info=(TextView)findViewById(R.id.game_info);
    img0=(ImageView)findViewById(R.id.img0);
    img1=(ImageView)findViewById(R.id.img1);
    img2=(ImageView)findViewById(R.id.img2);
    gameReset();
    img0.setOnClickListener(new View.OnClickListener(){
        @Override
        public void onClick(View v){
            isRight(v, 0);
        }
    });
    img1.setOnClickListener(new View.OnClickListener(){
        @Override
        public void onClick(View v){
            isRight(v, 1);
        }
    });
    img2.setOnClickListener(new View.OnClickListener(){
        @Override
        public void onClick(View v){
            isRight(v, 2);
        }
    });
    btn_again=(Button)findViewById(R.id.game_again);
    //为"再猜一次"按钮添加事件监听器
    btn_again.setOnClickListener(new View.OnClickListener(){
        @Override
        public void onClick(View v){
            gameReset();
            game_info.setText(R.string.game_info);
            img0.setImageAlpha(255);
```

```
            img1.setImageAlpha(255);
            img2.setImageAlpha(255);
            img0.setImageDrawable(getResources().getDrawable(R.drawable.glass
            _default));
            img1.setImageDrawable(getResources().getDrawable(R.drawable.glass
            _default));
            img2.setImageDrawable(getResources().getDrawable(R.drawable.glass
            _default));
        }
    });
}
//判断结果函数
public void isRight(View v, int index){
    img0.setImageAlpha(100);
    img1.setImageAlpha(100);
    img2.setImageAlpha(100);
    ImageView imgv=(ImageView)v;         //获取被单击的组件
    imgv.setImageDrawable(getResources().getDrawable(img[index]));
    imgv.setImageAlpha(255);
    if(img[index]==R.drawable.glass_story){
        game_info.setText("恭喜您,答对了!!");
    }else{
        game_info.setText("很抱歉,猜错了!!");
    }
}
//随机摆放玻璃瓶函数
public void gameReset(){
    for(int i=0; i<img.length; i++){
        int temp=img[i];
        int index=(int)(Math.random() * 2);
        img[i]=img[index];
        img[index]=temp;
    }
}
```

运行效果如图 3-25 所示。

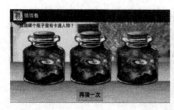

图 3-25 "猜猜看"游戏实例运行界面

3.8 本章小结

本章主要学习了 Android 游戏开发的 5 种布局方式和常用组件的使用方法。重点掌握页面切换和传递参数的 3 种方法，掌握 Android 事件处理机制，无论是应用系统开发还是游戏开发，这些知识都会经常用到。最后，通过 3 个综合实例贯穿本章的知识。其中涉及了 Java 程序设计基础、面向对象编程思想及软件架构等相关知识。

3.9 思考与练习

（1）尝试开发一个游戏注册界面和登录界面，注册后输入设置的用户名和密码可以跳转到登录界面，登录成功后有信息提示和用户名的界面传值。

（2）尝试再次拓展本章"BMI 计算器"实例，将目标用户改成美国人、英国人（使用英制计算身高（英尺、英寸）和体重（磅）的用户），由原本用一个文字编辑框输入身高改成两个文字编辑框（分别用于输入英尺、英寸）。

英制转换成公制的单位换算如下：

$$1ft = 12in$$
$$1in = 2.54cm$$
$$1lb = 0.45359kg$$

第 4 章

Android 游戏开发之图形界面

学习目标:
- 掌握消息类 Message 的应用。
- 掌握消息处理类 Handler 的应用。
- 掌握子线程中更新 UI 的方法。
- 掌握 View 和 SurfaceView 视图框架。
- 掌握 Canvas(画布)和 Paint(画笔)的应用。
- 掌握位图操作方法。
- 掌握剪切区域的运用。
- 掌握图像特效设置。
- 掌握动画。

本章导读:

本章内容是全书的重中之重,是 Android 游戏开发的核心知识,主要讲解 Android 中线程与消息处理机制、游戏开发的视图框架、绘画、位图操作、显示剪切区域、图像特效以及游戏动画等。主要用于处理游戏中的界面绘制与刷新、逻辑处理、地图生成、图像矩阵变换及状态机制等。

4.1 线程与消息处理

在 Android 游戏开发中,对于一些比较耗时的操作,通常会为其开辟一个单独的线程来执行,这样可以减少用户的等待时间。默认情况下,所有操作都在主线程中进行,这个线程负责管理与 UI 相关的事件,而在自己创建的子线程中,又不能对 UI 组件进行操作。因此,Android 提供了消息处理传递机制以解决这一问题。可以通过 Thread 类的构造方法创建线程对象,也可以采用实现 Runnable 接口的方式。无论用哪一种方式,都要重写 run 方法。

4.1.1 循环者类 Looper

Android 使用消息机制实现线程间的通信,线程通过 Looper 建立自己的消息循环,

MessageQueue(消息队列)是 FIFO 的消息队列,Looper 负责从 MessageQueue 中取出消息,并且分发到消息指定目标 Handler 对象。Handler 对象绑定到线程的局部变量 Looper,封装了发送消息和处理消息的接口。

默认情况下,Android 中新创建的线程是没有开启消息循环的。但主线程除外,系统会自动为主线程创建 Looper 对象,开启消息循环。所以在主线程中用 Handler myHandler=new Handler()创建 Handler 对象时不会出错,而在新创建的非主线程中应用就会产生异常。如果想要在非主线程中创建 Handler 对象,首先需要使用 Looper 类的 prepare()方法来初始化一个 Looper 对象,然后创建这个 Handler 对象,再使用 Looper 对象发送并处理消息。

4.1.2 Handler 消息传递机制

消息处理类(Handler)允许发送和处理 Message 或 Rannable 对象到其所在线程的消息队列中,主要有以下两个作用。

(1) 将 Message 或 Runnable 应用 post()方法或 sendMessage()方法发送到 MessageQueue 中,在发送时可以指定延迟时间、发送时间或者要携带的 Bundle 数据。当 MessageQueue 循环到该 Message 时,调用相应的 Handler 对象的 HandlerMessage()方法对其进行处理。

(2) 在子线程中与主线程进行通信,也就是在工作线程中与 UI 线程进行通信。

Handler 类提供的常用方法如下。

- void handleMessage(Message msg):处理消息方法,该方法通常用于被重写。
- final boolean hasMessages(int what):检查消息队列中是否包含 what 属性为指定值的消息。
- final boolean hasMessage(int what, Object object):检查消息队列中是否包含 what 属性为指定值且 Object 属性为指定对象的消息。
- Message obtainMessage():获取消息。
- sendEmptyMessage(int what):发送空消息。
- final boolean sendEmptyMessageDelayed(int what, long delayMillis):指定多少毫秒后发送空消息。
- final boolean sendMessage(Message msg):立即发送消息。
- final boolean sendMessageDelayed(Message msg, long delayMillis):指定多少毫秒之后发送消息。

★注意:在一个线程中,只能有一个 Looper 和 MessageQueue,但可以有多个 Handler,而且这些 Handler 可以共享一个 Looper 和 MessageQueue。

4.1.3 消息类 Message

消息类(Message)被存放在消息队列中,一个消息队列中可以包含多个 Message 对象,每个 Message 对象可以通过 Message.obtain()或 Handler.obtainMessage()方法获得。Message 类的属性如表 4-1 所示。

表 4-1　Message 类的属性

属　　性	类　　型	描　　述
arg1	int	存放整型数据
arg2	int	存放整型数据
obj	Object	存放发送给接收器的 Object 类型的对象
replyTo	Messenger	指定此 Message 发送到何处的可选 Message 对象
what	int	指定用户自定义的消息代码,使接收者可以了解这个消息的信息

在使用 Message 类时,需注意以下 3 点:

(1)尽管 Message 有 public 的默认构造方法,一般需要使用 Message.obtain()或 Handler.obtainMessage()方法从消息池中获得空消息对象,以节省资源。

(2)如果一个 Message 只需要携带简单的 int 类型信息,应优先使用 Message.arg1 和 Message.arg2 属性来传递信息,这比用 Bundle 更节省内存。

(3)尽可能使用 Message.what 来标识信息,以便用不同方式处理 Message。

4.1.4　基础实例:快乐舞者

本例实现一个不停跳舞的卡通人物,不间断地变换各种动作。

具体实现步骤如下。

步骤 1:新建项目 ProjectDanceBoy,准备游戏中所需要的图片资源,这里包括背景图片和人物序列图(如图 4-1 所示)。把图片资源复制到项目根目录下 res/drawable-mdpi 文件夹中。

　　bg.png　　　boy0.png　　boy1.png　　boy2.png　　boy3.png　　boy4.png

图 4-1　卡通人物图片素材

步骤 2:修改 res/layout 目录中的 XML 布局文件,添加背景和一个 ImageView 组件。代码如下:

```
<?xml version="1.0" encoding="utf-8"?>
<LinearLayout xmlns:android="http://schemas.android.com/apk/res/android"
    android:orientation="vertical"
    android:layout_width="fill_parent"
    android:layout_height="fill_parent"
    android:gravity="center"
    android:background="@drawable/bg">
    <ImageView
        android:id="@+id/myImageView"
```

```
        android:layout_width="fill_parent"
        android:layout_height="wrap_content"
        android:src="@drawable/boy0"
        android:gravity="center"/>
</LinearLayout>
```

步骤 3：在主活动文件中定义需要的变量。代码如下：

```
private ImageView myImageView;
private Handler myHandler;
//定义线程运行标识
private Boolean isDancing=true;
```

步骤 4：在主活动文件的 onCreate() 方法中获取组件，实现 Handler 类。代码如下：

```
setRequestedOrientation(ActivityInfo.SCREEN_ORIENTATION_LANDSCAPE);
                                       //设置屏幕为横向
myImageView= (ImageView)findViewById(R.id.myImageView);
myHandler=new Handler(){              //实例化 Handler 对象
    @Override
    public void handleMessage(Message msg){
        super.handleMessage(msg);
        switch(msg.what){
            case 0:
              myImageView.setImageResource(R.drawable.boy0);
              break;
            case 1:
              myImageView.setImageResource(R.drawable.boy1);
              break;
            case 2:
              myImageView.setImageResource(R.drawable.boy2);
              break;
            ...                        //此处省略部分代码
            case 18:
              myImageView.setImageResource(R.drawable.boy18);
              break;
        }
    }
};
new myThread().start();               //启动线程
```

步骤 5：在主活动文件中建立线程类，用于定时给 Handler 对象传递参数。代码如下：

```
class myThread extends Thread{
    @Override
    public void run(){
```

```
            int what=1;
            while(isDancing){
                myHandler.sendEmptyMessage((what++)%19);
                try {
                    Thread.sleep(100);
                } catch(InterruptedException e){
                    e.printStackTrace();
                }
            }
        }
    }
```

步骤 6：在 onDestroy()方法中结束线程。代码如下：

```
@Override
protected void onDestroy(){
    isDancing=false;
    super.onDestroy();
}
```

运行效果如图 4-2 所示。

4.1.5　基础实例：风中的气球

本例实现一个在屏幕上来回移动的气球动画。

具体实现步骤如下。

步骤 1：新建项目 MoveBall，准备游戏中所需要的图片资源，这里包括背景和气球图片（如图 4-3 所示）。把图片资源复制到项目根目录下 res/drawable-mdpi 文件夹中。

图 4-2　"快乐舞者"实例运行界面

图 4-3　"风中的气球"图片素材

步骤 2：修改 res/layout 目录中的 XML 布局文件，添加背景和一个 ImageView 组件。代码如下：

```
<FrameLayout xmlns:android="http://schemas.android.com/apk/res/android"
    xmlns:tools="http://schemas.android.com/tools"
    android:layout_width="fill_parent"
```

```xml
    android:layout_height="fill_parent"
    android:background="@drawable/background">
    <ImageView
        android:id="@+id/img_ball"
        android:layout_width="wrap_content"
        android:layout_height="wrap_content"
        android:src="@drawable/ball"
        android:contentDescription="@string/app_name"/>
</FrameLayout>
```

步骤 3：在主活动文件中声明需要的变量，获取相关组件，实现 Handler 类及线程类。代码如下：

```java
public class MainActivity extends Activity {
    private boolean flag=true;                        //标记变量
    private boolean flag_x=true;                      //为 true 表示从左向右
    private ImageView img_ball;                       //声明一个 ImageView 对象
    private Handler handler;                          //声明一个 Handler 对象
    private int x=50;
    private int y=100;
    @Override
    public void onCreate(Bundle savedInstanceState){
        super.onCreate(savedInstanceState);
        setContentView(R.layout.activity_main);
        img_ball=(ImageView)findViewById(R.id.img_ball);   //获取 ImageView 对象
        Display display=getWindowManager().getDefaultDisplay();
        Point point=new Point();
        display.getRealSize(point);
        final int mScreenW=point.x;                   //获取屏幕宽度
        handler=new Handler(){
            @Override
            public void handleMessage(Message msg){
                int move_x=0;
                if(msg.what==0x101){
                    move_x=msg.arg1;                  //获取移动的距离
                    if(x>(mScreenW - img_ball.getWidth())){
                        flag_x=false;
                    } else if(x<img_ball.getWidth()){
                        flag_x=true;
                    }
                    if(flag_x){
                        x+=move_x;
                    } else {
                        x -=move_x;
                    }
```

```
                if(flag){
                    y-=10;
                    flag=false;
                } else {
                    y+=10;
                    flag=true;
                }
                img_ball.setX(x);                    //设置气球图片 X 轴位置
                img_ball.setY(y);                    //设置气球图片 Y 轴位置
            }
                super.handleMessage(msg);
            }
        };
        Thread t=new Thread(new Runnable(){
            @Override
            public void run(){
                int index=0;                         //移动的距离
                while(!Thread.currentThread().isInterrupted()){
                    index=new Random().nextInt(20);  //产生一个随机数
                    Message m=handler.obtainMessage();//获取一个 Message
                    m.what=0x101;                    //设置消息标识
                    m.arg1=index;                    //保存移动的距离
                    handler.sendMessage(m);          //发送消息
                    try {
                        Thread.sleep(new Random().nextInt(500)+100);
                                                     //休眠一段时间
                    } catch(InterruptedException e){
                        e.printStackTrace();
                    }
                }
            }
        });
        t.start();                                   //开启线程
    }
}
```

在 Android 4.0 版本之前可以通过下面的方式获取屏幕的宽高。代码如下：

```
DisplayMetrics dm=new DisplayMetrics();
getWindowManager().getDefaultDisplay().getMetrics(dm);
    int mScreenW=dm.widthPixels;
    int mScreenH=dm.heightPixels;
```

采用以上方法获得的是屏幕的总宽和总高，包括状态栏的高度和导航栏的高度。但在 Android 4.0 以后，采用上面的方法获取的高度就是去掉导航栏高度外余下的高度值，需要采用下面的方法。代码如下：

```
Display display=getWindowManager().getDefaultDisplay();
    Point point=new Point();
    display.getRealSize(point);
    int mScreenW=point.x;
    int mScreenH=point.y;
```

运行效果如图4-4所示。

图4-4 "风中的气球"实例运行界面

4.2 Android 二维游戏开发视图

Android游戏开发中常用的3种视图是View、SurfaceView和GLSurfaceView。
- View：显示视图，内置画布，提供图形绘制函数、触屏事件、按键事件函数等。必须在UI主线程内更新画面，速度较慢。
- SurfaceView：基于View进行拓展的视图类，更适合2D游戏的开发。它是View的子类，类似使用双缓冲机制，在新的线程中更新画面，所以刷新速度比View快。
- GLSurfaceView：基于SurfaceView再次进行拓展的视图类，是专用于3D游戏开发的视图，是SurfaceView的子类（OpenGL专用）。

根据游戏特点，更新画面的类型一般分为以下两类。

（1）被动更新。画面依赖于onTouch来更新，例如棋类游戏，可以直接使用invalidate。因为这种情况下，这一次触摸和下一次触摸间隔的时间比较长，不会对操作产生影响。

（2）主动更新。需要一个单独的线程不停地重绘人的状态，例如一个人在一直跑动。这种情况下应避免阻塞主UI线程，所以显然View不合适，需要SurfaceView来控制。

游戏中的刷屏原理是：对于玩家来说，游戏是动态的；对于开发人员来说，游戏是静态的，是在不停地播放不同的画面，让玩家看到了动态效果。画笔（Paint）用于在画布（Canvas）上画各种图形图片等，视图用于将画布上的内容展现到手机屏幕上。

绘制在画布上的是静态图像，只有不停地展示不同的画布，才能实现动态效果。手机上的画布永远只是一张,不可能同时播放不同的画布,此时便需要对画布进行刷新来达到动态的效果,不断地刷新屏幕,重新绘制画布,如同使用一块橡皮擦擦去之前画布上的所有内容,然后重新绘制画布,如此反复,形成动态效果,这个过程就是"刷屏"。

4.2.1 View 框架

对于常规的游戏,在 View 中需要处理以下 3 种问题:控制事件、刷新 View 以及绘制 View。

(1) 处理按键事件 onKeyDown、屏幕触控 onTouchEvent 以及 Sensor 重力感应等方法。

(2) 刷新 View 的方法这里主要有 invalidate(int l,int t,int r,int b)刷新局部,4 个参数分别为左、上、右、下。整个 View 刷新使用 invalidate(),刷新一个矩形区域使用 invalidate(Rect dirty),刷新一个特性,例如 Drawable,使用 invalidateDrawable(Drawable drawable)。执行 invalidate 类的方法将会设置 View 为无效,最终导致 onDraw 方法被重新调用。除了使用 handler 方式外,可以在线程中直接使用 postInvalidate 方法来实现。

(3) 绘制 View 主要是在 onDraw()中通过形参 canvas 来处理,相关的绘制主要有 drawRect、drawLine、drawPath 等。

举例:调用自定义 View 视图

步骤 1:新建 MyView 类(继承 android.view.View 类),重写 onDraw(Canvas canvas)方法,实现游戏中的绘图操作。代码如下:

```
public class MyView extends View {
    private Paint paint;                              //定义画笔变量
    public MyView(Context context){
        super(context);
        setFocusable(true);                           //设置焦点
        paint=new Paint();                            //创建画笔实例
    }
    @Override
    protected void onDraw(Canvas canvas){
        super.onDraw(canvas);
        paint.setColor(Color.RED);                    //设置画笔颜色
        canvas.drawText("Hello", 150, 200, paint);    //在坐标点绘制文字
    }
}
```

在 onDraw()函数中最好不要创建对象,否则会提示警告信息,这是因为 onDraw()调用频繁,不断进行创建和垃圾回收会影响 UI 显示的性能。

★**注意**:手机屏幕是以最左上角的点为(0,0)点,水平向右是 X 轴的正方向,垂直向下是 Y 轴的正方向,最右下角以屏幕分辨率的最大宽度、高度为坐标值。

步骤 2:在主活动的 onCreate()方法中调用自定义的 MyView 类。代码如下:

```
public class MainActivity extends Activity{
    @Override
    protected void onCreate(Bundle savedInstanceState){
        super.onCreate(savedInstanceState);
        setContentView(new MyView(this));          //关联自定义的View
    }
```

以上代码只是在指定位置(X、Y)绘制文本,为了实现文本跟随手指移动的效果,需要重写触屏监听函数 onTouchEvent。此类实例中保存了玩家触屏的动作,常见的动作有按下、抬起、移动、屏幕压力、多点触屏等,另外也定义了很多动作的静态常量值。通过 event.getAction()方法获取玩家的动作与所需动作常量值匹配。代码如下:

```
@Override
public boolean onTouchEvent(MotionEvent event){
    int x=(int)event.getX();
    int y=(int)event.getY();
    if(event.getAction()==MotionEvent.ACTION_DOWN
            || event.getAction()==MotionEvent.ACTION_MOVE
            || event.getAction()==MotionEvent.ACTION_UP){
        textX=x;
        textY=y;
    }
    invalidate();
    return super.onTouchEvent(event);
}
```

调用 invalidate()方法,则重新绘图一次,也就是调用 onDraw()方法一次。如果在其他类中调用,则需要调用 postInvalidate()方法。

运行项目,用手指点击屏幕、离开屏幕都很正常,但当手指在屏幕中进行滑动时,文本的坐标没有跟随手指移动,也就是说系统无法获取 MotionEvent.ACTION_MOVE 的动作。其原因是:onTouchEvent()函数通常情况下会去执行 super.onTouchEvent()函数并传回布尔值,但 super.onTouchEvent()中的 super 有可能什么都没做,这样就会传回 false,导致后面的 event 动作可能接收不到值。所以,为了确保后面的 event 能顺利接收到布尔值,应该让触屏监听函数的返回值设为 true。修改触屏监听函数代码如下:

```
@Override
public boolean onTouchEvent(MotionEvent event){
    ...
    invalidate();
    return true;
}
```

针对当前的文本跟随用户手指的功能,在触屏监听函数中,其实没有必要获得用户的动作,因为不管用户是什么动作,开发人员需要的只是用户手指触摸在屏幕上的 X、Y

坐标位置，所以修改代码以使触屏监听函数简化。代码如下：

```
@Override
public boolean onTouchEvent(MotionEvent event){
    textX=(int)event.getX();
    textY=(int)event.getY();
    invalidate();
    return true;
}
```

在 Eclipse 中重写父类函数，操作步骤如下：选择主菜单的 Search 项，选中 Override/Implement Methods 选项（或在源代码处右击，选择快捷菜单中的相应命令），然后在出现的 Override/Implement Method 窗口中选中需要重写的函数单击"确定"按钮即可（Eclipse 中默认使用 Shift＋Alt＋S 组合键调出 Source 选项）。

4.2.2 SurfaceView 框架

SurfaceView 和 View 最本质的区别是：View 要在 UI 的主线程中更新画面，而 SurfaceView 是在一个新的单独线程中重新绘制画面，所以不会阻塞 UI 主线程。但这也带来事件同步的问题，也涉及线程同步。例如，每触屏一次，都需要 SurfaceView 中的 Thread 处理，一般就需要有一个事件队列的设计来保存触屏事件。SurfaceView 的原理如图 4-5 所示。

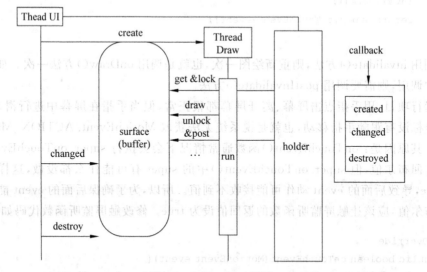

图 4-5　SurfaceView 原理

在游戏中，一般不会等用户每次触发了按键触屏事件才重绘画布，而是会定义一个时间去刷新画布，例如倒计时、动态的花草、流水等，这些游戏元素并不会跟玩家交互，但都是动态的。在游戏开发中，会有一个线程不停地重绘画布，实时更新游戏元素的状态。

游戏中除用画布给玩家最直接的动态展示外，还有很多逻辑需要不间断地更新，例如游戏中钱币的更新和 AI(人工智能)行为等。Android 中的 SurfaceView 类就是双缓冲机制,开发采用自动刷新屏幕的游戏时尽量使用 SurfaceView 类,这样效率较高,而且 SurfaceView 类的功能也更加完善。

举例:获取视图宽度和高度

步骤 1:新建项目 ProjectTestSurfaceView,自定义类 TestSurfaceView,继承 SurfaceView 类并实现 android.view.SurfaceHolder.Callback 接口。代码如下:

```java
public class TestSurfaceView extends SurfaceView implements Callback {
    private int screenW, screenH;                     //声明视图宽度和高度变量
    private SurfaceHolder sfh;                        //声明 SurfaceHolder 对象变量
    public TestSurfaceView(Context context){
        super(context);
        sfh=this.getHolder();                         //实例 SurfaceHolder
        sfh.addCallback(this);                        //为 SurfaceView 添加状态监听
        setFocusable(true);
    }
    /**
     * SurfaceView 视图状态发生改变,响应此函数
     */
    @Override
    public void surfaceChanged(SurfaceHolder holder, int format, int width, int height){
        //TODO: Auto-generated method stub
    }
    /**
     * SurfaceView 视图建立完成,响应此函数
     */
    @Override
    public void surfaceCreated(SurfaceHolder holder){
        screenW=this.getWidth();                      //获取视图宽度
        screenH=this.getHeight();                     //获取视图高度
        System.out.println("屏宽:"+screenW+" 屏高:"+screenH);
    }
    /**
     * SurfaceView 视图消亡时,响应此函数
     */
    @Override
    public void surfaceDestroyed(SurfaceHolder holder){
        //TODO: Auto-generated method stub
    }
}
```

步骤 2:在主活动的 onCreate 方法中调用自定义的 TestSurfaceView 类。代码如下:

```java
public class MainActivity extends Activity {
    @Override
    protected void onCreate(Bundle savedInstanceState){
        super.onCreate(savedInstanceState);
        setContentView(new TestSurfaceView(this));    //关联自定义的 SurfaceView
    }
}
```

在 SurfaceView 中获取视图的宽和高,一定要在视图创建之后获取,也就是在 surfaceCreated 函数执行之后获取,在此函数执行之前获取的永远是 0。

举例:跟随手指的文字

步骤 1:新建项目 ProjectSurfaceView,首先自定义类 MySurfaceView,继承自 SurfaceView 类并实现 android.view.SurfaceHolder.Callback 和 Runnable 接口。代码如下:

```java
public class MySurfaceView extends SurfaceView implements Callback, Runnable {
    private SurfaceHolder sfh;
    private boolean flag=true;                              //线程运行标志
    private Paint paint;
    private Canvas canvas;
    private Thread thread;
    private int textX=10, textY=10;
    public MySurfaceView(Context context){
        super(context);
        sfh=getHolder();
        sfh.addCallback(this);
        paint=new Paint();
        paint.setColor(Color.RED);
        setFocusable(true);
    }
    /**
     * 游戏绘制
     */
    public void myDraw(){
        try {
            canvas=sfh.lockCanvas();
            if(canvas !=null){
                canvas.drawRGB(255, 255, 255);
                canvas.drawText("Game", textX, textY, paint);
            }
        } catch(Exception e){
        } finally {
            if(canvas !=null){
                sfh.unlockCanvasAndPost(canvas);
            }
```

```
        }
    }
    @Override
    public void run(){
        while(flag){
            long startTime=System.currentTimeMillis();   //获取开始时间戳
            myDraw();
            long endTime=System.currentTimeMillis();     //获取结束时间戳
            if(endTime - startTime<50){
                try {
                    Thread.sleep(50 - (endTime - startTime));
                } catch(InterruptedException e){
                    e.printStackTrace();
                }
            }
        }
    }
    /**
     * 触屏事件监听
     */
    @Override
    public boolean onTouchEvent(MotionEvent event){
        textX=(int)event.getX();
        textY=(int)event.getY();
        return true;
    }
    @Override
    public void surfaceChanged(SurfaceHolder holder, int format, int width, int
    height){
        //TODO: Auto-generated method stub
    }
    @Override
    public void surfaceCreated(SurfaceHolder holder){
        thread=new Thread(this);
        thread.start();
    }
    @Override
    public void surfaceDestroyed(SurfaceHolder holder){
        flag=false;
    }
}
```

步骤 2：在主活动的 onCreate()方法中调用自定义的 MySurfaceView 类。代码如下：

```
public class MainActivity extends Activity {
    @Override
    protected void onCreate(Bundle savedInstanceState){
        super.onCreate(savedInstanceState);
        MySurfaceView mySurfaceView=new MySurfaceView(this);
        setContentView(mySurfaceView);
    }
}
```

本例需要注意以下 3 点。

(1) 线程标识位。

游戏开发中使用的线程一般都会在 run()函数中使用一个 while 死循环,在这个循环中调用绘图和逻辑函数,为了在游戏暂停或者游戏结束时便于销毁线程,在代码中声明一个布尔值,用于标识线程的状态。另外,单击 Android 手机上的 Back(返回)或 Home(主窗口)按钮时,默认会将当前程序切入到系统后台运行(程序中没有截获这两个按钮的前提下),这会造成 SurfaceView 视图的状态发生改变。当单击 Back(返回)按钮使当前程序切入后台,然后单击项目重新回到程序中,SurfaceView 的状态变化为 surfaceDestroyed→构造函数→surfaceCreated→surfaceChanged;单击 Home(主窗口)按钮使当前程序切入后台,单击项目重新回到程序中,SurfaceView 的状态变化为 surfaceDestroyed→surfaceCreated→surfaceChanged。

可以看到,单击 Back(返回)按钮并重新进入程序的过程要比单击 Home(主窗口)多执行一个构造函数,也就是说 SurfaceView 视图会被重新加载。所以,如果线程的初始化是在视图构造函数中或者在视图构造函数之前,那么线程启动也要放在视图构造函数中进行。如果把线程的初始化放在 surfaceCreated(视图创建)函数之前,而线程的启动放在 surfaceCreated 函数中,程序运行时玩家一旦单击 Home 按钮后再重新回到游戏时,程序就会抛出异常。异常是因为线程已经启动造成的,因为程序被 Home 按钮切入后台再从后台恢复时,会直接进入 surfaceCreated 函数中,又执行了一遍线程启动。

因此,线程的初始化与线程的启动都写在视图的 surfaceCreated 函数中,并且在视图销毁时将线程标识位的值改变为 false,这样既可以避免"线程已启动"的异常,又可以避免单击 Back 按钮无限增加线程数的问题。

(2) 异常处理。

当 SurfaceView 不可编辑或还没创建时,调用 lockCanvas 函数会返回 null,画布进行绘图时也会出现不可预知的问题,所以要对绘图函数使用 try…catch 异常处理。既然 lockCanvas()函数有可能获取为 null,为了避免出错,在使用画布开始绘制时,需要判断其是否为 null。

绘图的时候也会出现不可预知的问题,虽然使用 try 语句不会导致程序崩溃,但是一旦提交画布之前出错,那么解锁提交画布函数则无法被执行到,就会导致下次通过 lockCanvas()获取画布时由于画布上次没有解锁而抛出异常,所以将解锁提交的函数放入 finally 语句块中。另外,提交解锁之前要保证画布不为空,最好判断一下是否为空值。

（3）刷屏时间尽可能保证一致。

在线程循环中设计休眠时间，无法预料系统在处理逻辑时，时间的开销是否与上次相同，假设游戏线程的休眠时间为 X 毫秒，一般线程休眠的语句代码如下：

```
Thread.sleep(X);
```

优化后的语句代码如下：

```
long startTime=System.currentTimeMillis();     //通过系统函数获取开始时间戳
    ...                                         //此处为在线程中的绘图、逻辑等函数调用代码
long endTime=System.currentTimeMillis();       //通过系统函数获取结束时间戳
    if(endTime-startTime<X){                    //如果差值小于一定数值
        try{
            Thread.sleep(X-(endTime-startTime));  //设置休眠时间
        }catch(InterruptedException e){
            e.printStackTrace();
        }
    }
```

4.3 常用绘图类

在 Android 下进行 2D 绘图最常用的就是 Paint 类、Canvas 类、Bitmap 类和 BitmapFactory 类。其中 Paint 类代表画笔，Canvas 类代表画布。

4.3.1 Paint 类

Paint 是绘图的辅助类，一般是作为画布的参数来实现相应的效果，Paint 类中包含文字与位图的样式、颜色等属性信息。Paint 类常用的方法如表 4-2 所示。

表 4-2 Paint 类的常用方法

方　法	描　述
setARGB(int a,int r,int g,int b)	设置 ARGB 颜色，各参数均为 0～255 的整数，分别表示透明度、红色、绿色和蓝色值
setColor(int color)	设置颜色，参数 color 可以为 Color 类提供的颜色常量，也可以用 Color.rgb(int red,int green,int blue)的方法指定
setAlpha(int a)	设置透明度，值为 0～255 的整数
setAntiAlias(boolean aa)	设置是否使用抗锯齿功能（如果使用会使绘图速度变慢）
setDither(Boolean dither)	设置是否使用图像抖动处理，如果使用会使图像颜色更加平滑、饱满和清晰
setPathEffect(PathEffect effect)	设置绘制路径时的路径效果

续表

方法	描述
setShader(Shader shader)	设置Paint的填充效果,可以使用LinearGradient(线性渐变)、RadialGradient(径向渐变)和SweepGradient(角度渐变)
setShadowLayer(float radius,float dx,float dy,int color)	设置阴影,参数分别为阴影的角度、阴影在X轴和Y轴上的距离、阴影的颜色。如果参数radius的值为0将没有阴影
setStrokeCap(Paint.Cap cap)	当画笔的填充样式为STROKE或FILL_AND_STROKE时,设置笔刷的图形样式,参数值可以是Cap.BUTT、Cap.ROUND或Cap.SQUARE,主要体现在线的端点上
setStrokeJoin(Paint.Join join)	设置画笔转弯处的连接风格,参数值为Join.BEVEL、Join.MITER或Join.ROUND
setStrokeWidth(float width)	设置画笔的笔触宽度
setStyle(Paint.Style style)	设置画笔的填充风格,参数值为Style.FILL、Style.FILL_AND_STROKE或Style.STROKE
setTextSize(float textSize)	设置绘制文本时的文字大小
setFakeBoldText(Boolean fakeBoldText)	设置是否为粗体文字
setTextAlign(Paint.Align align)	设置绘制文本的对齐方式,参数值为Align.CENTER、Align.LEFT或Align.RIGHT

4.3.2 Canvas类

Canvas类提供了两个构造函数。
- Canvas():创建一个空的Canvas对象。
- Canvas(Bitmap bitmap):创建一个以bitmap(位图)为背景的画布。

Canvas类提供了很多drawXxx()方法,具有多种类型,可以画出点、线、矩形、圆形、椭圆、文字、位图等。Canvas类常用的方法如表4-3所示。

表4-3 Canvas类的常用方法

方法	描述
getWidth()	得到画布的宽度
getHeight()	得到画布的高度
drawBitmap(Bitmap bitmap,float left,float top,Paint paint)	在指定坐标绘制位图
drawLine(float startX,float startY,float stopX,float stopY,Paint paint)	在起始点和结束点之间绘制连线
drawPath(Path path,Paint paint)	根据给定的路径绘制连线
drawPoint(float x,float y,Paint paint)	根据给定的坐标绘制点
drawText(String text,int start,int end,Paint paint)	根据给定的坐标绘制文字

Canvas 可以绘制的对象有弧线（Arcs）、填充颜色（Argb 和 Color）、位图、圆（Circle 和 Oval）、点（Point）、线（Line）、矩形（Rect）、图片（Picture）、圆角矩形（RoundRect）、文本（Text）、顶点（Vertices）、路径（Path）。通过组合这些对象可以画出一些简单有趣的界面。另外，Canvas 类还提供了一些对画布位置进行变换的方法：rorate、scale、translate、skew（扭曲）等，而且允许通过获得它的转换矩阵对象（用 getMatrix 方法）直接操作它。

4.3.3 Bitmap 类

Bitmap 类代表位图，不仅可以获取图像文件信息，对图像进行剪切、旋转、缩放等操作，而且可以指定保存图像文件格式。Bitmap 类提供的常用方法如表 4-4 所示。

表 4-4 Bitmap 类的常用方法

方　　法	描　　述
createBitmap（Bitmap source, int x, int y, int width, int height, Matrix m, Boolean filter）	从源位图的指定坐标开始，取指定宽度和高度的一块图像来创建新的 Bitmap 对象，并按 Matrix 指定的规则进行变换
createBitmap（Bitmap source, int x, int y, int width, int height）	从源位图的指定坐标开始，取指定宽度和高度的一块图像来创建新的 Bitmap 对象
createBitmap(int width, int height, Bitmap.Config config)	创建一个指定宽度和高度的新的 Bitmap 对象
createBitmap(int[] colors, int width, int height, Bitmap.Config config)	使用颜色数组创建一个指定宽度和高度的新 Bitmap 对象（数组元素的个数为 width×height）
createBitmap(Bitmap src)	使用源位图创建一个新的 Bitmap 对象
createScaledBitmap(Bitmap src, int dstWidth, int dstHeight, Boolean filter)	将源位图缩放为一个指定宽度和高度的新的 Bitmap 对象
Compress（Bitmap.CompressFormat format, int qualite, OutputStream stream）	将 Bitmap 对象压缩为指定格式并保存到指定的文件输出流中
isRecycled()	判断 Bitmap 对象是否被回收
Recycle()	强制回收 Bitmap 对象

例如，创建一个包括 4 个像素，每个像素对应一种颜色的 Bitmap 对象，代码如下：

```
Bitmap bitmap=Bitmap.createBitmap(new int[] { Color.RED, Color.GREEN, Color.MAGENTA }, 4, 1, Config.RGB_565);
```

4.3.4 BitmapFactory 类

BitmapFactory 类为一个工具类，用于从不同的数据源解析和创建 Bitmap 对象。创建 Bitmap 对象的常用方法如表 4-5 所示。

表 4-5 BitmapFactory 类的常用方法

方　　法	描　　述
decodeFile(String pathName)	从指定路径的文件中解析并创建 Bitmap 对象
decodeFileDescriptor(FileDescriptor fd)	从 FileDescriptor 对应的文件中解析并创建 Bitmap 对象
decodeResource(Resources res,int id)	根据资源 ID 解析并创建 Bitmap 对象
decodeStream(InputStream is)	从指定的输入流中解析并创建 Bitmap 对象

例如，解析 SD 卡上的图片文件 game.jpg 并创建对应的 Bitmap 对象，代码如下：

```
String path="/sdcard/pictures/bccd/game.jpg";
Bitmap bitmap=BitmapFactory.decodeFile(path);
```

要解析 drawable 资源中保存的图片文件 game.jpg 创建对应的 Bitmap 对象，代码如下：

```
Bitmap bitmap=BitmapFactory.decodeResource(MainActivity.this.getResources(),R.drawable.game);
```

4.3.5　基础实例：游戏角色行走控制

本例实现对 RGP 类型的游戏人物行走进行控制。

具体实现步骤如下：

步骤 1：准备游戏中所需要的图片资源，这里包括背景图片和角色行走序列图片(如图 4-6 所示)。把图片资源复制到项目根目录下的 res/drawable-mdpi 文件夹中。

图 4-6　人物行走图片素材

步骤 2：建立内部类 GameView(继承自 android.view.View 类)，添加构造函数并实现其 onDraw 方法；实现 Runable 接口并重写其 run()方法。代码如下：

```
public class MyView extends View implements Runnable {
    private final static int RIGHT=1;            //标识向右走
    private final static int LEFT=2;             //标识向左走
    private int direction=1;                     //标识行走方向
    private Paint paint;                         //定义画笔
    private Bitmap gameBG;                       //定义背景
    private Bitmap roleRight[], roleLeft[];      //定义行走序列图片数组
    private float destX;                         //目标点 X 坐标
```

```java
private boolean isGame=true;                    //游戏是否结束
private boolean isMove=false;                   //是否开始行走
private int roleX=30, roleY=200;                //角色默认位置
private int frameIndex=0;                       //行走序列帧
private int speed=5;                            //行走速度
private Handler handler;                        //定义Handler对象
public MyView(Context context){
    super(context);
    paint=new Paint();
    paint.setAntiAlias(true);
    gameBG = BitmapFactory. decodeResource (getResources ( ), R. drawable.
    gamebg);
    roleRight=new Bitmap[4];                    //实例化向右走序列图片数组
    roleRight[0]=BitmapFactory.decodeResource(getResources(), R.drawable.
    role_right0);
    roleRight[1]=BitmapFactory.decodeResource(getResources(), R.drawable.
    role_right1);
    roleRight[2]=BitmapFactory.decodeResource(getResources(), R.drawable.
    role_right2);
    roleRight[3]=BitmapFactory.decodeResource(getResources(), R.drawable.
    role_right3);
    roleLeft=new Bitmap[4];                     //实例化向左走序列图片数组
    roleLeft[0]=BitmapFactory.decodeResource(getResources(), R.drawable.
    role_left0);
    roleLeft[1]=BitmapFactory.decodeResource(getResources(), R.drawable.
    role_left1);
    roleLeft[2]=BitmapFactory.decodeResource(getResources(), R.drawable.
    role_left2);
    roleLeft[3]=BitmapFactory.decodeResource(getResources(), R.drawable.
    role_left3);
    handler=new Handler(){
        @Override
        public void handleMessage(Message msg){
            super.handleMessage(msg);
            if(msg.what==1){
                invalidate();                   //刷新视图
            }
        }
    };
    new Thread(this).start();                   //启动线程
}
@Override
protected void onDraw(Canvas canvas){
    super.onDraw(canvas);
```

```
canvas.drawBitmap(gameBG, 0, 0, paint);
switch(direction){
case RIGHT:
    canvas.drawBitmap(roleRight[frameIndex], roleX, roleY, paint);
    break;
case LEFT:
    canvas.drawBitmap(roleLeft[frameIndex], roleX, roleY, paint);
    break;
}
}
@Override
public boolean onTouchEvent(MotionEvent event){
    destX=event.getX();                            //获取触屏 X 位置坐标
    isMove=true;                                   //角色开始行走
    return true;
}
@Override
public void run(){
    while(isGame){
        if(isMove){
            direction=destX>roleX ?RIGHT : LEFT;   //判断行走方向
            switch(direction){
            case RIGHT:                            //向右走
                roleX+=speed;
                break;
            case LEFT:                             //向左走
                roleX -=speed;
                break;
            }
            frameIndex=(frameIndex==3 ?0 : frameIndex+1);   //循环序列帧
            if(Math.abs(roleX -destX)<=5){         //如果到达位置
                frameIndex=0;                      //初始序列帧
                isMove=false;                      //角色停止行走
            }
            Message msg=handler.obtainMessage();   //获取 Handler 消息对象
            msg.what=1;
            handler.sendMessage(msg);              //发送消息
            try {
                Thread.sleep(100);
            } catch(InterruptedException e){
                e.printStackTrace();
            }
        }
    }
}
```

    }
}
```

**步骤 3**：在主活动的 onCreate()方法中,设置屏幕属性并调用 MyView 视图。代码如下：

```
@Override
protected void onCreate(Bundle savedInstanceState){
    super.onCreate(savedInstanceState);
    setRequestedOrientation(ActivityInfo.SCREEN_ORIENTATION_LANDSCAPE);
                            //设置横屏
    ……                      //此处省略隐藏标题栏、全屏及保持屏幕亮度的代码
    MyView gameView=new MyView(this);
    gameView.setFocusable(true);
    setContentView(gameView);
}
```

运行效果如图 4-7 所示。

图 4-7 人物行走实例运行界面

4.4 绘制 2D 图像

Android 提供了强大的二维图形库用于绘制 2D 图像,常用的是绘制几何图形、文本、路径和图片等。

4.4.1 绘制文本

在开发游戏的过程中,特别是 RPG(角色)类游戏时,显示文字信息多用绘制文本的方式实现。Canvas 类提供了 3 个常用的绘制文本的方法,如表 4-6 所示。
```

表 4-6　常用的绘制文本的方法

方　　法	描　　述
drawText(String text, float x, float y, Paint paint)	在画布的指定位置绘制文字
drawPosText(String text, float [ ] pos, Paint paint)	使用该方法绘制字符串时,要为每个字符指定位置
drawTextOnPath(String text, Path path, float hOffset, float vOffset, Paint paint)	沿路径绘制字符串

举例：游戏中的对话界面

步骤 1：新建项目,修改 res/layout 目录下的 XML 布局文件,设置背景图片。代码如下：

```
<?xml version="1.0" encoding="utf-8"?>
<FrameLayout xmlns:android="http://schemas.android.com/apk/res/android"
    android:id="@+id/myLayout "
    android:layout_width="fill_parent"
    android:layout_height="fill_parent"
    android:background="@drawable/background"
    android:orientation="vertical">
</FrameLayout>
```

步骤 2：在主活动文件中创建 MyView 内部类(继承自 android.view.View 类),添加构造方法并重写 onDraw(Canvas canvas)方法。代码如下：

```
public class MainActivity extends Activity {
    @Override
    public void onCreate(Bundle savedInstanceState){
        super.onCreate(savedInstanceState);
        setContentView(R.layout.main);
        FrameLayout ll=(FrameLayout)findViewById(R.id.myLayout);   //获取帧布局
        ll.addView(new MyView(this));
    }
    public class MyView extends View{
        public MyView(Context context){
            super(context);
        }
        @Override
        protected void onDraw(Canvas canvas){
            Paint paintText=new Paint();                    //创建一个采用默认设置的画笔
            paintText.setColor(Color.BLACK);                //设置画笔颜色
            paintText.setTextAlign(Align.LEFT);             //设置文字左对齐
            paintText.setTextSize(12);                      //设置文字大小
            paintText.setAntiAlias(true);                   //使用抗锯齿功能
            canvas.drawText("这里是百花谷!", 245, 40, paintText);   //绘制文字
```

```
        float[] pos=new float[]{ 185, 140, 200, 140, 215, 140, 230, 140, 185,
        155, 205, 155, 220, 155};                             //定义文字位置数组
        canvas.drawPosText("这是什么地方?", pos, paintText);   //绘制文字
        super.onDraw(canvas);
    }
  }
}
```

运行效果如图 4-8 所示。

图 4-8　游戏中的对话界面

4.4.2　绘制几何图形

Canvas 类提供了丰富的绘制几何图形的方法，包括点、线、弧、圆形、矩形等。常用的绘制几何图形的方法如表 4-7 所示。

表 4-7　常用的绘制几何图形的方法

方　　法	描　　述
drawPoint(float x,float y,Paint paint)	在指定坐标处绘制一个点
drawPoint(float[] pts,Paint paint)	根据数组坐标绘制多个点
drawLine(float startX, float startY, float stopX, float stopY,Paint paint)	根据定义的起点和终点坐标绘制一条线
drawLines(float[] pts,Paint paint)	根据数据坐标绘制多条线
drawOval(RectF oval,Paint paint)	在定义的矩形框内绘制椭圆
drawRect(float left, float top, float right, float bottom, Paint paint)	绘制矩形
drawRoundRect(RectF rect, float rx, float ry, Paint paint)	绘制圆角矩形

举例：绘制奥运五环。

步骤 1：新建项目，修改 res/layout 目录下的 XML 布局文件，设置背景图片。代码如下：

```xml
<?xml version="1.0" encoding="utf-8"?>
<FrameLayout xmlns:android="http://schemas.android.com/apk/res/android"
    android:id="@+id/myLayout "
    android:layout_width="fill_parent"
    android:layout_height="fill_parent"
    android:orientation="vertical">
</FrameLayout>
```

步骤2：在主活动文件中创建 MyView 内部类（继承自 android.view.View 类），添加构造方法并重写 onDraw(Canvas canvas)方法。代码如下：

```java
public class MainActivity extends Activity {
    @Override
    public void onCreate(Bundle savedInstanceState){
        super.onCreate(savedInstanceState);
        setContentView(R.layout.main);
        FrameLayout ll=(FrameLayout)findViewById(R.id.myLayout);
                                                        //获取布局管理器
        ll.addView(new MyView(this));    //添加自定义的MyView视图到帧布局管理中
    }
    public class MyView extends View{
        public MyView(Context context){
            super(context);
        }
        @Override
        protected void onDraw(Canvas canvas){
            canvas.drawColor(Color.WHITE);
            Paint paint=new Paint();                //创建采用默认设置的画笔
            paint.setAntiAlias(true);               //使用抗锯齿功能
            paint.setStrokeWidth(3);                //设置笔触的宽度
            paint.setStyle(Style.STROKE);           //设置填充样式为描边
            paint.setColor(Color.BLUE);
            canvas.drawCircle(50, 50, 30, paint);   //绘制蓝色的圆形
            paint.setColor(Color.YELLOW);
            canvas.drawCircle(100, 50, 30, paint);  //绘制黄色的圆形
            paint.setColor(Color.BLACK);
            canvas.drawCircle(150, 50, 30, paint);  //绘制黑色的圆形
            paint.setColor(Color.GREEN);
            canvas.drawCircle(75, 90, 30, paint);   //绘制绿色的圆形
            paint.setColor(Color.RED);
            canvas.drawCircle(125, 90, 30, paint);  //绘制红色的圆形
            super.onDraw(canvas);
        }
    }
}
```

运行效果如图 4-9 所示。

图 4-9 奥运五环

4.4.3 绘制路径

Android 中绘制路径有创建路径和绘制定义好的路径两种方式。创建路径可以使用 android.graphics.Path 类实现。Path 类的常用方法如表 4-8 所示。

表 4-8 Path 类的常用方法

方 法	描 述
addArc(RectF oval,float startAngle,float sweepAngle)	添加弧形路径
addCircle(float x,float y,float radius,Path.Direction dir)	添加圆形路径
addOval(RectF oval,Path.Direction dir)	添加椭圆路径
addRect	添加矩形路径
addRoundRect(RectF rect, float rx, float ry, Path.Direction dir)	添加圆角矩形路径
moveTo(float x,float y)	移动到指定点
lineTo(float x,float y)	在 moveTo 方法设置的起点与本方法指定的终点之间画一条直线,如果没有使用 moveTo 方法,则从(0,0)点开始绘制直线
quadTo(float x1,float y1,float x2,float y2)	根据指定的参数绘制一条线段轨迹
Close()	闭合路径

Path.direction 常量的可选值为 Path.Direction.CW(顺时针)和 Path.Direction.CCW(逆时针)。

在 onDraw 方法中,创建画笔,并设置画笔的相关属性,然后分别绘制一个圆形路径、折线路径、三角形路径和圆形路径文字。代码如下:

```
//绘制圆形路径
Path pathCircle=new Path();                          //创建并实例化一个 path 对象
pathCircle.addCircle(70, 70, 40, Path.Direction.CCW);  //添加逆时针的圆形路径
canvas.drawPath(pathCircle, paint);                  //绘制路径
//绘制折线路径
Path pathLine=new Path();                            //创建并实例化一个 Path 对象
```

```
pathLine.moveTo(150, 100);              //设置起始点
pathLine.lineTo(200, 45);               //设置第一段直线的结束点
pathLine.lineTo(250, 100);              //设置第二段直线的结束点
pathLine.lineTo(300, 80);               //设置第三段直线的结束点
canvas.drawPath(pathLine, paint);       //绘制路径
//绘制三角形路径
Path pathTr=new Path();                 //创建并实例化一个path对象
pathTr.moveTo(350, 80);                 //设置起始点
pathTr.lineTo(400, 30);                 //设置第一条边的结束点,也是第二条边的起始点
pathTr.lineTo(450, 80);                 //设置第二条边的结束点,也是第三条边的起始点
pathTr.close();                         //闭合路径
canvas.drawPath(pathTr, paint);         //绘制路径
//绘制沿圆形路径排列的环形文字
String str="朝辞白帝彩云间,千里江陵一日还";
Path path=new Path();                   //创建并实例化一个path对象
path.addCircle(550, 100, 48, Path.Direction.CW);    //添加顺时针的圆形路径
paint.setStyle(Style.FILL);             //设置画笔的填充方式
canvas.drawTextOnPath(str, path, 0, -18, paint);    //绘制沿圆形路径排列的文字
```

4.4.4 绘制图片

在 Android 中,Canvas 类绘制图片的常用方法如表 4-9 所示。

表 4-9 Canvas 类绘制图片的常用方法

方法	描述
drawBitmap(Bitmap bitmap, Rect src, RectF dst, Paint paint)	从指定点绘制从源图中挖取的部分图像
drawBitmap(Bitmap bitmap, float left, float top, Paint paint)	在指定点绘制图像
drawBitmap(Bitmap bitmap, Rect src, Rect dst, Paint paint)	从指定点绘制从源图中挖取的部分图像

举例:绘制 SD 卡指定图像

步骤 1:新建项目,修改 res/layout 目录下的 XML 布局文件,设置背景图片。代码如下:

```
<?xml version="1.0" encoding="utf-8"?>
<FrameLayout xmlns:android="http://schemas.android.com/apk/res/android"
    android:id="@+id/myLayout"
    android:layout_width="fill_parent"
    android:layout_height="fill_parent"
    android:orientation="vertical">
    <ImageView
        android:id="@+id/imgView"
        android:layout_width="100px"
```

```
            android:paddingTop="5px"
            android:layout_height="25px"/></FrameLayout>
```

步骤2：在主活动文件中创建 MyView 内部类（继承自 android.view.View 类），添加构造方法并重写 onDraw(Canvas canvas) 方法。代码如下：

```
public class MainActivity extends Activity {
    private ImageView imgView;
    @Override
    public void onCreate(Bundle savedInstanceState){
        super.onCreate(savedInstanceState);
        setContentView(R.layout.main);
        FrameLayout ll=(FrameLayout)findViewById(R.id.myLayout);
                                              //获取布局管理器
        ll.addView(new MyView(this));         //添加自定义视图到帧布局管理器
        imgView=(ImageView)findViewById(R.id.imageView1);
    }
    public class MyView extends View {
        public MyView(Context context){
            super(context);
        }
        @Override
        protected void onDraw(Canvas canvas){
            Paint paint=new Paint();
            String path="/sdcard/pictures/bccd/img01.png";
                                              //指定图片文件的路径
            Bitmap bm=BitmapFactory.decodeFile(path);
                                              //获取图片文件对应的 Bitmap 对象
            canvas.drawBitmap(bm, 0, 30, paint);
                                              //在画布指定位置绘制获取的 Bitmap 对象
            Rect src=new Rect(95, 150, 175, 240);   //设置挖取的区域
            Rect dst=new Rect(420, 30, 500, 120);   //设置绘制的区域
            canvas.drawBitmap(bm, src, dst, paint); //绘制挖取到的图像
            Bitmap bitmap=Bitmap.createBitmap(new int[] { Color.RED,
                Color.GREEN, Color.BLUE, Color.MAGENTA }, 4, 1,
                Config.RGB_565);              //使用颜色数组创建一个 Bitmap 对象
            iv.setImageBitmap(bitmap);        //为 ImageView 指定要显示的位图
            super.onDraw(canvas);
        }
    }
}
```

步骤3：重写 onDestroy() 方法，回收 ImageView 组件中使用的 Bitmap 资源。代码如下：

```
@Override
```

```
protected void onDestroy(){
    BitmapDrawable b=(BitmapDrawable)imgView.getDrawable();
    if(b !=null && !b.getBitmap().isRecycled()){
        b.getBitmap().recycle();                    //回收资源
    }
    super.onDestroy();
}
```

4.5 图像特效

在 Android 中可以为图像添加旋转、缩放、倾斜、平移和渲染等特效。在 AndroidAPI 中提供 setXXX()、postXXX() 和 preXXX() 3 种方法。其中 setXXX() 方法用于直接设置 Matrix 的值,每使用一次,整个 Matrix 都会改变;postXXX() 方法采用后乘的方式设置 Matrix 的值,可以连续多次使用这种方法以完成多个变换;postXXX() 方法采用前乘的方式设置 Matrix 的值,设置的操作最先发生。

4.5.1 旋转图像

对图像进行旋转操作,可以使用 android.graphics.Matrix 类提供的 setRotate()、postRotate() 和 preRotate() 3 个方法。

举例:应用 Matrix 旋转图像

步骤 1:新建项目,修改 res/layout 目录下的 XML 布局文件,将默认的布局管理器与组件删除,添加一个帧布局管理器,用于在主活动文件的 onCreate() 方法中显示自定义的绘图类。

步骤 2:创建 MyView 内部类(继承自 android.view.View 类),添加构造方法并重写 onDraw(Canvas canvas)方法。主要代码如下:

```
Paint paint=new Paint();
paint.setAntiAlias(true);
Bitmap bitmap_bg=BitmapFactory.decodeResource(MainActivity.this.getResources
(), R.drawable.background);
canvas.drawBitmap(bitmap_bg, 0, 0, paint);          //绘制背景图像
Bitmap bitmap _ rabbit = BitmapFactory. decodeResource (MainActivity. this.
getResources(), R.drawable.rabbit);
canvas.drawBitmap(bitmap_rabbit, 0, 0, paint);      //绘制原图
//应用 setRotate(float degrees)方法旋转图像
Matrix matrix=new Matrix();
matrix.setRotate(30);                               //以(0, 0)点为轴心旋转 30°
canvas.drawBitmap(bitmap_rabbit, matrix, paint);    //绘制图像并应用 matrix 的变换
//应用 setRotate(float degrees, float px, float py)方法旋转图像
Matrix m=new Matrix();
```

```
m.setRotate(90, 87, 87);                    //以(87,87)点为轴心旋转 90°
canvas.drawBitmap(bitmap_rabbit, m, paint);  //绘制图像并应用 Matrix 的变换
```

4.5.2 缩放图像

对图像进行缩放操作,可以使用 android.graphics.Matrix 类提供的 setScale()、postScale()和 preScale()3 个方法。

举例:应用 Matrix 缩放图像

步骤 1:新建项目,修改 res/layout 目录下的 XML 布局文件,将默认的布局管理器与组件删除,添加一个帧布局管理器,用于在主活动文件的 onCreate()方法中显示自定义的绘图类。

步骤 2:创建 MyView 内部类(继承自 android.view.View 类),添加构造方法并重写 onDraw(Canvas canvas)方法。主要代码如下:

```
Paint paint=new Paint();
paint.setAntiAlias(true);
Bitmap bitmap_bg=BitmapFactory.decodeResource(MainActivity.this.getResources
(), R.drawable.background);
canvas.drawBitmap(bitmap_bg, 0, 0, paint);          //绘制背景
Bitmap bitmap _ rabbit = BitmapFactory. decodeResource (MainActivity. this.
getResources(), R.drawable.rabbit);
//应用 setScale(float sx, float sy)方法缩放图像
Matrix matrix=new Matrix();
matrix.setScale(2f, 2f);          //以(0,0)点为轴心将图像在 X 轴和 Y 轴均缩放为 200%
canvas.drawBitmap(bitmap_rabbit, matrix, paint);   //绘制图像并应用 matrix 的变换
//应用 setScale(float sx, float sy, float px, float py)方法缩放图像
Matrix m=new Matrix();
m.setScale(0.8f, 0.8f, 156, 156);
                    //以(156,156)点为轴心将图像在 X 轴和 Y 轴均缩放为 80%
canvas.drawBitmap(bitmap_rabbit, m, paint);        //绘制图像并应用 matrix 的变换
canvas.drawBitmap(bitmap_rabbit, 0, 0, paint);     //绘制原图
```

4.5.3 倾斜图像

对图像进行倾斜操作,可以使用 android.graphics.Matrix 类提供的 setSkew()、postSkew()和 preSkew()3 个方法。

举例:应用 Matrix 倾斜图像

步骤 1:新建项目,修改 res/layout 目录下的 XML 布局文件,将默认的布局管理器与组件删除,添加一个帧布局管理器,用于在主活动文件的 onCreate()方法中显示自定义的绘图类。

步骤 2:创建 MyView 内部类(继承自 android.view.View 类),添加构造方法并重写 onDraw(Canvas canvas)方法。主要代码如下:

```
Paint paint=new Paint();
paint.setAntiAlias(true);
Bitmap bitmap_bg=BitmapFactory.decodeResource(MainActivity.this.getResources
(), R.drawable.background);
canvas.drawBitmap(bitmap_bg, 0, 0, paint);            //绘制背景
Bitmap bitmap _ rabbit = BitmapFactory. decodeResource (MainActivity. this.
getResources(), R.drawable.rabbit);
//应用setSkew(float sx, float sy)方法倾斜图像
Matrix matrix=new Matrix();
matrix.setSkew(2f, 1f);
                    //以(0, 0)点为轴心将图像在X轴上倾斜量为2,在Y轴上倾斜量为1
canvas.drawBitmap(bitmap_rabbit, matrix, paint);      //绘制图像并应用matrix的变换
//应用setSkew(float sx, float sy, float px, float py)方法倾斜图像
Matrix m=new Matrix();
m.setSkew(-0.5f, 0f, 78, 69);         //以(78, 69)点为轴心将图像在X轴上倾斜量为-0.5
canvas.drawBitmap(bitmap_rabbit, m, paint);           //绘制图像并应用matrix的变换
canvas.drawBitmap(bitmap_rabbit, 0, 0, paint);        //绘制原图
```

4.5.4 平移图像

对图像进行平移操作,可以使用android.graphics.Matrix类提供的setTranslate()、postTranslate()和preTranslate()3个方法。

举例:应用Matrix平移图像

步骤1:新建项目,修改res/layout目录下的XML布局文件,将默认的布局管理器与组件删除,添加一个帧布局管理器,用于在主活动文件的onCreate()方法中显示自定义的绘图类。

步骤2:创建MyView内部类(继承自android.view.View类),添加构造方法并重写onDraw(Canvas canvas)方法。主要代码如下:

```
Paint paint=new Paint();
paint.setAntiAlias(true);
Bitmap bitmap_bg=BitmapFactory.decodeResource(MainActivity.this.getResources
(), R.drawable.background);
canvas.drawBitmap(bitmap_bg, 0, 0, paint);            //绘制背景
Bitmap bitmap _ rabbit = BitmapFactory. decodeResource (MainActivity. this.
getResources(), R.drawable.rabbit);
canvas.drawBitmap(bitmap_rabbit, 0, 0, paint);        //绘制原图
Matrix matrix=new Matrix();                           //创建一个Matrix的对象
matrix.setRotate(30);                                 //将matrix旋转30°
matrix.postTranslate(100, 50);                        //将matrix平移到(100, 50)的位置
canvas.drawBitmap(bitmap_rabbit, matrix, paint);      //绘制图像并应用matrix的变换
```

4.5.5 渲染图像

Android 中渲染图像主要应用 BitmapShader 类,创建该类对象可通过以下的构造方法实现:

```
BitmapShader(Bitmap bitmap, Shader.TileMode tileX, Shader.TileMode tileY)
```

其中,bitmap 参数用于指定一个位图对象,tileX 参数用于指定水平方向图像的重复方式,tileY 参数用于指定垂直方向图像的重复方式。

★注意:Shader.TileMode 类型的参数包括 CLAMP(用边界颜色填充剩余空间)、MIRROR(镜像方式)和 REPEAT(重复方式)3 个可选值。

举例:应用 BitmapShader 渲染图像

步骤 1:新建项目,修改 res/layout 目录下的 XML 布局文件,将默认的布局管理器与组件删除,添加一个帧布局管理器。

步骤 2:定义视图宽度和高度变量,在主活动文件的 onCreate()方法中显示自定义的绘图类。代码如下:

```
private int view_width;
private int view_height;
@Override
public void onCreate(Bundle savedInstanceState){
    super.onCreate(savedInstanceState);
    setContentView(R.layout.main);
    FrameLayout ll=(FrameLayout)findViewById(R.id.myLayout);
                                                    //获取帧布局管理器
    ll.addView(new MyView(this));                   //添加自定义视图到帧布局管理器
}
```

步骤 3:创建 MyView 内部类(继承自 android.view.View 类),添加构造方法并重写 onDraw(Canvas canvas)方法。主要代码如下:

```
public class MyView extends View {
    public MyView(Context context){
        super(context);
        view_width=context.getResources().getDisplayMetrics().widthPixels;
                                                    //屏宽
        view_height=context.getResources().getDisplayMetrics().heightPixels;
                                                    //屏高
    }
    @Override
    protected void onDraw(Canvas canvas){
        Paint paint=new Paint();
        paint.setAntiAlias(true);
        Bitmap bitmap_bg = BitmapFactory.decodeResource (MainActivity.this.
```

```
        getResources(), R.drawable.android);
//创建一个在水平和垂直方向都重复的 BitmapShader 对象
BitmapShader bitmapshader=new BitmapShader(bitmap_bg, TileMode.REPEAT,
TileMode.REPEAT);
paint.setShader(bitmapshader);            //设置渲染对象
canvas.drawRect(0, 0, view_width, view_height, paint);
                                          //绘制要渲染的矩形
Bitmap bm=BitmapFactory.decodeResource(MainActivity.this.getResources
(), R.drawable.img02);
//创建一个在水平方向上重复，在垂直方向上镜像的 BitmapShader 对象
BitmapShader bs = new BitmapShader (bm, TileMode. REPEAT, TileMode.
MIRROR);
    paint.setShader(bs);                  //设置渲染对象
RectF oval=new RectF(0, 0, 280, 180);
canvas.translate(40, 20);      //将画面在 X 轴上平移 40 像素，在 Y 轴上平移 20 像素
canvas.drawOval(oval, paint);  //绘制一个使用 BitmapShader 渲染的椭圆形
super.onDraw(canvas);
    }
}
```

4.6 剪切区域

4.6.1 剪切区域原理

剪切区域也称可视区域，是由画布进行设置的，指的是在画布上设置一块区域，设置了可视区域以后，将看不见区域以外绘制的任何内容。Android 提供的设置可视区域的方法如表 4-10 所示。

表 4-10 剪切区域的常用方法

方　　法	描　　述
clipRect（int left，int top，int right，int bottom）	剪切区域函数，前两个参数是可视区域的左上角坐标，后两个参数为可视区域的右下角坐标
clipPath(Path path)	利用 Path 来设置可视区域的形状
clipRegion(Region region)	利用 Region 来对画布设置可视区域（Region 类表示区域的集合）；常用函数 op(Rect rect,Op op)，参数 op 表示区域块的显示方式：Region. Op. UNION 为区域全部显示，Region. Op. INTERSECT 为区域的交集显示，Region. Op. XOR 为不显示交集区域

4.6.2 基础实例：RPG 游戏地图生成

本例实现二维 RGP 类型游戏中地图的编辑与生成。

具体实现步骤如下。

步骤 1：准备游戏中所需要的图片资源（如图 4-10 所示，原图为顺时针旋转 90°）。把图片复制到项目根目录下的 res/drawable-mdpi 文件夹中。

图 4-10　RPG 游戏地图素材

步骤 2：下载二维地图编辑器软件 MapWin，新建 10 行×15 列的地图文件（设置单元格宽 32 像素，高 32 像素），创建地图并导出图元位置二维数组。MapWin 软件界面如图 4-11 所示。

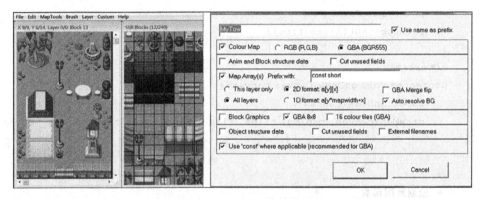

图 4-11　MapWin 软件界面

步骤 3：建立内部类 GameView（继承自 android.view.View 类），添加构造函数并实现其 onDraw() 方法。代码如下：

```
public class MyView extends View {
    private Bitmap mapImg;
    private Paint paint;
    private int tileW, tileH;                    //定义单元格的宽和高
    private int tileNum;
    private int mapX, mapY;                      //定义地图位移
    private short map_map0[][]={                 //地图位置
        { 117, 126, 119, 128, 1, 1, 1, 4, 1, 20 },
        { 125, 134, 127, 136, 1, 4, 11, 23, 24, 28 },
        { 133, 142, 143, 144, 1, 4, 4, 31, 32, 20 },
        { 141, 150, 151, 152, 97, 1, 4, 4, 12, 28 },
        { 12, 1, 1, 1, 105, 1, 1, 1, 1 },
        { 1, 1, 1, 1, 113, 56, 6, 7, 56, 13 },
```

```java
        { 1, 185, 186, 187, 1, 16, 14, 15, 16, 13 },
        { 1, 193, 194, 195, 1, 1, 1, 1, 17, 1 },
        { 1, 201, 202, 203, 1, 1, 1, 1, 25, 3 },
        { 1, 209, 210, 211, 97, 1, 1, 123, 124, 4 },
        { 12, 217, 218, 219, 105, 1, 1, 131, 132, 1 },
        { 4, 225, 226, 227, 113, 1, 1, 139, 140, 1 },
        { 12, 233, 234, 235, 5, 1, 1, 147, 148, 1 },
        { 106, 106, 106, 106, 1, 1, 106, 106, 106, 106 },
        { 114, 114, 114, 114, 64, 64, 114, 114, 114, 114 }
    };
    public MyView(Context context){
        super(context);
        tileW=32;                                         //单元格宽度
        tileH=32;                                         //单元格高度
        mapImg=BitmapFactory.decodeResource(getResources(), R.drawable.map);
        tileNum=mapImg.getWidth()/tileW;        //每一行的单元格数量
        paint=new Paint();
        paint.setAntiAlias(true);
    }
    @Override
    protected void onDraw(Canvas canvas){
        super.onDraw(canvas);
        drawMap(canvas, mapImg, tileW, tileH, mapX, mapY, tileNum, map_map0);
                                                          //绘制地图
    }
    /**
     * 绘制地图函数
     * @param canvas
     * @param mapImg   地图原图图片
     * @param tileW    每个tile的宽
     * @param tileH    每个tile的高
     * @param mapX     地图的X位置
     * @param mapY     地图的Y位置
     * @param tileNum  地图原图图片上的一行tile的总个数
     * @param map      地图数组
     */
    void drawMap(Canvas canvas, Bitmap mapImg, int tileW, int tileH, int mapX,
        int mapY, int tileNum, short map[][]){
        for(int i=0; i<map.length; i++){
            for(int j=0; j<map[i].length; j++){
                canvas.save();
                canvas.clipRect(j * tileW+mapX, i * tileH+mapY, (j+1)
                    * tileW+mapX, i * tileH+tileH+mapY);
                canvas.drawBitmap(mapImg, j * tileW+mapX
```

```
                    -((map[i][j]-1)%tileNum)* tileW, i * tileH+mapY
                    -((map[i][j]-1)/tileNum)* tileH, paint);
            canvas.restore();
        }
    }
}
```

步骤4：在主活动文件 MainActivity.java 的 onCreate()方法中，设置屏幕属性并调用 MyView 视图。代码如下：

```
@Override
protected void onCreate(Bundle savedInstanceState){
    super.onCreate(savedInstanceState);
    setRequestedOrientation(ActivityInfo.SCREEN_ORIENTATION_LANDSCAPE);
                            //设置横屏
    …                       //此处省略隐藏标题栏、全屏及保持屏幕亮度的代码
    MyView gameView=new MyView(this);
    gameView.setFocusable(true);
    setContentView(gameView);
}
```

运行效果如图 4-12 所示。

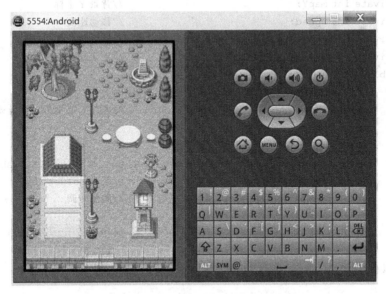

图 4-12 RPG 游戏地图生成实例运行界面

4.6.3 基础实例：游戏中的自动滚屏

本例实现射击类游戏或过关类游戏中背景的自动滚动。

具体实现步骤如下。

步骤1：准备游戏中所需要的背景图片资源（如图4-13所示，原图为顺时针旋转90°）；把图片复制到项目根目录下的res/drawable-mdpi文件夹中。

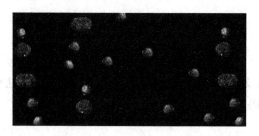

图4-13 自动滚屏背景素材

步骤2：建立内部类GameSurfaceView（继承自android.view.SurfaceView类），添加构造函数并实现Callback接口（android.view.SurfaceHolder.Callback）。实现Runable接口并重写其run()方法。代码如下：

```java
public class GameSurfaceView extends SurfaceView implements Callback, Runnable {
    private SurfaceHolder holder;              //定义SurfaceHolder对象
    private Bitmap bgMap;                       //定义Bitmap对象
    private Paint paint;                        //定义画笔
    private Canvas canvas;                      //定义画布
    private int mapY;                           //背景Y坐标
    private int speed;                          //移动速度
    private boolean isGame=true;                //标识游戏是否进行
    private MainActivity activity;              //主活动对象变量
    public MyView(Context context){
        super(context);
        activity=(MainActivity)context;         //通过上下文传入主活动对象
        paint=new Paint();
        paint.setAntiAlias(true);
        bgMap=BitmapFactory.decodeResource(getResources(), R.drawable.bg);
        holder=getHolder();                     //获取Holder对象
        holder.addCallback(this);               //添加回调
        speed=3;
    }
    /**
     * 背景移动
     */
    private void move(){
        mapY+=speed;                            //移动背景
        if(mapY>=this.getHeight())              //如果背景图Y坐标大于屏幕高度
            //回到初始位置(屏幕高度 - 背景图高度)
            mapY=this.getHeight()-bgMap.getHeight();
    }
```

```java
/**
 * 图片绘制
 */
    private void render(){
        canvas=holder.lockCanvas();              //锁定画布
        if(mapY<0)                                //如果背景图Y坐标小于0
            canvas.drawBitmap(bgMap, 0, mapY, paint);
        else {
            canvas.drawBitmap(bgMap, 0, mapY -bgMap.getHeight(), paint);
            canvas.drawBitmap(bgMap, 0, mapY, paint);
        }
        holder.unlockCanvasAndPost(canvas);      //将画布解锁
    }
    @Override
    public void surfaceChanged(SurfaceHolder holder, int format, int width, int height){
        //TODO: Auto-generated method stub
    }
    @Override
    public void surfaceCreated(SurfaceHolder holder){
        mapY=this.getHeight()-bgMap.getHeight();  //背景初始位置(屏高 -背景图高)
        isGame=true;                              //游戏开始
        new Thread(this).start();                 //启动线程
    }
    @Override
    public void surfaceDestroyed(SurfaceHolder holder){
        isGame=false;                             //游戏结束
        activity.finish();                        //同时关闭主活动
    }
    @Override
    public void run(){
        while(isGame){
            move();                               //调用背景移动函数
            render();                             //调用背景绘制函数
            try {
                Thread.sleep(100);
            } catch(InterruptedException e){
                e.printStackTrace();
            }
        }
    }
}
```

步骤3：在主活动的 onCreate 方法中，设置屏幕属性并调用 MyView 视图。代码

如下:

```
@Override
protected void onCreate(Bundle savedInstanceState){
    super.onCreate(savedInstanceState);
    setRequestedOrientation(ActivityInfo.SCREEN_ORIENTATION_LANDSCAPE);
        …                          //此处省略隐藏标题栏、全屏及保持屏幕亮度的代码
    GameSurfaceView gameView=new GameSurfaceView(this);
    gameView.setFocusable(true);
    setContentView(gameView);
}
```

运行效果如图 4-14 所示。

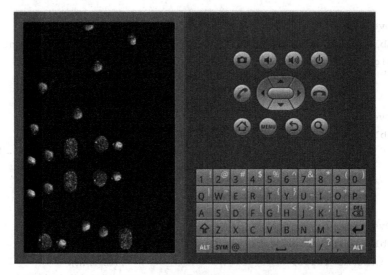

图 4-14　自动滚屏实例运行界面

4.7　游戏动画

在 Android 中提供了逐帧动画和补间动画两种动画类型,均可以在 XML 文件中定义动画资源文件。另外,也可以通过重写 Animation 的 applyTransformation()函数实现自定义动画效果。

4.7.1　逐帧动画

逐帧动画就是按顺序播放静态图像,先定义一组生成动画的图片资源。代码如下:

```
<?xml version="1.0" encoding="utf-8"?>
<animation-list xmlns:android="http://schemas.android.com/apk/res/android"
    android:oneshot="true|false">
```

```
<item android:drawable=" @ drawable/图片资源名称 1" android: duration =
    "integer"/>
… <!--此处省略部分<item></item>标记 -->
<item android: drawable = " @ drawable/图片资源名称 1" android: duration =
    "integer"/>
</animation-list>
```

在上面的代码中,android:oneshot 属性用于设置是否循环播放(默认值为 true),android:drawable 属性用于指定要显示的图片资源,android:duration 属性指定图片资源持续的时间。

举例:奔跑的小动物

步骤 1:新建项目,准备表现奔跑小动物的序列图片(如图 4-15 所示),并把图片资源复制到根目录下的 res/drawable-mdpi 文件夹中。

图 4-15 小动物序列图片素材

步骤 2:在 res 目录中创建名称为 anim 的文件夹,并在该文件夹中添加一个名称为 run_lionet.xml 的 XML 动画资源文件。代码如下:

```
<?xml version="1.0" encoding="utf-8"?>
<animation-list xmlns:android="http://schemas.android.com/apk/res/android">
    <item android:drawable="@drawable/lionet01" android:duration="50"/>
    <item android:drawable="@drawable/lionet02" android:duration="50"/>
    <item android:drawable="@drawable/lionet02" android:duration="50"/>
    <item android:drawable="@drawable/lionet04" android:duration="50"/>
</animation-list>
```

步骤 3:修改 res/layout 目录下的 XML 布局文件,设置背景为步骤 2 中创建的动画资源文件。代码如下:

```
<LinearLayout xmlns:android="http://schemas.android.com/apk/res/android"
    xmlns:tools="http://schemas.android.com/tools"
    android:id="@+id/myLayout"
    android:background="@anim/run_lionet"
    android:layout_width="fill_parent"
    android:layout_height="fill_parent"
    android:orientation="vertical">
</LinearLayout>
```

步骤 4:在默认的主活动文件中获取布局文件和 AnimationDrawable 对象,完成动画控制。代码如下:

```java
public class MainActivity extends Activity {
    private boolean flag=true;
    @Override
    protected void onCreate(Bundle savedInstanceState){
        super.onCreate(savedInstanceState);
        setContentView(R.layout.activity_main);
        LinearLayout myLayout= (LinearLayout)findViewById(R.id.myLayout);
        final AnimationDrawable anim= (AnimationDrawable)myLayout.getBackground();
        myLayout.setOnClickListener(new View.OnClickListener(){
            @Override
            public void onClick(View v){
                if(flag){
                    anim.start();
                    flag=false;
                }else{
                    anim.stop();
                    flag=true;
                }
            }
        });
    }
}
```

要在布局文件和组件中使用上面定义的动画资源,通常可以将其设置为背景。如果是在 Java 代码中创建逐帧动画,首先创建 AnimationDrawable 对象,然后调用 addFrame() 方法向动画中添加帧,每调用一个 addFrame() 方法,可以添加一个帧。

4.7.2 补间动画

补间动画是通过对场景里的对象不断做图像变换(平移、缩放、旋转)来产生动画效果。在实现补间动画时,只需定义动画开始和结束的关键帧。在 Android 中提供了透明度渐变动画、旋转动画、缩放动画和平移动画 4 种补间动画。这 4 种补间动画都具有的常用属性如表 4-11 所示。

表 4-11　4 种补间动画的常用属性

属　　性	描　　述
android:repeatMode	设置动画的重复方式,可选值为 reverse(反向)或 restart(重新开始)
android:repeatCount	设置动画的重复次数,属性可以是代表次数的数值或 infinite(无限循环)
android:duration	设置动画的持续时间(单位:毫秒)

1. 透明度渐变动画

通过视图组件透明度的变化来实现渐隐渐显效果,主要为动画指定开始时的透明度和结束时的透明度。定义透明度渐变动画的常用属性如表 4-12 所示。

表 4-12 定义透明度渐变动画的常用属性

属　　性	描　　述
android:fromAlpha	设置动画开始时的透明度(值为 0.0 代表完全透明,值为 1.0 代表完全不透明)
android:toAlpha	设置动画结束时的透明度(值为 0.0 代表完全透明,值为 1.0 代表完全不透明)
android:interpolator	@android:anim/linear_interpolator　一直做匀速改变
	@android:anim/accelerate_interpolator　动画开始时改变速度慢,后加速
	@android:anim/decelerate_interpolator　动画开始时改变速度快,后减速
	@android:anim/accelerate_decelerate_interpolator　动画开始和结束时改变速度慢,中间加速
	@android:anim/cycle_interpolator　动画播放特定次数,变化速度按正弦曲线改变
	@android:anim/bounce_interpolator　动画结束的地方采用弹球效果
	@android:anim/overshoot_interpolator　动画快速到终点并超出一部分,最后回到结束地方
	@android:anim/anticipate_interpolator　动画开始的地方先向后退一部分,最后快速到动画结束地方
	@android:anim/anticipate_overshoot_interpolator　动画开始的地方先向后退一部分,再开始动画,到结束的地方再超出一部分,最后回到动画结束的地方

例如,定义一个视图组件从完全不透明到完全透明、持续时间为 2s 的动画,代码如下：

```
<?xml version="1.0" encoding="utf-8"?>
<set xmlns:android="http://schemas.android.com/apk/res/android">
    <alpha
        android:duration="2000"
        android:fromAlpha="1"
        android:toAlpha="0"/>
</set>
```

2．旋转动画

通过为动画指定开始时的旋转角度、结束时的旋转角度以及持续时间来创建动画。在旋转时可以通过指定轴心点坐标的方式改变旋转中心。定义旋转动画时常用的属性如表 4-13 所示。

表 4-13 定义旋转动画的常用属性

属 性	描 述
android:pivotX	设置轴心点 X 轴坐标
android:pivotY	设置轴心点 Y 轴坐标
android:fromDegrees	设置动画开始时的旋转角度
android:toDegrees	设置动画结束时的旋转角度
android:interpolator	设置动画的变化速度(匀速、加速、减速或抛物线式变速等)

例如,定义一个使图片从 0°转到 360°、持续时间为 2s、以图片中心为旋转中心点的动画,代码如下:

```
<?xml version="1.0" encoding="utf-8"?>
<set xmlns:android="http://schemas.android.com/apk/res/android">
    <rotate
        android:duration="2000"
        android:fromDegrees="0"
        android:pivotX="50%"
        android:pivotY="50%"
        android:toDegrees="360"/>
</set>
```

3. 缩放动画

通过为动画指定开始时的缩放系数、结束时的缩放系数和持续时间来创建动画。在缩放时可以通过指定轴心点坐标的方式改变缩放中心。定义缩放动画时常用的属性如表 4-14 所示。

表 4-14 定义缩放动画的常用属性

属 性	描 述
android:pivotX	设置轴心点 X 轴坐标
android:pivotY	设置轴心点 Y 轴坐标
android:fromXScale	设置动画开始时水平方向的缩放系数(值为 1.0 表示不变化)
android:toXScale	设置动画结束时水平方向的缩放系数(值为 1.0 表示不变化)
android:fromYScale	设置动画开始时垂直方向的缩放系数(值为 1.0 表示不变化)
android:toYScale	设置动画结束时垂直方向的缩放系数(值为 1.0 表示不变化)
android:interpolator	设置动画的变化速度(匀速、加速、减速或抛物线式变速等)

例如,定义一个将图片放大 2 倍、持续时间为 2s、以图片中心为缩放中心点的动画,代码如下:

```
<?xml version="1.0" encoding="utf-8"?>
<set xmlns:android="http://schemas.android.com/apk/res/android">
    <scale
        android:duration="2000"
```

```
            android:fromXScale="1"
            android:fromYScale="1"
            android:pivotX="50%"
            android:pivotY="50%"
            android:toXScale="2.0"
            android:toYScale="2.0"/>
</set>
```

4. 平移动画

通过为动画指定开始时的位置、结束时的位置和持续时间来创建动画。定义平移动画时常用的属性如表 4-15 所示。

表 4-15　定义平移动画的常用属性

属　　性	描　　述
android:fromXDelta	设置动画开始时水平方向上的起始位置
android:toXDelta	设置动画结束时水平方向上的起始位置
android:fromYDelta	设置动画开始时垂直方向上的起始位置
android:toYDelta	设置动画结束时垂直方向上的起始位置
android:interpolator	设置动画的变化速度(匀速、加速、减速或抛物线式变速等)

例如,定义一个让图片从(0,0)坐标点到(200,200)坐标点、持续时间为 2s 的动画,代码如下:

```
<?xml version="1.0" encoding="utf-8"?>
<set xmlns:android="http://schemas.android.com/apk/res/android">
    <translate
        android:duration="2000"
        android:fromXDelta="0"
        android:fromYDelta="0"
        android:toXDelta="200"
        android:toYDelta="200"/>
</set>
```

举例：补间动画实例

步骤 1：新建项目,在 res 目录下新建名称为 anim 的文件夹,在该文件夹内创建实现透明度动画、旋转动画、缩放动画和平移动画的资源文件。

创建名称为 anim_alpha.xml 的透明度动画资源文件。该动画持续时间为 2s,实现从完全不透明到完全透明,再到完全不透明的动画效果。代码如下:

```
<?xml version="1.0" encoding="utf-8"?>
<set xmlns:android="http://schemas.android.com/apk/res/android">
    <alpha
        android:duration="2000"
        android:fillAfter="true"
```

```
        android:fromAlpha="1"
        android:repeatCount="1"
        android:repeatMode="reverse"
        android:toAlpha="0"/>
</set>
```

创建名称为 anim_rotate.xml 的旋转动画资源文件。该动画持续时间为 2s，实现从 0°转到 720°，再从 360°转到 0°的动画效果。代码如下：

```
<?xml version="1.0" encoding="utf-8"?>
<set xmlns:android="http://schemas.android.com/apk/res/android">
    <rotate
        android:duration="2000"
        android:fromDegrees="0"
        android:interpolator="@android:anim/accelerate_interpolator"
        android:pivotX="50%"
        android:pivotY="50%"
        android:toDegrees="720"/>
    <rotate
        android:duration="2000"
        android:fromDegrees="360"
        android:interpolator="@android:anim/accelerate_interpolator"
        android:pivotX="50%"
        android:pivotY="50%"
        android:startOffset="2000"
        android:toDegrees="0"/>
</set>
```

创建名称为 anim_scale.xml 的缩放动画资源文件。该动画持续时间为 2s，实现将元素放大 2 倍，再逐渐收缩为原尺寸的动画效果。代码如下：

```
<?xml version="1.0" encoding="utf-8"?>
<set xmlns:android="http://schemas.android.com/apk/res/android">
    <scale
        android:duration="2000"
        android:fillAfter="true"
        android:fromXScale="1"
        android:fromYScale="1"
        android:interpolator="@android:anim/decelerate_interpolator"
        android:pivotX="50%"
        android:pivotY="50%"
        android:repeatCount="1"
        android:repeatMode="reverse"
        android:toXScale="2.0"
        android:toYScale="2.0"/>
</set>
```

创建名称为 anim_translate.xml 的平移动画资源文件。该动画持续时间为 2s，实现从左侧移动到右侧，再从右侧移动到左侧的动画效果。代码如下：

```xml
<?xml version="1.0" encoding="utf-8"?>
<set xmlns:android="http://schemas.android.com/apk/res/android">
    <translate
        android:duration="2000"
        android:fillAfter="true"
        android:fromXDelta="0"
        android:fromYDelta="0"
        android:repeatCount="1"
        android:repeatMode="reverse"
        android:toXDelta="860"
        android:toYDelta="0">
    </translate>
</set>
```

步骤 2：修改 res/layout 目录下的 XML 布局文件，采用线性布局方式，添加一个 ImageView 组件和 4 个按钮组件。代码如下：

```xml
<LinearLayout xmlns:android="http://schemas.android.com/apk/res/android"
    xmlns:tools="http://schemas.android.com/tools"
    android:layout_width="fill_parent"
    android:layout_height="fill_parent"
    android:orientation="vertical">
    <ImageView
        android:id="@+id/img_superman"
        android:layout_width="wrap_content"
        android:layout_height="wrap_content"
        android:src="@drawable/superman"/>
    <LinearLayout
        android:layout_width="match_parent"
        android:layout_height="wrap_content">
        <Button
            android:id="@+id/btn_alpha"
            android:layout_width="wrap_content"
            android:layout_height="wrap_content"
            android:text="透明动画"
            android:layout_weight="1"/>
        <!--省略部分代码 -->
        ...
    </LinearLayout>
</LinearLayout>
```

步骤 3：打开默认创建的主活动文件，在 onCreate() 方法中完成动画调用。代码

如下：

```
public class MainActivity extends Activity {
    private ImageView img_superman;
    private Button btn_alpha;
    @Override
    protected void onCreate(Bundle savedInstanceState){
        super.onCreate(savedInstanceState);
        setContentView(R.layout.activity_main);
        final Animation animation_alpha=AnimationUtils.loadAnimation(this, R.anim.anim_alpha);                    //获取透明度变化动画资源
        final Animation animation_rotate=AnimationUtils.loadAnimation(this, R.anim.anim_rotate);                  //获取旋转动画资源
        final Animation animation_scale=AnimationUtils.loadAnimation(this, R.anim.anim_scale);                    //获取缩放动画资源
        final Animation animation_translate = AnimationUtils. loadAnimation(this, R.anim.anim_translate);         //获取平移动画资源
        img_superman= (ImageView)findViewById(R.id.img_superman);
        btn_alpha= (Button)findViewById(R.id.btn_alpha);
        btn_alpha.setOnClickListener(new View.OnClickListener(){
            @Override
            public void onClick(View v){
                img_superman.startAnimation(animation_alpha);
                                                    //播放透明度变化动画
            }
        });
        …                        //此处省略旋转、缩放、平移3个功能按钮的事件代码
    }
}
```

运行效果如图4-16所示。

4.7.3　自定义动画

自定义动画要重写Animation的applyTransformation()函数，然后通常要实现initialize()函数，这是一个回调函数，告诉Animation目标View的大小，可以初始化一些相关的参数。在绘制动画的过程中会反复调用applyTransformation()函数，参数interpolatedTime的值（0～1）在每次调用中都会变化。

通过参数Transformation来获取变换的矩阵（matrix），通过改变矩阵就可以实现各种复杂的效果。代码如下：

图4-16　补间动画实例运行界面

```java
public class MyAnimation extends Animation {
int mCenterX, mCenterY;
    public MyAnimation(){
    }
@Override
protected void applyTransformation(float interpolatedTime, Transformation t){
    //通过 Matrix.setScale 函数来缩放,该函数的两个参数代表 X、Y 轴缩放因子
    //由于 interpolatedTime 是从 0 到 1 变化,所以在这里实现的效果就是控件从最小逐渐
      变化到最大
    Matrix matrix=t.getMatrix();
    matrix.setScale(interpolatedTime, interpolatedTime);
    //Matrix 可以实现各种复杂的变换
    //preTranslate 函数是在缩放前移动而 postTranslate 是在缩放完成后移动。
    matrix.preTranslate(-mCenterX, -mCenterY);
    matrix.postTranslate(mCenterX, mCenterY);
}
@Override
public void initialize(int width, int height, int parentWidth, int parentHeight)
{
    super.initialize(width, height, parentWidth, parentHeight);
    //初始化中间坐标
    mCenterX=width/2;
    mCenterY=height/2;
    //设置变换持续的时间 2500ms,然后设置 Interpolator 为 LinearInterpolator
    //并设置 FillAfter 为 true,这样可以在动画结束的时候保持动画的完整性
    setDuration(2000);
    setFillAfter(true);
    setInterpolator(new LinearInterpolator());
    }
}
```

接口 applyTransformation 的作用就是提供一个变换矩阵来对 View 进行变换,需要使用的 Matrix 的功能(如位移、大小等)都有响应的接口可以直接使用。

举例:自定义动画实例

步骤 1:为了方便定义动画的状态(也可以称为"关键帧"),简单定义一个类。代码如下:

```java
public class AnimationStatus {
    public float X=0;
    public float Y=0;
    public float Scale_X=1;
    public float Scale_Y=1;
    public float Alpha=1;
    public float Time=0;
```

}

步骤 2：编写继承自 Animation 的 ExtAnimation 对象。主要的接口和代码如下：

```
public class ExtAnimation extends Animation {
private List mAniList=new ArrayList();         //声明用于保存动画关键帧的列表
public ViewSwitchAnimation(){
    this.setFillAfter(true);
    this.setFillEnabled(true);
    this.setDuration(1000);
}
 //添加一个关键帧(添加时 Status 请按 Time 排序)
 public void addStatus(AnimationStatus status, float timepoint){
    keyStatus.Time=timepoint;
    mAniList.add(status);
 }
@Override
public void initialize(int width, int height, int parentWidth, int parentHeight)
{
    super.initialize(width, height, parentWidth, parentHeight);
 }
//最重要的方法
@Override
protected void applyTransformation(float interpolatedTime, Transformation t){
    if(mAniList.size()<1)
        return;
    int index1=0;
    int index2=1;
    for(int i=0; i<mAniList.size()-1;++i){
        if(interpolatedTime >= mAniList.get(i).Time && interpolatedTime <=
        mAniList.get(i+1).Time){
            index1=i;
            index2=i+1; break;
        }
    }
//在 keyStatus1 和 keyStatus2 指定的状态之间做动画
AnimationStatus keyStatus1=mAniList.get(index1);
AnimationStatus keyStatus2=mAniList.get(index2);
    if(keyStatus1==null || keyStatus2==null)
        return;
    Matrix matrix=t.getMatrix();                    //取得当前变换矩阵
    float alpha1=keyStatus1.Alpha;                  //调整透明度
    float alpha2=keyStatus2.Alpha;
    t.setAlpha(alpha1+ (alpha2 -alpha1) * (offset_time)/timeTotal);
    float x1=keyStatus1.X;                          //调整位移
```

```
            float y1=keyStatus1.Y;
            float x2=keyStatus2.X;
            float y2=keyStatus2.Y;
            matrix.preTranslate(x1+(x2 - x1) * offset_time/timeTotal, y1+(y2 - y1) *
            offset_time/timeTotal);
            float sx1=keyStatus1.Scale_X;         //调整大小
            float sy1=keyStatus1.Scale_Y;
            float sx2=keyStatus2.Scale_X;
            float sy2=keyStatus2.Scale_Y;
            matrix.preScale(sx1+(sx2 - sx1) * offset_time/timeTotal, sy1+(sy2 - sy1)
            * offset_time/timeTotal);
        }
}
```

以上就是一个最简单的自定义动画,可以连续加入任意多个 AnimationStatus 来指定动画的关键状态,比如(0,0,1,1,1)到(10,50,1.5f,1.5f,0.5f)到(10,−50,2,2,0)表示:从初始状态,首先移动到(10,50)坐标处,并且长宽变为1.5倍,Alpha 变成0.5;然后再次移动到(10,−50)坐标处,并且长宽变为2倍,Alpha 变为0。这样实现的好处在于:与两个 AnimationSet 顺序执行相比,消除了 AniamtionSet 切换时的"卡顿"现象,整个动画流程将流畅很多。

上述的实现中,在每两个关键帧之间的动画采取的只是简单的线性函数,即时间变量(interpolatedTime)的一元一次方程,也可以采用更复杂的计算函数(变量为 interpolatedTime)来计算每一次的位移等,从而实现任何复杂的动画曲线。

4.8 综合实例一: 小小弹球

4.8.1 功能描述

本例实现一款简单的弹球游戏,利用触屏控制底部弹板弹起小球,游戏失败后,再次触屏可以重启游戏。

4.8.2 关键技术

本例实现的关键是 Timer 类的使用和小球与边界、弹板是否接触的计算。代码如下:

```
final Timer timer=new Timer();
    timer.schedule(new TimerTask(){
        @Override
        public void run(){
            if(ballX<=0 || ballX>=(screen_width -ball_size)){
                xSpeed=-xSpeed;
```

```
                    }
                    if(ballY<=0 || ballY>=(rectY-ball_size)
                            && ballX>=rectX && ballX<=(rectX+rect_width)){
                        ySpeed=-ySpeed;
                    } else if(ballY>=(rectY-ball_size)&&(ballX<rectX
                            || ballX>(rectX+rect_width))){
                        isGameOver=true;
                        timer.cancel();
                    }
                    ballX+=xSpeed;
                    ballY+=ySpeed;
                    Message msg=new Message();
                    msg.what=1;
                    handler.sendMessage(msg);
                }
            }, 0, 300);
            new Thread(this).start();
        }
```

4.8.3 实现过程

步骤1：新建项目MyPinBall，定义游戏戏中需要的常量和变量。代码如下：

```
private static final int MOVE=0;                    //弹板移动信号
private static final int LEFT=1;                    //弹板向左移动
private static final int RIGHT=2;                   //弹板向右移动
private int direction=3;                            //弹板的移动方向
private int screenW, screenH;
Random rand=new Random();
private int ballX=rand.nextInt(150)+20;             //小球的X坐标
private int ballY=rand.nextInt(20)+10;              //小球的Y坐标
private int rectX=60;                               //定义弹板的X坐标
private int rectY=420;                              //定义弹板的Y坐标
private int rect_width=150;                         //定义弹板的宽度
private int rect_height=20;
private int ball_size=15;                           //定义小球尺寸
private double xyRate=rand.nextDouble()-0.5;        //定义小球运动的速率
private int ySpeed=30;                              //定义小球在Y轴方向上的速度
private int xSpeed=(int)(ySpeed * xyRate * 2);      //定义小球在X轴方向上的速度
private boolean isLose=false;                       //标识游戏是否失败
private boolean isMove=false;                       //标识弹板是否移动
private int rect_move_speed=5;                      //定义弹板移动速度
private int destX=10;                               //接收触屏X坐标值
private Handler handler;                            //定义Handler对象
```

```
private Timer timer=new Timer();                    //定义 Timer 对象
```

步骤 2：建立内部类 GameView（继承自 android.view.View 类），添加构造函数并实现其 onDraw()方法。代码如下：

```
public class GameView extends View {
    public GameView(Context context){
        super(context);
        setFocusable(true);                         //设置视图获取焦点
    }
    @Override
    protected void onDraw(Canvas canvas){
        super.onDraw(canvas);
        Paint paint=new Paint();
        paint.setAntiAlias(true);
        paint.setStyle(Paint.Style.FILL);
        if(isLose){                                 //如果游戏结束
            paint.setColor(Color.RED);
            paint.setTextSize(24);
            canvas.drawText("游戏结束!", 50, 20, paint);
        } else {
            paint.setColor(Color.rgb(60, 200, 20));
            canvas.drawCircle(ballX, ballY, ball_size, paint);
                                                    //绘制小球
            canvas.drawRect(rectX, rectY, rectX+rect_width, rectY+rect_height,
            paint);                                 //绘制弹板
        }
    }
}
```

步骤 3：在主活动文件的 onCreate()方法中设置屏幕、调用内部类度设置弹板和小球的初始位置。代码如下：

```
@Override
protected void onCreate(Bundle savedInstanceState){
    super.onCreate(savedInstanceState);
    ...                         //此处省略隐藏标题栏、全屏及保持屏幕亮度的代码
    final GameView gameView=new GameView(this);
    setContentView(gameView);
    Display display=getWindowManager().getDefaultDisplay();
    Point point=new Point();
    display.getRealSize(point);
    screenW=point.x;                                //获取屏幕宽度
    screenH=point.y;                                //获取屏幕高度
    rectX=(screenW - rect_width)/2;                 //设置弹板的 X 坐标
    rectY=screenH - 30;                             //设置弹板的 Y 坐标
```

```
ballX=rand.nextInt(screenW-40)+20;        //设置小球的 X 坐标
ballY=20;                                 //设置小球的 Y 坐标
if(rand.nextBoolean())                    //小球出现时的左右随机方向
{
    xSpeed=-xSpeed;
}
}
```

步骤 4：应用 Timer 类实现小球在屏幕内的运动规则，定义并实现 Handler 实例，接收线程消息并判断弹板移动的方向。代码如下：

```
handler=new Handler(){
    @Override
    public void handleMessage(Message msg){
        super.handleMessage(msg);
        if(msg.what==1){
            gameView.invalidate();
        }
    }
};
timer.schedule(new TimerTask(){
    @Override
    public void run(){
        //如果小球碰到右边框或左边框
        if(ballX>=screenW-ball_size || ballX<=ball_size){
            xSpeed=-xSpeed;
        }
        //当小球碰到顶部，小球在弹板范围内
        if(ballY<=ball_size||(ballY>(rectY-ball_size)
                    && ballX>rectX && ballX<rectX+rect_width)){
            ySpeed=-ySpeed;
        }
        //小球超出弹板范围
        if(ballY>=rectY-ball_size
                    &&(ballX<=rectX || ballX>(rectX+rect_width))){
            timer.cancel();
            isLose=true;
        }
        ballX=ballX+xSpeed;                //小球向 X 方向移动
        ballY=ballY+ySpeed;                //小球向 Y 方向移动
        handler.sendEmptyMessage(1);       //发送消息
    }
}, 0, 300);
```

步骤 5：重写当前主活动的 onTouchEvent(MotionEvent event)方法，获取触屏位

置。代码如下:

```java
@Override
public boolean onTouchEvent(MotionEvent event){
    if(event.getAction()==MotionEvent.ACTION_DOWN){
        if((int)event.getX()>=screenH - rect_width){
            destX=screenW - rect_width;
        } else {
            destX= (int)event.getX();
        }
        isMove=true;                              //弹板可以移动
    }
    return true;
}
```

步骤6:在当前主活动中实现 Runable 接口并重写其 run() 方法,实现触屏后弹板的左右滑动。代码如下:

```java
@Override
public void run(){
    while(isLose==false){                         //当游戏没有失败
        if(isMove){                               //当移动信号为真时
            //根据触屏的 X 坐标设置弹板移动方向
            direction=destX< rectX ? LEFT : RIGHT;
            switch(direction){
            case LEFT:                            //向左移动
                rectX -=rect_move_speed;  //如果弹板的 X 坐标大于或等于触屏的 X 坐标
                if(rectX<=destX){         //如果弹板的 X 坐标小于或等于触屏的 X 坐标
                    isMove=false;                 //弹板停止移动
                }
                break;
            case RIGHT:                           //向右移动
                rectX+=rect_move_speed;
                if(rectX>=destX){         //如果弹板的 X 坐标大于或等于触屏的 X 坐标
                    isMove=false;                 //弹板停止移动
                }
                break;
            }
            handler.sendEmptyMessage(MOVE);
        }
        try {
            Thread.sleep(100);
        } catch(InterruptedException e){
            e.printStackTrace();
        }
    }
```

            }
    }

修改 Handler 对象接收消息的条件。代码如下:

```
if(msg.what==1 || msg.what==MOVE){
    gameView.invalidate();
}
```

在主活动中启动线程。代码如下:

```
new Thread(this).start();
```

步骤 7:编写重新启动游戏函数。代码如下:

```
public void restartGame()
    {
        ballX=rand.nextInt(screenW-40)+20;
        ballY=rand.nextInt(20)+20;
        timer.purge();                          //移除定时器
        timer=null;
        timer=new Timer();
        timer.schedule(new TimerTask()
        {
            @Override
            public void run()
            {
                //如果小球碰到右边框或左边框
                if(ballX>=screenW -ball_size || ballX<=ball_size)
                {
                    xSpeed=-xSpeed;
                }
                //当小球碰到顶部,小球在弹板范围内
                if(ballY<=ball_size||(ballY>(rectY -ball_size)
                    && ballX>rectX && ballX< rectX  +rect_width))
                {
                    ySpeed=-ySpeed;
                }
                //小球超出弹板范围
                if(ballY>=rectY -ball_size
                    &&(ballX<=rectX || ballX>(rectX+rect_width)))
                {
                    timer.cancel();
                    isLose=true;
                }
                ballX=ballX+xSpeed;
                ballY=ballY+ySpeed;
```

```
            handler.sendEmptyMessage(1);
        }
    }, 0, 300);
    new Thread(this).start();
}
```

在 onTouchEvent 方法的触屏判断中添加对游戏重启函数的调用。代码如下：

```
if(event.getAction()==MotionEvent.ACTION_DOWN){
    if((int)event.getX()>=screenH-rect_width){
        …                              //此处省略部分代码
    if(isLose)
    {
        isLose=false;
        restartGame();
    }
}
```

4.8.4 实例拓展

为游戏添加背景图，用图片资源替换游戏中采用绘图函数实现的小球和弹板，用重力传感器控制弹板移动。

步骤 1：准备游戏中所需要的图片资源，这里包括游戏背景、小球及弹板 3 张图片（如图 4-17 所示）。并把图片资源复制到项目根目录下的 res/drawable-mdpi 文件夹中。

设置游戏背景，把游戏中的小球和弹板换成素材图片。修改游戏视图类 MyView 中的 onDraw()函数。代码如下：

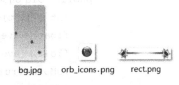

图 4-17 弹球图片素材

```
@Override
public void draw(Canvas canvas){
    super.draw(canvas);
    Paint paint=new Paint();
    paint.setAntiAlias(true);
    paint.setStyle(Paint.Style.FILL);
    Bitmap bg=BitmapFactory.decodeResource(getResources(), R.drawable.bg);
    float sx=screenW/(float)bg.getWidth();
    float sy=screenH/(float)bg.getHeight();
    Matrix m=new Matrix();
    m.postScale(sx, sy, 0, 0);
    canvas.drawBitmap(bg, m, paint);
    if(isLose){                          //如果游戏结束
        paint.setColor(Color.RED);
```

```
            paint.setTextSize(24);
            canvas.drawText("游戏结束!", 50, 20, paint);
        }else{
            paint.setColor(Color.rgb(200, 60, 50));
            Bitmap bm=BitmapFactory.decodeResource(getResources(), R.drawable.orb
            _icons);
            canvas.drawBitmap(bm, ballX, ballY, paint);
            Bitmap rm= BitmapFactory. decodeResource (getResources (), R. drawable.
            rect);
            rect_width=rm.getWidth();
            canvas.drawBitmap(rm, rectX, rectY, paint);
        }
    }
```

步骤 2：编写 changeDirection() 函数，应用重力传感器控制弹板左右移动。代码如下：

```
    public void changeDirection(){
        SensorManager sm= (SensorManager)getSystemService(SENSOR_SERVICE);
        Sensor sr=sm.getDefaultSensor(Sensor.TYPE_ACCELEROMETER);
        SensorEventListener se=new SensorEventListener(){
            @Override
            public void onSensorChanged(SensorEvent event){
                int x=(int)event.values[SensorManager.DATA_X];
                float y=event.values[SensorManager.DATA_Y];
                float z=event.values[SensorManager.DATA_Z];
                isMove=true;
                if(x>0){
                    direction=LEFT;
                } else if(x<0){
                    direction=RIGHT;
                } else {
                    isMove=false;
                }
            }
            @Override
            public void onAccuracyChanged(Sensor sensor, int accuracy){
                //TODO: Auto-generated method stub
            }
        };
        sm.registerListener(se, sr, SensorManager.SENSOR_DELAY_GAME);
    }
```

步骤 3：去除重写线程 run()方法中用于判断弹板移动方向的三元运算代码 (direction＝destX＜rectX？LEFT：RIGHT；)，在自定义视图 MyView 的构造函数

GameView(Context context)和重启游戏函数 restartGame()中调用 changeDirection() 函数。

运行效果如图 4-18 所示。

图 4-18 弹球实例运行界面

★提示：传感器的知识与用法见第 5 章。

4.9 综合实例二：动态游戏导航界面

4.9.1 功能描述

本例实现一个图形化的动态游戏菜单界面。

4.9.2 关键技术

本例实现的关键是根据游戏不同的状态标识绘制不同的界面。代码如下：

```
private void drawMenu(Canvas canvas){
        canvas.drawBitmap(menu[0], 0, 0, paint);              //绘制菜单界面背景
        canvas.clipRect (240, 260, 240 + menu [1]. getWidth ()/6, 260 + menu [1].
        getHeight());                                         //剪切菜单项区域
        canvas.drawBitmap(menu[1], 240 - menuIndex * menu[1].getWidth()/6, 260,
        paint);                                               //绘制当前菜单项
        canvas.clipRect(205, 350, 234, 370, Op.UNION);        //剪切左切换按钮区域
        canvas.drawBitmap(menu[2], 205, 350, paint);          //绘制左切换按钮
        canvas.clipRect(290, 350, 319, 370, Op.UNION);        //剪切右切换按钮区域
        canvas.drawBitmap(menu[2], 290, 330, paint);          //绘制右切换按钮
        //剪切花瓣区域
        canvas.clipRect(flowerX, flowerY, flowerX+15, flowerY+15, Op.UNION);
        canvas.drawBitmap(flower[frameIndex], flowerX, flowerY, paint);
                                                              //绘制花瓣
    }
```

4.9.3 实现过程

步骤 1：新建项目 ProjectDynamicMenu，准备游戏中所需要的背景图片资源（如图 4-19 所示）。把图片复制到项目根目录下的 res/drawable-mdpi 文件夹中。

图 4-19 动态导航图片素材

步骤 2：建立内部类 GameSurfaceView（继承自 android.view.SurfaceView 类），添加构造函数并实现 Callback 接口（android.view.SurfaceHolder.Callback）；实现 Runable 接口并重写其 run()方法。代码如下：

```
public class GameSurfaceView extends SurfaceView implements Callback,
Runnable {
    private final int LOGO=0;            //标识闪屏游戏状态
    private final int MENU=1;            //标识菜单游戏状态
    private final int GAME=2;            //标识进入游戏状态
    private int gameState=LOGO;          //标识游戏所处状态(默认为闪屏状态)
    private SurfaceHolder holder;
    private Canvas canvas;
    private Paint paint;
    private boolean isGame;
    private Bitmap logo[];               //闪屏状态界面图像数组
    private int timeCount;               //延时计数
    private Bitmap menu[];               //菜单状态界面图像数组
    private int menuIndex=0;             //菜单项位置索引
    private Bitmap flower[];             //花瓣序列图像数组
    private int flowerX=300, flowerY;    //花瓣初始位置
    private int frameIndex=1;            //花瓣切换帧
    private MainActivity activity;
    public GameSurfaceView(Context context){
        super(context);
        activity=(MainActivity)context;
        holder=getHolder();
        paint=new Paint();
        paint.setAntiAlias(true);
```

```java
        holder.addCallback(this);
    logo=new Bitmap[3];                    //实例化闪屏界面图像数组
        logo[0]=BitmapFactory.decodeResource(getResources(), R.drawable.
            toast);
        logo[1]=BitmapFactory.decodeResource(getResources(), R.drawable.
            splash1);
        logo[2]=BitmapFactory.decodeResource(getResources(), R.drawable.
            splash2);
        menu=new Bitmap[3];                //实例化菜单界面图像数组
        menu[0]=BitmapFactory.decodeResource(getResources(), R.drawable.
            menubg);
        menu[1]=BitmapFactory.decodeResource(getResources(), R.drawable.
            menu);
        menu[2]=BitmapFactory.decodeResource(getResources(), R.drawable.
            arrow);
        flower=new Bitmap[3];              //实例化花瓣图像数组
        flower[0]=BitmapFactory.decodeResource(getResources(),R.drawable.
            flower0);
        flower[1]=BitmapFactory.decodeResource(getResources(), R.drawable.
            flower1);
        flower[2]=BitmapFactory.decodeResource(getResources(), R.drawable.
            flower2);
    }
    @Override
    public void surfaceChanged(SurfaceHolder holder, int format, int width,
            int height){
        //TODO: Auto-generated method stub
    }
    @Override
    public void surfaceCreated(SurfaceHolder holder){
        isGame=true;
        new Thread(this).start();
    }
    @Override
    public void surfaceDestroyed(SurfaceHolder holder){
        isGame=false;
        activity.finish();
    }
    @Override
    public boolean onTouchEvent(MotionEvent event){
        switch(event.getAction()){
        case MotionEvent.ACTION_DOWN:
            switch(gameState){                    //判断游戏状态
            case LOGO:                            //闪屏状态
```

```
                    gameState=MENU;                     //切换到菜单状态
                    frameIndex=0;                       //花瓣图像数组帧索引
                    break;
                case MENU:                              //菜单状态
                    float pointX=event.getX();          //获取触屏的X坐标
                    float pointY=event.getY();          //获取触屏的Y坐标
                    if(pointX>=205 && pointX<=234 && pointY>=350
                            && pointY<=370)             //判断触屏位置
                        menuIndex= (menuIndex>0 ?menuIndex -1 : 5);
                                                        //计算菜单索引
                    else if(pointX>=290 && pointX<=319 && pointY>=350
                            && pointY<=370)
                        menuIndex= (menuIndex>4 ?0 : menuIndex+1);
                    else if(pointX>=240
                            && pointX<=240+menu[1].getWidth()/6
                            && pointY>=260 && pointY<=260+menu[1].getHeight()){
                        switch(menuIndex){              //判断菜单图像位置索引
                        case 0:
                            gameState=GAME;
                            break;
                        case 5:
                            isGame=false;
                            activity.finish();
                            break;
                        }
                    }
                    break;
                }
                break;
            }
            return true;
        }
        /**
         * 绘制闪屏状态下的游戏界面
         * @param canvas 当前画布
         */
        private void drawLogo(Canvas canvas){
            canvas.drawBitmap(logo[frameIndex], 0, 0, paint);  //绘制闪屏状态游戏背景
            canvas.drawBitmap(logo[0], getWidth()/2 - logo[0].getWidth()/2,
                    getHeight()-50, paint);             //绘制触屏直接进入游戏提示图片
        }
        /**
         * 绘制菜单状态下的游戏界面
         * @param canvas 当前画布
```

```java
     */
    private void drawMenu(Canvas canvas){
        canvas.drawBitmap(menu[0], 0, 0, paint);              //绘制菜单界面背景
        canvas.clipRect(240, 260, 240+menu[1].getWidth()/6, 260+menu[1].
        getHeight());                                          //剪切菜单项区域
        canvas.drawBitmap(menu[1], 240-menuIndex * menu[1].getWidth()/6,
        260, paint);                                           //绘制当前菜单项
        canvas.clipRect(205, 350, 234, 370, Op.UNION);         //剪切左切换按钮区域
        canvas.drawBitmap(menu[2], 205, 350, paint);           //绘制左切换按钮
        canvas.clipRect(290, 350, 319, 370, Op.UNION);         //剪切右切换按钮区域
        canvas.drawBitmap(menu[2], 290, 330, paint);           //绘制右切换按钮
        //剪切花瓣区域
        canvas.clipRect(flowerX, flowerY, flowerX+15, flowerY+15, Op.UNION);
        canvas.drawBitmap(flower[frameIndex], flowerX, flowerY, paint);
                                                               //绘制花瓣
    }
    /**
     * 渲染游戏画面
     */
    private void render(){
        canvas=holder.lockCanvas();                            //锁定画布
        switch(gameState){                                     //判断游戏状态
        case LOGO:                                             //闪屏状态
            drawLogo(canvas);                                  //调用闪屏状态绘制函数
            break;
        case MENU:                                             //菜单状态
            drawMenu(canvas);                                  //调用菜单状态绘制函数
            break;
        case GAME:
            break;
        }
        holder.unlockCanvasAndPost(canvas);                    //解锁画布
    }
    @Override
    public void run(){
        while(isGame){                                         //当游戏开始
            switch(gameState){                                 //判断游戏状态
            case LOGO:                                         //闪屏状态
                timeCount++;                                   //延时操作计数
                if(timeCount %15==0){                          //对15取余
                    if(frameIndex<2)                           //判断闪屏图像数组索引
                        frameIndex++;
                    else {
                        gameState=MENU;                        //切换到菜单状态
```

```
                    frameIndex=0;
                }
            }
            break;
        case MENU:                                    //菜单状态
            frameIndex=(frameIndex==2?0:frameIndex+1);
                                                      //获取花瓣索引
            flowerX-=5;                               //花瓣的X坐标
            flowerY+=8;                               //花瓣的Y坐标
            if(flowerY>getHeight()){                  //判断花瓣是否超出屏高位置
                flowerX=300;
                flowerY=0;
            }
            break;
        case GAME:                                    //游戏状态
            break;
        }
        render();
        try {
            Thread.sleep(100);
        } catch(InterruptedException e){
            e.printStackTrace();
        }
    }
}
```

步骤3：在主活动的onCreate()方法中，设置屏幕属性并调用MyView视图。代码如下：

```
@Override
protected void onCreate(Bundle savedInstanceState){
    super.onCreate(savedInstanceState);
    setRequestedOrientation(ActivityInfo.SCREEN_ORIENTATION_LANDSCAPE);
        …                       //此处省略隐藏标题栏、全屏和保持屏幕亮度的代码
    GameSurfaceView gameView=new GameSurfaceView(this);
    gameView.setFocusable(true);
    setContentView(gameView);
}
```

运行效果如图4-20所示。

4.9.4　实例拓展

实现游戏处于菜单状态下大量花瓣不断飘落的效果。

图 4-20　游戏动态导航运行界面

步骤 1：修改有关花瓣的部分变量，定义花瓣对象池二维数数组，并初始化对象池中每个花瓣的宽度和高度。代码如下：

```
private Bitmap flowerImg[];
private int flower[][];              //每一行表示一个花瓣对象，行数表示对象池的初始化大小
//private int flowerX=300, flowerY;   //花瓣初始位置
Random random;
```

步骤 2：在构造方法中修改花瓣图像资源获取部分代码，添加花瓣对象池。代码如下：

```
flowerImg=new Bitmap[3];                //实例化花瓣图像数组
flowerImg[0] = BitmapFactory.decodeResource(getResources(), R.drawable.flower0);
flowerImg[1] = BitmapFactory.decodeResource(getResources(), R.drawable.flower1);
flowerImg[2] = BitmapFactory.decodeResource(getResources(), R.drawable.flower2);
random=new Random();
//列属性：visible(0 为不可见，1 为可见), x, y, speedX, speedY, width, height,
  frameIndex
flower=new int[10][8];
for(int i=0; i<flower.length; i++){
    flower[i][5]=flowerImg[0].getWidth();
    flower[i][6]=flowerImg[0].getHeight();
}
```

步骤 3：分别编写生成花瓣函数 creatFlower()、移动花瓣函数 moveFlower()、绘制花瓣函数 paintFlower() 和设置花瓣是否可见函数 setFlowerVisible()。代码如下：

```
/**
 * 生成花瓣函数
```

```java
 */
private void creatFlower(){
//列属性：visible（0 为不可见，1 为可见），x，y，speedX，speedY，width，height，
  frameIndex
    for(int i=0; i<flower.length; i++){
        if(flower[i][0]==0){
            flower[i][0]=1;
            flower[i][1]=Math.abs(random.nextInt()%161)+160;
            flower[i][2]=-flower[i][6];
            flower[i][3]=-(Math.abs(random.nextInt()%10)+5);
            flower[i][4]=Math.abs(random.nextInt()%16)+10;
            break;
        }
    }
}
/**
 * 移动花瓣函数
 */
private void moveFlower(){
//列属性：visible（0 为不可见，1 为可见），x，y，speedX，speedY，width，
  height，frameIndex
    for(int i=0; i<flower.length; i++){
        if(flower[i][0]==1){
            flower[i][1]+=flower[i][3];
            flower[i][2]+=flower[i][4];
            flower[i][7]=(flower[i][7]==2 ? 0 : flower[i][7]+1);
        }
    }
}
/**
 * 绘制花瓣函数
 */
private void paintFlower(Canvas canvas, Paint paint){
//列属性：visible（0 为不可见，1 为可见），x，y，speedX，speedY，width，height，
  frameIndex
    for(int i=0; i<flower.length; i++){
        if(flower[i][0]==1){
            canvas.save();
            canvas.clipRect(flower[i][1], flower[i][2], flower[i][1]+flower[i]
              [5], flower[i][2]+flower[i][6], Op.REPLACE);
            canvas.drawBitmap(flowerImg[flower[i][7]], flower[i][1], flower[i]
              [2], paint);
            canvas.restore();
        }
```

    }
}
/**
 * 设置花瓣是否可见函数
 */
void setFlowerVisible(){
//列属性:visible (0 为不可见, 1 为可见), x, y, speedX, speedY, width, height,
    frameIndex
    for(int i=0; i<flower.length; i++){
        if(flower[i][0]==1){
            if(flower[i][1]+flower[i][5]<=0|| flower[i][2]>=getHeight())
                flower[i][0]=0;
        }
    }
}
```

**步骤 4**：调用 paintFlower()函数，替换菜单状态界面绘制函数中绘制花瓣的代码。代码如下：

```
private void drawMenu(Canvas canvas){
    ...                                          //此处省略绘制背景、菜单项等部分代码
    //canvas.clipRect(flowerX, flowerY, flowerX+15, flowerY+15, Op.UNION);
    //canvas.drawBitmap(flower[frameIndex], flowerX, flowerY, paint);
                                                 //绘制花瓣
    paintFlower(canvas, paint);
}
```

步骤 5：修改线程 run()方法中菜单状态下的代码如下所示：

```
@Override
public void run(){
    while(isGame){                               //当游戏开始
        switch(gameState){                       //判断游戏状态
            case LOGO:                           //闪屏状态
                ...                              //此处省略闪屏状态下的操作代码
                break;
            case MENU:                           //菜单状态
                frameIndex= (frameIndex==2 ? 0 : frameIndex+1);
                timeCount++;
            if(timeCount %5==0)
                creatFlower();                   //调用花瓣生成函数
                moveFlower();                    //调用移动花瓣函数
                setFlowerVisible();              //调用设置花瓣是否可见函数
                break;
            case GAME:                           //游戏状态
                break;
```
```

```
        }
        render();
        ...                                   //此处省略线程休眠部分的代码
    }
}
```

运行效果如图 4-21 所示。

图 4-21 游戏动态导航花瓣效果

4.10 综合实例三：打地鼠

4.10.1 功能描述

本例实现一个打地鼠游戏。在一个有多个"洞穴"的场景中，每个"洞穴"随机显示地鼠，用户可以用手触摸出现的地鼠。如果触摸到则该地鼠不再显示，同时在屏幕上通过消息提示框显示打到了几只地鼠。

4.10.2 关键技术

本例实现的关键是如何在指定的位置随机显示地鼠，这里主要是通过线程与消息处理进行控制。首先使用 Thread 对象记录地鼠出现的位置，然后通过 Handler 消息控制地鼠的出现。使用 Thread 对象记录地鼠出现位置的关键代码如下：

```
Thread t=new Thread(new Runnable(){
    @Override
    public void run(){
        int index=0;                                  //创建一个记录地鼠位置的索引值
        while(!Thread.currentThread().isInterrupted()){
            index=new Random().nextInt(position.length);
                                                      //产生一个随机数
            Message m=handler.obtainMessage();    //获取一个 Message
```

```
            m.what=0x101;                          //设置消息标识
            m.arg1=index;                          //保存地鼠标位置的索引值
            handler.sendMessage(m);                //发送消息
            try {
                Thread.sleep(new Random().nextInt(500)+500);
                                                   //休眠一段时间
            } catch(InterruptedException e){
                e.printStackTrace();
            }
        }
    }
});
t.start();                                          //开启线程
```

通过 Handler 消息控制地鼠出现的关键代码如下：

```
handler=new Handler(){
    @Override
    public void handleMessage(Message msg){
        int index=0;
        if(msg.what==0x101){
            index=msg.arg1;                        //获取位置索引值
            mouse.setX(position[index][0]);        //设置 X 轴位置
            mouse.setY(position[index][1]);        //设置 Y 轴位置
            mouse.setVisibility(View.VISIBLE);     //设置地鼠显示
        }
        super.handleMessage(msg);
    }
};
```

4.10.3 实现过程

步骤1：新建项目 MyMouse(Android SDK 22.2.1，Target SDK 4.3)，准备游戏中需要的图片资源，这里包括地鼠、背景图片及 Logo 图标 3 张图片(如图 4-22 所示)。并把图片资源复制到项目根目录下的 res/drawable-mdpi 文件夹中。

mouse.png

bg.png

ic_launcher.png

图 4-22 打地鼠游戏图片素材

步骤2：修改 res/layout 目录中的 XML 布局文件，将默认生成的 TextView 组件删除，添加一个 ImageView 组件，设置该组件的背景为地鼠图片。代码如下：

```xml
<RelativeLayout xmlns:android="http://schemas.android.com/apk/res/android"
    xmlns:tools="http://schemas.android.com/tools"
    android:layout_width="fill_parent"
    android:layout_height="fill_parent"
    android:background="@drawable/bg">
    <ImageView
        android:id="@+id/img_mouse"
        android:layout_width="wrap_content"
        android:layout_height="wrap_content"
        android:src="@drawable/mouse"/>
</RelativeLayout>
```

步骤 3：在主活动文件中声明所需要的变量；创建一个新的线程，实现地鼠的随机位置并发送消息；创建一个 Handler 对象，获取线程传过来的消息中的位置坐标，并设置地鼠的显示位置。代码如下：

```java
private ImageView img_mouse;
private int i=0;                                    //记录打到的数量
private int position[][]=new int[][] { { 90, 132 }, { 230, 175 },
        { 179, 145 }, { 238, 118 }, { 227, 97 }, { 320, 92 }, { 320, 145 },
        { 345, 149 } };                             //位置数组，根据屏幕分辨率进行设置
private static Handler handler=null;                //创建 Handler 对象
    @Override
    protected void onCreate(Bundle savedInstanceState){
        super.onCreate(savedInstanceState);
        setContentView(R.layout.activity_main);
        setRequestedOrientation(ActivityInfo.SCREEN_ORIENTATION_LANDSCAPE);
        img_mouse= (ImageView)findViewById(R.id.img_mouse);
        img_mouse.setOnTouchListener(new OnTouchListener(){
                                                    //添加触摸监听
            @Override
            public boolean onTouch(View v, MotionEvent event){
                i++;
                Toast.makeText(MainActivity.this,"打到["+i+"]只地鼠!",
                        Toast.LENGTH_SHORT).show();
                                                    //显示打到的数量
                return false;
            }
        });
        handler=new Handler(){
            @Override
            public void handleMessage(Message msg){
                super.handleMessage(msg);
                if(msg.arg1==0x666){
                    int index=msg.what;             //获取传值
```

```java
            img_mouse.setVisibility(View.INVISIBLE);   //设置不可见
            img_mouse.setX(position[index][0]);         //设置 X 轴位置
            img_mouse.setY(position[index][1]);         //设置 Y 轴位置
            img_mouse.setVisibility(View.VISIBLE);      //设置可见
        }
    }
    };
    new myThread().start();                             //启动线程
}
@Override
protected void onDestroy(){
    super.onDestroy();
    Thread.currentThread().interrupt();                 //中断线程
}
class myThread extends Thread {
@Override
public void run(){
    super.run();
    int index=0;
    while(!Thread.currentThread().isInterrupted()){
        index=new Random().nextInt(position.length);
        Message msg=handler.obtainMessage();            //获取一个 Message 对象
        msg.what=index;                                 //保存位置索引值
        msg.arg1=0x666;                                 //保存消息标识
        handler.sendMessage(msg);                       //发送消息
        try {
            Thread.sleep(new Random().nextInt(500)+300); //线程休眠
        } catch(InterruptedException e){
            e.printStackTrace();
        }
    }
}
}
```

运行效果如图 4-23 所示。

图 4-23　打地鼠游戏运行界面

★**注意**：Android 中 UI 应用的开发中经常会使用 view.setVisibility() 来设置控件的可见性，该函数有 3 个可选值 VISIBLE(可见)、INVISIBLE(不可见)和 GONE(隐藏)。INVISIBLE 和 GONE 的主要区别是：当控件 visibility 属性为 INVISIBLE 时，界面保留了 View 控件所占有的空间；而控件属性为 GONE 时，界面则不保留 View 控件所占有的空间。

4.11 综合实例四：游戏中的瞄准镜

4.11.1 功能描述

本例实现射击类游戏中瞄准镜随触屏手指移动的效果。

4.11.2 关键技术

实现本例的关键是对放大图像的局部抠取及平移放大图像时矩阵位置的计算。创建圆形图像的关键代码如下：

```
shader=new BitmapShader(bitmap_bg_big, TileMode.CLAMP, TileMode.CLAMP);
shapeDrawable=new ShapeDrawable(new OvalShape());           //创建圆形的Shape
shapeDrawable.setBounds(0, 0, RADIUS * 2, RADIUS * 2);      //设置圆的外切矩形
shapeDrawable.getPaint().setShader(shader);                 //设置画笔形状
```

平移矩阵新位置计算的代码如下：

```
matrix.setTranslate(RADIUS - x * FACTOR, RADIUS - y * FACTOR);   //平移矩阵位置
shapeDrawable.getPaint().getShader().setLocalMatrix(matrix);
shapeDrawable.setBounds(x -RADIUS, y -RADIUS, x+RADIUS, y+RADIUS);   //圆外切矩形
```

4.11.3 实现过程

步骤 1：新建项目 Projectmagnifier，把游戏中所需要的图片资源，包括游戏原地图、游戏放大地图及瞄准镜 3 张图片(如图 4-24 所示)复制到根目录下的 res/drawable-mdpi 文件夹中。

图 4-24　瞄准镜实例图片素材

步骤 2：在主活动文件中创建 MyView 内部类，该类继承自 android.view.View 类，

并添加构造方法和重写 onDraw(Canvas canvas)方法,然后在 onCreate()方法中关联自定义的 MyView 视图。代码如下:

```
public class MainActivity extends Activity {
    @Override
    protected void onCreate(Bundle savedInstanceState){
        super.onCreate(savedInstanceState);
        setContentView(new MyView(this));              //关联自定义视图
    }
    public class MyView extends View {
        public MyView(Context context){
            super(context);
        }
        @Override
        protected void onDraw(Canvas canvas){
            super.onDraw(canvas);
        }
    }
}
```

步骤 3:在内部类 MyView 中,完成图像控制及触屏事件。代码如下:

```
public class MyView extends View {
    private Bitmap bitmap_bg_small, bitmap_bg_big, bitmap_magnifier;
                                                    //图像变量
    private int m_left=0;                           //瞄准镜默认左边距
    private int m_top=0;                            //瞄准镜默认上边距
    private static final int RADIUS=50;             //瞄准镜的半径
    private static final int FACTOR=2;              //瞄准镜的放大倍数
    private BitmapShader shader;                    //定义 BitmapShader 对象变量
    private ShapeDrawable shapeDrawable;            //定义 ShapeDrawable 对象变量
    private Matrix matrix;                          //定义矩阵变量
    public MyView(Context context){
        super(context);
        bitmap_bg_small = BitmapFactory.decodeResource(getResources(), R.
        drawable.tankmap_small);                    //获取背景图像
        bitmap_bg_big = BitmapFactory.decodeResource(getResources(), R.
        drawable.tankmap);                          //获取大背景图像
        bitmap_magnifier = BitmapFactory.decodeResource(getResources(), R.
        drawable.magnifier);                        //获取瞄准镜图像
        shader= new BitmapShader(bitmap_bg_big, TileMode.CLAMP, TileMode.
        CLAMP);                                     //创建 BitmapShader 对象
        shapeDrawable=new ShapeDrawable(new OvalShape());
                                                    //创建圆形的 Shape
        shapeDrawable.setBounds(0, 0, RADIUS * 2, RADIUS * 2);
```

```
        shapeDrawable.getPaint().setShader(shader);   //设置画笔形状
        matrix=new Matrix();                          //实例化矩阵对象
    }
    @Override
    protected void onDraw(Canvas canvas){
        super.onDraw(canvas);
        canvas.drawBitmap(bitmap_bg_small, 0, 0, null);   //绘制背景图像
        shapeDrawable.draw(canvas);                       //绘制瞄准镜
        canvas.drawBitmap(bitmap_magnifier, m_left, m_top, null);
                                                          //绘制大图像
    }
    @Override
    public boolean onTouchEvent(MotionEvent event){
        int x=(int)event.getX();                      //获取当前触摸点的 X 轴坐标
        int y=(int)event.getY();                      //获取当前触摸点的 Y 轴坐标
        matrix.setTranslate(RADIUS-x * FACTOR, RADIUS-y * FACTOR);
        shapeDrawable.getPaint().getShader().setLocalMatrix(matrix);
        shapeDrawable.setBounds(x-RADIUS, y-RADIUS, x+RADIUS, y+RADIUS);
            m_left=x-bitmap_magnifier.getWidth()/2;   //瞄准镜的左边距
            m_top=y-bitmap_magnifier.getHeight()/2;   //瞄准镜的上边距
            invalidate();
            return true;
    }
}
```

运行效果如图 4-25 所示。

图 4-25 瞄准镜实例运行界面

4.12 综合实例五：发疯的小猪

4.12.1 功能描述

本例实现一只小猪在围栏内左右两个方向来回奔跑的动画效果，折返条件是碰到围

栏的边缘。

4.12.2 关键技术

在实现本例的过程中,最关键的技术就是通过动画形式显示小猪的奔跑状态,这里主要用到 Animation 对象获取动画资源,通过重写该对象的 onAnimationEnd()方法并调用 startAnimation()方法实现小猪奔跑状态的切换及动画的播放。关键代码如下:

```
final Animation translateright = AnimationUtils.loadAnimation(this, R.anim.
    translateright);                                            //获取"向右奔跑"的动画资源
final Animation translateleft=AnimationUtils.loadAnimation(this,
    R.anim.translateleft);                                      //获取"向左奔跑"的动画资源
anim=(AnimationDrawable)iv.getBackground();                     //获取应用的帧动画
...
translateleft.setAnimationListener(new AnimationListener(){
    @Override
    public void onAnimationStart(Animation animation){}
    @Override
    public void onAnimationRepeat(Animation animation){}
    @Override
    public void onAnimationEnd(Animation animation){
        iv.setBackgroundResource(R.anim.motionright);
                                                                //重新设置ImageView应用的帧动画
        iv.startAnimation(translateright);                      //播放"向右奔跑"的动画
        anim=(AnimationDrawable)iv.getBackground();             //获取应用的帧动画
        anim.start();                                           //开始播放帧动画
    }
});
```

4.12.3 实现过程

步骤 1:新建项目 PigRun,把游戏中所需要的图片资源,包括背景图片、Logo 图标及小猪左右跑动状态 6 张图片(如图 4-26 所示)复制到根目录下的 res/drawable-mdpi 文件夹中。

图 4-26 发疯小猪系列图片素材

步骤 2:在新建项目的 res 目录中新建 anim 目录,并在该目录中创建实现小猪左右奔跑动作的逐帧动画资源文件。

创建名称为 motionright.xml 的 XML 资源文件,在该文件中定义小猪向右奔跑动作

的动画,该动画由两帧组成,也就是由两个预先定义好的图片组成。代码如下:

```xml
<?xml version="1.0" encoding="utf-8"?>
<animation-list xmlns:android="http://schemas.android.com/apk/res/android">
    <item android:drawable="@drawable/pig1" android:duration="40"/>
    <item android:drawable="@drawable/pig2" android:duration="40"/>
</animation-list>
```

创建名称为 translateright.xml 的 XML 资源文件,在该文件中定义小猪向右奔跑的补间动画,设置水平方向上向右平移 850 像素,持续时间为 3s。代码如下:

```xml
<?xml version="1.0" encoding="utf-8"?>
<set xmlns:android="http://schemas.android.com/apk/res/android">
    <translate
        android:fromXDelta="0"
        android:toXDelta="850"
        android:fromYDelta="0"
        android:toYDelta="0"
        android:duration="3000">
    </translate></set>
```

同理,完成小猪向左奔跑动作的动画 motionleft.xml 和向左奔跑的补间动画 translateleft.xml。

步骤 3:修改项目 res/layout 目录下的布局文件 main.xml,将默认生成的 TextView 组件删除,添加一个 ImageView 组件,设置该组件的背景为逐帧动画资源 motionright。代码如下:

```xml
<?xml version="1.0" encoding="utf-8"?>
<LinearLayout xmlns:android="http://schemas.android.com/apk/res/android"
    android:id="@+id/linearLayout1"
    android:background="@drawable/background"
    android:layout_width="fill_parent"
    android:layout_height="fill_parent"
    android:orientation="vertical">
    <ImageView
        android:id="@+id/imageView1"
        android:layout_width="wrap_content"
        android:layout_height="wrap_content"
        android:background="@anim/motionright"
        android:layout_marginTop="280px"
        android:layout_marginLeft="30px"/>
</LinearLayout>
```

步骤 4:在主文件 MainActivity.java 中,首先获取应用动画效果的 ImgeView 组件,并获取小猪向左右奔跑的动画资源、ImageView 组件应用的逐帧动画以及线性布局管理器,并显示一个提示信息,再为线性布局管理器添加触摸监听器。

重写 onTouch()方法,开始播放逐帧动画及补间动画,最后为"向左奔跑"、"向右奔跑"动画添加动画监听器,在重写的 onAnimationEnd()方法中改变要使用的逐帧动画和补间动画并播放,从而实现小猪来回奔跑的动画效果。代码如下:

```java
public class MainActivity extends Activity {
    private AnimationDrawable anim;
    @Override
    public void onCreate(Bundle savedInstanceState){
        super.onCreate(savedInstanceState);
        setContentView(R.layout.main);
        final ImageView iv=(ImageView)findViewById(R.id.imageView1);
        final Animation translateright=AnimationUtils.loadAnimation(this, R.anim.translateright);          //获取"向右奔跑"的动画资源
        final Animation translateleft=AnimationUtils.loadAnimation(this, R.anim.translateleft);            //获取"向左奔跑"的动画资源
        anim=(AnimationDrawable)iv.getBackground();    //获取应用的帧动画
        LinearLayout ll=(LinearLayout)findViewById(R.id.linearLayout1);
                                                        //获取布局
        Toast.makeText(this, "触摸屏幕开始播放...", Toast.LENGTH_SHORT).show();
        ll.setOnTouchListener(new OnTouchListener(){
            @Override
            public boolean onTouch(View v, MotionEvent event){
                anim.start();                      //开始播放帧动画
                iv.startAnimation(translateright);   //播放"向右奔跑"的动画
                return false;
            }
        });
        translateright.setAnimationListener(new AnimationListener(){
            @Override
            public void onAnimationStart(Animation animation){}
            @Override
            public void onAnimationRepeat(Animation animation){}
            @Override
            public void onAnimationEnd(Animation animation){
                iv.setBackgroundResource(R.anim.motionleft);
                iv.startAnimation(translateleft);    //播放"向左奔跑"的动画
                anim=(AnimationDrawable)iv.getBackground();
                                                   //获取应用的帧动画
                anim.start();                      //开始播放帧动画
            }
        });
        translateleft.setAnimationListener(new AnimationListener(){
            @Override
            public void onAnimationStart(Animation animation){}
```

```
            @Override
            public void onAnimationRepeat(Animation animation){}
            @Override
            public void onAnimationEnd(Animation animation){
                iv.setBackgroundResource(R.anim.motionright);
                iv.startAnimation(translateright);        //播放"向右奔跑"的动画
                anim=(AnimationDrawable)iv.getBackground();
                                                          //获取应用的帧动画
                anim.start();                             //开始播放帧动画
            }
        });
    }
}
```

运行效果如图 4-27 所示。

图 4-27　发疯的小猪实例运行界面

4.13　综合实例六：开心涂鸦

4.13.1　功能描述

本例实现一个代表白板的空白区域，用户可以通过菜单选择画笔在白板上随意绘制（文字和各种图案等），并能够将绘制的内容保存到 SD 卡中。

4.13.2　关键技术

在实现过程中，当主要用到的是保存画板内容时，可以调用 Bitmap 类的 compress() 方法将绘图内容压缩为 PNG 格式输出到文件输出流对象中。关键代码如下：

```
public void saveBitmap(String fileName)throws IOException {
    File file=new File("/sdcard/pictures/"+fileName+".png");
                                                      //创建文件对象
    file.createNewFile();                             //创建一个新文件
```

```
        FileOutputStream fileOS=new FileOutputStream(file);    //创建一个文件输出流对象
        //将绘图内容压缩为 PNG 格式输出到输出流对象中
        cacheBitmap.compress(Bitmap.CompressFormat.PNG, 100, fileOS);
            fileOS.flush();                         //将缓冲区中的数据全部写出到输出流中
            fileOS.close();                                 //关闭文件输出流对象
    }
```

4.13.3 实现过程

步骤 1：新建项目 MyDraw，在 res 目录中创建一个 menu 目录，并在该目录中创建菜单资源文件 toolsmenu.xml，在该件中编写程序所需要的功能菜单。代码如下：

```
<?xml version="1.0" encoding="utf-8"?>
<menu xmlns:android="http://schemas.android.com/apk/res/android">
    <item android:title="@string/color">
        <menu>
            <!--定义一组单选菜单项 -->
            <group android:checkableBehavior="single">
                <!--定义子菜单 -->
                <item android:id="@+id/red" android:title="@string/color_
                    red"/>
                <item android:id="@+id/green" android:title="@string/color_
                    green"/>
                <item android:id="@+id/blue" android:title="@string/color_
                    blue"/>
            </group>
        </menu>
    </item>
    <item android:title="@string/width">
        <menu>
            <!--定义子菜单 -->
            <group>
                <item android:id="@+id/width_1" android:title="@string/width_1"/>
                <item android:id="@+id/width_2" android:title="@string/width_2"/>
                <item android:id="@+id/width_3" android:title="@string/width_3"/>
            </group>
        </menu>
    </item>
    <item android:id="@+id/clear" android:title="@string/clear"/>
    <item android:id="@+id/save" android:title="@string/save"/></menu>
```

步骤 2：创建一个名称为 DrawView 的类（继承自 android.view.View），在该类中定义程序所需的属性，然后添加构造方法并重写 onDraw(Canvas canvas)方法。代码如下：

```
public class DrawView extends View {
```

```java
        private int view_width=0;              //屏幕的宽度
        private int view_height=0;             //屏幕的高度
        private float preX;                    //起始点的 X 坐标值
        private float preY;                    //起始点的 Y 坐标值
        private Path path;                     //定义路径变量
        public Paint paint=null;               //定义画笔
        Bitmap cacheBitmap=null;               //定义一个内存中的图片,该图片将作为缓冲区
        Canvas cacheCanvas=null;               //定义 cacheBitmap 上的 Canvas 对象
        public DrawView(Context context, AttributeSet set){
            super(context, set);
            view_width=context.getResources().getDisplayMetrics().widthPixels;
            view_height=context.getResources().getDisplayMetrics().heightPixels;
            System.out.println(view_width+" * "+view_height);
            //创建一个与该 View 相同大小的缓存区
            cacheBitmap=Bitmap.createBitmap(view_width, view_height, Config.ARGB_8888);
            cacheCanvas=new Canvas();
            path=new Path();
            cacheCanvas.setBitmap(cacheBitmap);  //在 cacheCanvas 上绘制 cacheBitmap
            paint=new Paint(Paint.DITHER_FLAG);
            paint.setColor(Color.RED);           //设置默认的画笔颜色
            paint.setStyle(Paint.Style.STROKE);  //设置填充方式为描边
            paint.setStrokeJoin(Paint.Join.ROUND); //设置笔刷的图形样式
            paint.setStrokeCap(Paint.Cap.ROUND);   //设置画笔转弯处的连接风格
            paint.setStrokeWidth(1);             //设置默认笔触的宽度为 1 像素
            paint.setAntiAlias(true);            //使用抗锯齿功能
            paint.setDither(true);               //使用抖动效果
        }
        @Override
        public void onDraw(Canvas canvas){
            canvas.drawColor(0xFFFFFFFF);        //设置背景颜色
            Paint bmpPaint=new Paint();          //采用默认设置创建一个画笔
            canvas.drawBitmap(cacheBitmap, 0, 0, bmpPaint);
                                                 //绘制 cacheBitmap
            canvas.drawPath(path, paint);        //绘制路径
            canvas.save(Canvas.ALL_SAVE_FLAG);   //保存画布的状态
            canvas.restore();                    //恢复画布之前保存的状态,防止影响后续绘制
        }
        @Override
        public boolean onTouchEvent(MotionEvent event){
            //获取触摸事件的发生位置
            float x=event.getX();
            float y=event.getY();
            switch(event.getAction()){
```

```java
            case MotionEvent.ACTION_DOWN:
                path.moveTo(x, y);                  //将绘图的起始点移到(x, y)坐标点的位置
                preX=x;
                preY=y;
                break;
            case MotionEvent.ACTION_MOVE:
                float dx=Math.abs(x-preX);
                float dy=Math.abs(y-preY);
                if(dx>=5 || dy>=5){                 //判断是否在允许的范围内
                    path.quadTo(preX, preY, (x+preX)/2, (y+preY)/2);
                    preX=x;
                    preY=y;
                }
                break;
            case MotionEvent.ACTION_UP:
                cacheCanvas.drawPath(path, paint);  //绘制路径
                path.reset();
                break;
        }
        invalidate();
        return true;                                //返回true表明处理方法已经处理该事件
    }
    public void clear(){
        paint.setXfermode(new PorterDuffXfermode(PorterDuff.Mode.CLEAR));
        paint.setStrokeWidth(50);                   //设置笔触的宽度
    }
    public void save(){
        try {
            saveBitmap("myPicture");
        } catch(IOException e){
            e.printStackTrace();
        }
    }
    //保存绘制好的位图
    public void saveBitmap(String fileName)throws IOException {
        File file=new File("/sdcard/pictures/"+fileName+".png");
                                                    //创建文件对象
        file.createNewFile();                       //创建一个新文件
        FileOutputStream fileOS=new FileOutputStream(file);
                                                    //创建文件输出流对象
        //将绘图内容压缩为PNG格式输出到输出流对象中
        cacheBitmap.compress(Bitmap.CompressFormat.PNG,100,fileOS);
        fileOS.flush();                             //将缓冲区中的数据全部写出到输出流中
        fileOS.close();                             //关闭文件输出流对象
```

    }
}

步骤 3：修改 res/layout 目录的 XML 布局文件，将默认生成的 TextView 组件删除，添加一个帧布局管理器，并加创建的自定义视图。代码所下：

```xml
<?xml version="1.0" encoding="utf-8"?>
<FrameLayout xmlns:android="http://schemas.android.com/apk/res/android"
    android:layout_width="fill_parent"
    android:layout_height="fill_parent"
    android:orientation="vertical">
    <com.mingrisoft.DrawView
        android:id="@+id/drawView1"
        android:layout_width="match_parent"
        android:layout_height="match_parent"/>
</FrameLayout>
```

步骤 4：在主程序文件中，为实例添加选项菜单。首先重写 onCreateOptionsMenu 方法，在该方法中实例化一个 MenuInflater 对象，并调用该对象的 inflate()方法解析菜单文件。代码如下：

```java
public class DrawActivity extends Activity {
    @Override
    public void onCreate(Bundle savedInstanceState){
        super.onCreate(savedInstanceState);
        setContentView(R.layout.main);
    }
    //创建选项菜单
    @Override
    public boolean onCreateOptionsMenu(Menu menu){
        MenuInflater inflator=new MenuInflater(this);    //实例化 MenuInflater 对象
        inflator.inflate(R.menu.toolsmenu, menu);        //解析菜单文件
        return super.onCreateOptionsMenu(menu);
    }
    //当菜单项被选择时，作出相应的处理
    @Override
    public boolean onOptionsItemSelected(MenuItem item){
        DrawView dv=(DrawView)findViewById(R.id.drawView1);
                                                         //获取自定义的绘图视图
        dv.paint.setXfermode(null);                      //取消擦除效果
        dv.paint.setStrokeWidth(1);                      //初始化画笔的宽度
        switch(item.getItemId()){
            case R.id.red:
                dv.paint.setColor(Color.RED);            //设置画笔的颜色为红色
                item.setChecked(true);
```

```
            break;
        case R.id.green:
            dv.paint.setColor(Color.GREEN);      //设置画笔的颜色为绿色
            item.setChecked(true);
            break;
        case R.id.blue:
            dv.paint.setColor(Color.BLUE);       //设置画笔的颜色为蓝色
            item.setChecked(true);
            break;
        case R.id.width_1:
            dv.paint.setStrokeWidth(1);          //设置笔触的宽度为 1 像素
            break;
        case R.id.width_2:
            dv.paint.setStrokeWidth(5);          //设置笔触的宽度为 5 像素
            break;
        case R.id.width_3:
            dv.paint.setStrokeWidth(10);         //设置笔触的宽度为 10 像素
            break;
        case R.id.clear:
            dv.clear();                          //擦除绘画
            break;
        case R.id.save:
            dv.save();                           //保存绘画
            break;
        }
        return true;
    }
}
```

步骤 5：在 androidManifest.xml 文件中添加向 SD 卡上保存文件所需要的权限。代码如下：

```
<uses-permission android:name="android.permission.MOUNT_
UNMOUNT_FILESYSTEMS"/>
<uses-permission android:name="android.permission.WRITE_
EXTERNAL_STORAGE"/>
```

图 4-28　开心涂鸦实例运行界面

运行效果如图 4-28 所示。

4.14　本章小结

本章讲解了 Android 游戏开发中的多线程技术。由于在 Android 中不能在子线程（也称工作线程）中更新主线程（也称 UI 线程）中的组件，因此引入消息传递机制。另外，重点讲解了游戏开发中的 View、SurfaceView 两个视图框架的使用以及 Canvas 对象、

Paint 对象、剪切区域、位图操作、游戏动画等图像和动画处理技术。这些是游戏开发的常用知识,需要重点掌握。

4.15 思考与练习

(1) 应用 Android 的绘制函数绘制一个 Android 机器人图案。

(2) 尝试设计并开发一款打气球游戏。可以通过触摸打破屏幕上随机出现的气球,有记分显示功能。在打破一定数量的气球时播放不同的鼓励动画。

(3) 尝试设计并开发一款开心动物园游戏。在游戏场景中有能做各种动作的小动物,触摸不同的小动物会有不同的动作或行为。

(4) 尝试设计并开发一个绘画板。可以对绘制的内容进行保存,并能对保存的绘制内容进行回放。

第 5 章

Android 多媒体与传感器

学习目标：
- 掌握用 Camera 进行图像采集。
- 了解 Android 支持音频和视频格式。
- 掌握用 MediaPlayer 播放音频。
- 掌握用 SoundPool 播放音频。
- 掌握用 VideoView 组件播放视频。
- 掌握用 MediaPlay 和 SurfaceView 播放视频。
- 了解 Android 传感器的框架。
- 掌握方向、重力等常用 Android 传感器的使用。

本章导读：

本章主要讲解 Android 中的音频及视频等多媒体应用知识，另外介绍了传感器的应用方法。主要用于游戏中的各类音效控制和使用方向、重力等传感器操作游戏元素。

5.1 Camera 图像采集

在 Android 中的 Camera 类(位于 android.hardware 包中)用于处理相机相关事件。Camera 类没有构造方法，是通过其提供的一系列方法对相机进行设置与操作。Camera 类常用的方法如表 5-1 所示。

表 5-1　Camera 类的常用方法

方　　法	描　　述
getParameters	获取相机参数
Camera.open()	打开相机
Release()	释放相机资源
setParameters(Camera.Parameters params)	设置相机的拍照参数
setPreviewDisplay(SurfaceHolder holder)	为相机指定一个用来显示相机预览画面的 SurfaceView

续表

方　　法	描　　述
startPreview()	开始预览画面
takePicture（Camera. ShutterCallback shutter, Camera. PictureCallback raw,Camera. PictureCallback jpeg）	进行拍照
stopPreview()	停止预览画面

举例：调用系统相机

步骤 1：修改 res/layout 目录下的 XML 布局文件,添加一个 ImageView 组件和一个按钮组件。代码如下：

```xml
<RelativeLayout xmlns:android="http://schemas.android.com/apk/res/android"
    xmlns:tools="http://schemas.android.com/tools"
    android:layout_width="match_parent"
    android:layout_height="match_parent">
    <ImageView
        android:id="@+id/photo_img"
        android:layout_width="fill_parent"
        android:layout_height="300dp"
        android:layout_alignParentLeft="true"
        android:src="@drawable/ic_launcher"/>
    <Button
        android:id="@+id/btn_takephoto"
        android:layout_width="wrap_content"
        android:layout_height="wrap_content"
        android:layout_alignParentBottom="true"
        android:layout_centerHorizontal="true"
        android:layout_marginBottom="15dp"
        android:text="拍照"/>
</RelativeLayout>
```

步骤 2：在 AndroidManifest.xml 文件中添加访问 SD 卡和控制相机的权限。代码如下：

```xml
<!--授予程序可以向SD卡中保存文件的权限 -->
<uses-permission android:name="android.permission.MOUNT_UNMOUNT_FILESYSTEMS"/>
<uses-permission android:name="android.permission.WRITE_EXTERNAL_STORAGE"/>
<!--授予程序使用摄像头的权限 -->
<uses-permission android:name="android.permission.CAMERA"/>
<uses-feature android:name="android.hardware.camera"/>
<uses-feature android:name="android.hardware.camera.autofocus"/>
```

步骤 3：在主活动的 onCreate()方法中获取相关组件,调用系统相机。代码如下：

```java
public class MainActivity extends Activity {
    private Button btn_takephoto;
    private ImageView view;
    @Override
    public void onCreate(Bundle savedInstanceState){
        super.onCreate(savedInstanceState);
        setContentView(R.layout.activity_main);
        btn_takephoto= (Button)findViewById(R.id.btn_takephoto);
        view= (ImageView)findViewById(R.id.photo_img);
        btn_takephoto.setOnClickListener(new OnClickListener(){
            @Override
            public void onClick(View v){
                Intent intent=new Intent(MediaStore.ACTION_IMAGE_CAPTURE);
                startActivityForResult(intent, 1);
            }
        });
    }
```

步骤4：重写主活动的 onActivityResult()方法，接收由系统相机传回的图像。代码如下：

```java
@Override
protected void onActivityResult(int requestCode, int resultCode, Intent data){
    super.onActivityResult(requestCode, resultCode, data);
    if(resultCode==Activity.RESULT_OK){
        String sdStatus=Environment.getExternalStorageState();
        if(!sdStatus.equals(Environment.MEDIA_MOUNTED)){    //检测SD卡是否可用
            Toast.makeText(this, "请安装SD卡!", Toast.LENGTH_SHORT).show();
            return;
        }
        String name=DateFormat.format("yyyyMMdd_hhmmss", Calendar.getInstance
            (Locale.CHINA))+".jpg";
        Toast.makeText(this, name, Toast.LENGTH_LONG).show();
        Bundle bundle=data.getExtras();
        Bitmap bitmap= (Bitmap)bundle.get("data");
                            //获取相机返回的数据，并转为Bitmap格式
        FileOutputStream b=null;
        File file=new File("/sdcard/Image/");
        file.mkdirs();                                      //创建文件夹
        String fileName="/sdcard/Image/"+name;
        try {
            b=new FileOutputStream(fileName);
            bitmap.compress(Bitmap.CompressFormat.JPEG, 100, b);
                                                            //把数据写入文件
        } catch(FileNotFoundException e){
```

```
            e.printStackTrace();
    } finally {
        try {
            b.flush();                      //将缓冲区中的数据全部写到输出流中
            b.close();                      //关闭文件输出流对象
        } catch(IOException e){
            e.printStackTrace();
        }
    }
    try {
        view.setImageBitmap(bitmap);        //将图片显示在 ImageView 里
    } catch(Exception e){
        Log.e("error", e.getMessage());
    }
    }
}
```

5.2 游戏音乐与音效

Android 支持常用的音频和视频格式,音频格式包括 MP3(.mp3)、3GPP(.3gp)、OGG(.ogg)和 WAVE(.ave)等。在游戏开发中,一般情况下,播放游戏背景音乐的类是 MediaPlayer,而用于游戏音效的则是 SoundPool 类。MediaPlayer 与 SoundPool 的优劣分析如表 5-2 所示。

表 5-2 MediaPlayer 与 SoundPool 的优劣分析

类 名	优 点	缺 点
MediaPlayer	支持很大的音乐文件播放,而且不会同 SoundPool 一样需要加载准备一段时间,MediaPlayer 能及时播放音乐。适合播放游戏背景音乐	资源占用量较高,延迟时间较长,不支持多个音频同时播放等。除此之外,使用 MediaPlayer 播放音乐时,尤其是在快速连续播放声音(比如连续猛点按钮)时,会非常明显地出现 1~3s 的延迟;当然此问题可以通过使用 MediaPlayer.seekTo()方法来解决
SoundPool	支持多个音乐文件同时播放。游戏音效的播放采用 SoundPool 更好,游戏中肯定会出现多个音效同时播放的情况	①最大只能申请 1MB 的内存空间,用户只能使用一些很短的声音片段,而不能用它来播放歌曲或者游戏背景音乐。②提供了 pause 和 stop 方法,但可能会导致程序莫名其妙地终止。③音频格式建议使用 OGG 格式。如果使用 WAVE 格式的音频文件,在播放的情况下有时会出现异常关闭的情况。④播放音乐文件时,如果在构造中就调用播放函数来播放音乐,其效果则是没有声音。不是因为函数没有执行,而是 SoundPool 需要加载准备时间。当然这个准备时间很短,不会影响使用,只是程序刚运行时会没有声音

5.2.1 MediaPlayer 类

MediaPlayer 类通过调用静态方法 create(Context context, intresid) 得到。MediaPlayer 类常用的方法如表 5-3 所示。

表 5-3 MediaPlayer 类的常用方法

方　　法	描　　述
prepare()	为播放音乐文件做准备工作
start()	播放音乐
pause()	暂停音乐播放
stop()	停止音乐播放
setLooping(boolean looping)	设置音乐是否循环播放(true 表示循环,false 表示不循环)
seekTo(int msec)	将音乐播放跳转到某一时间点(以毫秒为单位)
getDuration()	获取播放的音乐文件总时间长度
getCurrentPosition()	得到当前播放音乐的时间点

暂停音乐播放后,可继续播放,再次调用 start() 函数即可;停止音乐播放后,无法继续播放,必须重新调用 prepare() 做播放音乐的准备工作,然后再调用 start() 函数播放音乐。

除此之外,音乐管理类 AudioManager 提供了获取当前音乐大小以及最大音量等方法。AudioManager 类的常用方法如表 5-4 所示。

表 5-4 AudioManager 类的常用方法

方　　法	描　　述
setStreamVolume(int streamType, int index, int flags)	设置音量大小;参数 streamType 为音量类型,参数 index 为音量大小,参数 flags 为设置一个或者多个标识
getStreamVolume(int streamType)	获取当前音量值
getStreamMaxVolume(int streamType)	获取当前音量最大值
Activity.setVolumeControlStream(int streamType)	设置控制音量的类型

使用 MediaPlayer 类控制音频,只需创建该类对象并为其指定要播放的音频文件,然后调用相应的控制音频的方法即可。

举例:音乐播放器

步骤 1:修改 XML 布局文件,添加"播放"、"暂停/继续"和"停止"3 个按钮。

步骤 2:将要播放的音频文件复制到 SD 卡的根目录。

步骤 3:在主活动文件中定义所需变量,重写 onCreate() 方法并添加各个控制按钮的处理事件。代码如下:

```
public class MainActivity extends Activity {
    private MediaPlayer player;                    //声明一个 MediaPlayer 对象
```

```java
    private boolean isPause=false;                    //标识是否暂停
    private File file;                                //声明音频文件
    private TextView txt_hint;
    private Button btn_play, btn_pauorcon, btn_stop;
    @Override
    public void onCreate(Bundle savedInstanceState){
        super.onCreate(savedInstanceState);
        setContentView(R.layout.main);
        btn_play= (Button)findViewById(R.id.btn_play);
        btn_pauorcon= (Button)findViewById(R.id.btn_pauorcon);
        btn_stop= (Button)findViewById(R.id.btn_stop);
        txt_hint=(TextView)findViewById(R.id.txt_hint);
        String path=Environment.getExternalStorageDirectory().getPath();
        File file=new File(path+File.separator+"music.mp3");
                                                      //获取 SD 卡音频文件
        if(file.exists()){                            //如果文件存在
            //创建 MediaPlayer 对象
            player=MediaPlayer.create(this, Uri.parse(file.getAbsolutePath()));
        } else {
            txt_hint.setText("要播放的音频文件不存在!");
            btn_play.setEnabled(false);
            return;
        }
        //为 MediaPlayer 对象添加完成事件监听器
        player.setOnCompletionListener(new OnCompletionListener(){
            @Override
            public void onCompletion(MediaPlayer mp){
                play();                               //重新开始播放
            }
        });
        //为"播放"按钮添加单击事件监听器
        btn_play.setOnClickListener(new OnClickListener(){
            @Override
            public void onClick(View v){
                play();                               //开始播放音乐
                if(isPause){
                    btn_pauorcon.setText("暂停");
                    isPause=false;                    //设置暂停标记变量的值为 false
                }
                btn_pauorcon.setEnabled(true);        //"暂停/继续"按钮可用
                btn_stop.setEnabled(true);            //"停止"按钮可用
                btn_play.setEnabled(false);           //"播放"按钮不可用
            }
        });
```

```java
//为"暂停/继续"按钮添加单击事件监听器
btn_pauorcon.setOnClickListener(new OnClickListener(){
    @Override
    public void onClick(View v){
        if(player.isPlaying()&& !isPause){
            player.pause();                    //暂停播放；
            isPause=true;
            ((Button)v).setText("继续");
            txt_hint.setText("暂停播放音频...");
            btn_play.setEnabled(true);         //"播放"按钮可用
        } else {
            player.start();                    //继续播放
            ((Button)v).setText("暂停");
            txt_hint.setText("继续播放音频...");
            isPause=false;
            btn_play.setEnabled(false);        //"播放"按钮不可用
        }
    }
});
//为"停止"按钮添加单击事件监听器
Btn_stop.setOnClickListener(new OnClickListener(){
    @Override
    public void onClick(View v){
        player.stop();                         //停止播放；
        txt_hint.setText("停止播放音频...");
        btn_pauorcon.setEnabled(false);        //"暂停/继续"按钮不可用
        btn_stop.setEnabled(false);            //"停止"按钮不可用
        btn_play.setEnabled(true);             //"播放"按钮可用
    }
});
}
//播放音乐的方法
private void play(){
    try {
        player.reset();
        player.setDataSource(file.getAbsolutePath());
                                               //重新设置要播放的音频
        player.prepare();                      //预加载音频
        player.start();                        //开始播放
        txt_hint.setText("正在播放音频...");
    } catch(Exception e){
        e.printStackTrace();                   //输出异常信息
    }
}
```

```
    @Override
    protected void onDestroy(){
        if(player.isPlaying()){
            player.stop();                    //停止音频的播放
        }
        player.release();                     //释放资源
        super.onDestroy();
    }
}
```

5.2.2 SoundPool 类

SoundPool 类就是音频池,可以同时播放多个短促的音频,而且占用的资源少,适合在游戏中播放消息提示音、按键音、枪声、爆炸音等密集而短暂的声音。SoundPool 类提供了一个构造方法,用来创建 SoundPool 对象,该构造方法的语法格式如下:

```
SoundPool(int maxStreams, int streamType, int srcQuality)
```

在构造函数中,参数 maxStreams 用于指定可以容纳多少个音频,参数 streamType 用于指定音频类型(可以通过 AudioManager 类提供的常量,通常使用 STREAM_MUSIC),参数 srcQuality 用于指定音频的品质(0 为默认值)。SoundPool 类提供的常用方法如表 5-5 所示。

表 5-5 SoundPool 类的常用方法

方法	描述
load(Context context, int resId, int priority)	加载音频文件,返回音频 ID。参数 priority 用于标识优先考虑的声音
play(int soundID,float leftVolume,float rightVolume, int priority, int loop, float rate)	播放音频。参数 soundID 为加载后得到的音频文件 ID;参数 leftVolume 为左声道的音量(范围:0.0～1.0);参数 rightVolume 为右声道的音量,范围同上;参数 priority 为音频流的优先级(0 是最低优先级);参数 loop 为音频的播放次数(−1 为无限循环,0 为正常一次,大于 0 的数表示循环次数);参数 rate 为播放速率(取值范围为 0.5～2.0,1.0 为正常播放)
pause(int streamID)	暂停音频播放。参数 streamID 为音乐文件加载后的流 ID
stop(int streamID)	结束音频播放
release()	释放 SoundPool 资源
setLoop(int streamID,int loop)	设置循环次数。参数 loop 为循环次数
setRatee(int stream,float rate)	设置播放速率
setVolume(int streamID, float leftVolume,float rightVolume)	设置音量大小
setPriority(int streamID,int priority)	设置流的优先级

举例：通过 SoundPool 类播放音频

步骤 1：修改 XML 布局文件，添加 5 个控制按钮。

步骤 2：在 res 目录下新建 raw 文件夹，将要播放的 4 个 wav 音频文件复制到该文件夹中。

步骤 3：在主活动文件中定义所需变量，重写 onCreate()方法并添加控制按钮的处理事件。代码如下：

```java
public class MainActivity extends Activity {
    private SoundPool soundpool;                        //声明一个 SoundPool 对象
    //创建一个 HashMap 对象
    private HashMap<Integer, Integer> soundmap=new HashMap<Integer, Integer>();
    private Button btn1, btn2, btn3, btn4;
    @Override
    public void onCreate(Bundle savedInstanceState){
        super.onCreate(savedInstanceState);
        setContentView(R.layout.main);
        btn1=(Button)findViewById(R.id.button1);
        btn2=(Button)findViewById(R.id.button2);
        btn3=(Button)findViewById(R.id.button3);
        btn4=(Button)findViewById(R.id.button4);
        //创建一个 SoundPool 对象，该对象可以容纳 5 个音频流
        soundpool=new SoundPool(5, AudioManager.STREAM_SYSTEM, 0);
        //将要播放的音频流保存到 HashMap 对象中
        soundmap.put(1, soundpool.load(this, R.raw.ding1, 1));
        soundmap.put(2, soundpool.load(this, R.raw.ding2, 1));
        soundmap.put(3, soundpool.load(this, R.raw.ding3, 1));
        soundmap.put(4, soundpool.load(this, R.raw.ding4, 1));
        soundmap.put(5, soundpool.load(this, R.raw.ding5, 1));
        //为各按钮添加单击事件监听器
        Btn1.setOnClickListener(new OnClickListener(){
            @Override
            public void onClick(View v){
                soundpool.play(soundmap.get(1), 1, 1, 0, 0, 1);
                                                        //播放指定的音频
            }
        });
        …                                               //此处省略 3 个按钮的事件
    }
    //重写键被按下的事件
    @Override
    public boolean onKeyDown(int keyCode, KeyEvent event){
        soundpool.play(soundmap.get(5), 1, 1, 0, 0, 1);    //播放按键音
        return true;
```

        }
    }

5.2.3 基础实例：游戏音效

本例实现为游戏界面添加背景音乐和按键音效。

具体实现步骤如下。

步骤 1：新建项目 MyGameMusic；在 res 文件夹中创建名称为 raw 的文件夹，加入所需要的 mp3 背景音乐文件和 wav 音效文件（如图 5-1 所示）。

修改 res/layout 中的 XML 布局文件。代码如下：

图 5-1　音频文件目录

```xml
<?xml version="1.0" encoding="utf-8"?>
<FrameLayout xmlns:android="http://schemas.android.com/apk/res/android"
    android:layout_width="match_parent"
    android:layout_height="match_parent"
    android:background="@drawable/background">
    <ImageView
        android:id="@+id/rabbit"
        android:layout_width="wrap_content"
        android:layout_height="wrap_content"
        android:src="@drawable/rabbit"/>
</FrameLayout>
```

步骤 2：打开主程序文件，做如下操作。

（1）创建程序中所需要的成员变量。代码如下：

```
private SoundPool soundpool;                         //声明一个 SoundPool 对象
//创建一个 HashMap 对象
private HashMap<Integer,Integer> soundmap=new HashMap<Integer,Integer>();
private ImageView rabbit;
private int x=0;                                     //兔子在 X 轴的位置
private int y=0;                                     //兔子在 Y 轴的位置
private int width=0;                                 //屏幕的宽度
private int height=0;                                //屏幕的高度
```

（2）在 onCreate()方法中，首先实例化 SoundPool 对象，并将要播放的全部音频保存在 HashMap 对象中，然后获取布局管理器中添加的图片组件，设置默认位置。代码如下：

```
soundpool=new SoundPool(5,AudioManager.STREAM_SYSTEM,0);
                    //创建一个 SoundPool 对象，该对象可以容纳 5 个音频流
                    //将要播放的音频流保存到 HashMap 对象中
soundmap.put(1,soundpool.load(this,R.raw.chimes,1));
soundmap.put(2,soundpool.load(this,R.raw.enter,1));
```

```
soundmap.put(3, soundpool.load(this, R.raw.notify, 1));
soundmap.put(4, soundpool.load(this, R.raw.ringout, 1));
soundmap.put(5, soundpool.load(this, R.raw.ding, 1));
rabbit= (ImageView)findViewById(R.id.rabbit);
width=MainActivity.this.getResources().getDisplayMetrics().widthPixels;
height=MainActivity.this.getResources().getDisplayMetrics().heightPixels;
x=width/2-44;                                    //计算兔子在 X 轴的位置
y=height/2-35;                                   //计算兔子在 Y 轴的位置
rabbit.setX(x);                                  //设置兔子在 X 轴的位置
rabbit.setY(y);                                  //设置兔子在 Y 轴的位置
```

(3) 重写键盘 onKeyDown()方法,根据按键设置音效,并控制图片组件的移动。代码如下:

```
@Override
public boolean onKeyDown(int keyCode, KeyEvent event){
switch(keyCode){
case KeyEvent.KEYCODE_DPAD_LEFT:                 //向左方向键
     soundpool.play(soundmap.get(1), 1, 1, 0, 0, 1);   //播放指定的音频
     if(x>0){
         x-=10;
         rabbit.setX(x);                         //移动小兔子
     }
     break;
   case KeyEvent.KEYCODE_DPAD_RIGHT:             //向右方向键
     soundpool.play(soundmap.get(2), 1, 1, 0, 0, 1);   //播放指定的音频
     if(x<width-88){
         x+=10;
         rabbit.setX(x);                         //移动小兔子
     }
     break;
   case KeyEvent.KEYCODE_DPAD_UP:                //向上方向键
     soundpool.play(soundmap.get(3), 1, 1, 0, 0, 1);   //播放指定的音频
     if(y>0){
         y-=10;
         rabbit.setY(y);                         //移动小兔子
     }
     break;
   case KeyEvent.KEYCODE_DPAD_DOWN:              //向下方向键
     soundpool.play(soundmap.get(4), 1, 1, 0, 0, 1);   //播放指定的音频
     if(y<height-70){
         y+=10;
         rabbit.setY(y);                         //移动小兔子
     }
     break;
```

```
            default:
                soundpool.play(soundmap.get(5), 1, 1, 0, 0, 1);  //播放默认按键音
        }
        return super.onKeyDown(keyCode, event);
    }
```

步骤 3：在 res 目录下新建一个 menu 文件夹，并在该文件夹中创建一个名称为 setting.xml 的菜单资源，在该文件中添加一个控制是否播放背景音乐的多选菜单组，默认为选中状态。代码如下：

```xml
<?xml version="1.0" encoding="utf-8"?>
<menu xmlns:android="http://schemas.android.com/apk/res/android">
    <group android:id="@+id/setting"
        android:checkableBehavior="all">
        <item android:id="@+id/bgsound"
            android:title="播放背景音乐"
            android:checked="true">
        </item>
    </group>
</menu>
```

步骤 4：重写 MainActivity.java 中的 onCreateOptionsMenu 和 onOptionsItemSelected 方法，添加菜单并对选取状态进行处理，用于根据菜单项的选取状态控制是否播放背景音乐。代码如下：

```java
@Override
public boolean onCreateOptionsMenu(Menu menu){
    MenuInflater inflater=new MenuInflater(this);
                                              //实例化一个 MenuInflater 对象
    inflater.inflate(R.menu.setting, menu);   //解析菜单文件
    return super.onCreateOptionsMenu(menu);
}
@Override
public boolean onOptionsItemSelected(MenuItem item){
    if(item.getGroupId()==R.id.setting){      //判断是否选择了参数设置菜单组
        if(item.isChecked()){                 //当菜单项已经被选中
            item.setChecked(false);           //设置菜单项不被选中
            Music.stop(this);
        }else{
            item.setChecked(true);            //设置菜单项被选中
            Music.play(this, R.raw.jasmine);
        }
    }
    return true;
```

步骤 5：创建 Music.java 类文件，用于控制背景音乐的播放。代码如下：

```java
public class Music {
    private static MediaPlayer mp=null;                  //声明一个 MediaPlayer 对象
    public static void play(Context context, int resource){
        stop(context);
        //获取选项菜单存储的首选值判断是否播放背景音乐
        if(SettingsActivity.getBgSound(context)){
            mp=MediaPlayer.create(context, resource);
            mp.setLooping(true);                         //是否循环播放
            mp.start();                                  //开始播放
        }
    }
    public static void stop(Context context){
        if(mp !=null){
            mp.stop();                                   //停止播放
            mp.release();                                //释放资源
            mp=null;
        }
    }
}
```

步骤 6：创建 SettingsActivity.java 类（继承 PreferenceActivity 类），用于实现自动存储首选项的值。首先重写 onCreate()方法，调用 addPreferencesFromResource()方法加载首选项资源文件，然后在 getBgSound()方法中获取是否播放背景音乐的首选项的值。代码如下：

```java
public class SettingsActivity extends PreferenceActivity {
    @Override
    protected void onCreate(Bundle savedInstanceState){
        super.onCreate(savedInstanceState);
        addPreferencesFromResource(R.xml.setting);
    }
    //获取是否播放背景音乐的首选项的值
    public static boolean getBgSound(Context context){
        return PreferenceManager.getDefaultSharedPreferences(context)
            .getBoolean("bgsound", true);
    }
}
```

★**提示**：PreferenceActivity 类用于实现对程序设置参数的存储。在该活动中，设置参数的存储是完全自动的，不需要手动保存。

步骤 7：在主程序文件中，重写 onPause()和 onResume()方法，实现返回、进入界面

时音乐停止和播放。代码如下:

```
@Override
protected void onPause(){
    Music.stop(this);                                    //停止播放背景音乐
    super.onPause();
}
@Override
protected void onResume(){
    Music.play(this, R.raw.jasmine);                     //播放背景音乐
    super.onResume();
}
```

5.2.4 基础实例:游戏开场动画

本例完成游戏开场动画视频的制作。
具体实现步骤如下。

步骤 1:新建项目 MyGameFlash,在 res 文件夹中创建名称为 raw 的文件夹,加入所需要的背景音乐文件,复制图片素材到 res/drawable-mdpi 文件夹(如图 5-2 所示)。

步骤 2:修改 res/layout 中的布局文件 main.xml,添加 Imageview 组件并设置帧布局管理器的背景图片。代码如下:

图 5-2 图片及声音素材目录

```
<?xml version="1.0" encoding="utf-8"?>
<FrameLayout xmlns:android="http://schemas.android.com/apk/res/android"
    android:layout_width="match_parent"
    android:layout_height="match_parent"
    android:background="@drawable/background">
    <ImageView
        android:id="@+id/rabbit"
        android:layout_width="wrap_content"
        android:layout_height="wrap_content"
        android:src="@drawable/rabbit"/>
</FrameLayout>
```

步骤 3:创建 StartActivity.java,重写 onCreate()方法,首先获取 VideoView 组件,并获取要播放的文件对应的 URL;然后为 VideoView 组件指定要播放的视频,并让其获得焦点;再调用 start()方法开始播放视频;最后为 VideoView 组件添加完成事件监听器,重写 onCompletion()方法并调用 startMain()自定义函数进入游戏画面。代码如下:

```
public class StartActivity extends Activity {
    private VideoView video;                             //声明 VideoView 对象
    @Override
```

```
public void onCreate(Bundle savedInstanceState){
    super.onCreate(savedInstanceState);
    setContentView(R.layout.start);
    video=(VideoView)findViewById(R.id.video);      //获取 VideoView 组件
    Uri uri=Uri.parse("android.resource://com.yctc/"+R.raw.ycmusic);
                                                     //获取 URI
    video.setVideoURI(uri);                          //指定要播放的视频
    video.requestFocus();                            //让 VideoView 获得焦点
    try {
        video.start();                               //开始播放视频
    } catch(Exception e){
        e.printStackTrace();                         //输出异常信息
    }
    //为 VideoView 添加完成事件监听器
    video.setOnCompletionListener(new OnCompletionListener(){
        @Override
        public void onCompletion(MediaPlayer mp){
            startMain();                             //进入游戏主界面
        }
    });
}
//进入游戏主界面
private void startMain(){
    Intent intent = new Intent(StartActivity.this, MainActivity.class);
    startActivity(intent);                           //启动新的活动
    StartActivity.this.finish();                     //结束当前活动
}
```

5.3 播放视频

Android 支持的视频格式有 3GPP(.3gp)和 MPEG-4(.mp4)等。在 Android 中提供了一个 VideoView 组件,用于播放视频文件。首先需要在布局文件中创建该组件,然后在活动中获取组件,并使用其 setVideoPath()方法或 setVideoURL()方法加载视频文件,最后调用 start()、stop()和 pause()方法控制视频。在 Android 中还提供了一个 MediaController 组件,用于通过图形界面方式控制视频的播放。

举例：应用组件播放视频

步骤 1：修改 res/layout 目录下的 XML 布局文件,添加一个 VideoView 组件。代码如下：

```
<VideoView
    android:id="@+id/video"
```

```
            android:layout_width="match_parent"
            android:layout_height="wrap_content"
            android:layout_gravity="center"/>
```

步骤 2：将要播放的视频文件复制到 SD 卡的根目录（或自定义目录）。

步骤 3：在主活动文件中定义所需变量，重写 onCreate()方法并添加控制按钮的处理事件。代码如下：

```java
public class MainActivity extends Activity {
    private VideoView video;                              //声明 VideoView 对象
    @Override
    public void onCreate(Bundle savedInstanceState){
        super.onCreate(savedInstanceState);
        setContentView(R.layout.main);
        video= (VideoView)findViewById(R.id.video);       //获取 VideoView 组件
        String path=Environment.getExternalStorageDirectory().getPath();
        File file=new File(path+File.separator+"Falling.3gp");
                                                          //获取 SD 卡上视频
        MediaController mc=new MediaController(MainActivity.this);
        if(file.exists()){                                //判断视频文件是否存在
            video.setVideoPath(file.getAbsolutePath());
                                                          //指定要播放的视频
            video.setMediaController(mc);
                                //设置 VideoView 与 MediaController 相关联
            video.requestFocus();                         //让 VideoView 获得焦点
            try {
                video.start();                            //开始播放视频
            } catch(Exception e){
                e.printStackTrace();                      //输出异常信息
            }
            //为 VideoView 添加完成事件监听器
            video.setOnCompletionListener(new OnCompletionListener(){
                @Override
                public void onCompletion(MediaPlayer mp){
                    Toast.makeText(MainActivity.this,"视频播放完毕!",
                    Toast.LENGTH_SHORT).show();           //弹出消息提示框显示播放完毕
                }
            });
        }else{
            Toast.makeText(this,"要播放的视频文件不存在",Toast.LENGTH_SHORT).
            show();
        }
    }
}
```

也可以使用 MediaPlayer 播放视频,但需要使用 SurfaceView 组件来显示视频图像。一般分 4 个步骤进行:定义 SurfaceView 组件,创建 MediaPlayer 对象,将视频画面输出到 SurfaceView,以及调用 MediaPlayer 对象的相应方法控制视频。

举例:应用视图播放视频

步骤 1:修改 XML 布局文件,添加一个 SurfaceView 组件和 3 个控制按钮。代码如下:

```xml
<VideoView
    android:id="@+id/video"
    android:layout_width="match_parent"
    android:layout_height="wrap_content"
    android:layout_gravity="center"/>
```

步骤 2:将要播放的视频文件复制到 SD 卡的根目录(或自定义目录)。

步骤 3:在主活动文件中定义所需变量,重写 onCreate()方法并添加控制按钮的处理事件。代码如下:

```java
public class MainActivity extends Activity {
    private MediaPlayer mp;                             //声明 MediaPlayer 对象
    private SurfaceView video_view;                     //声明 SurfaceView 对象
    private Button btn_play, btn_pauorcon, btn_stop;
    @Override
    public void onCreate(Bundle savedInstanceState){
        super.onCreate(savedInstanceState);
        setContentView(R.layout.main);
        mp=new MediaPlayer();                           //实例化 MediaPlayer 对象
        video_view= (SurfaceView)findViewById(R.id.video_view);
        btn_play= (Button)findViewById(R.id.btn_play);
        btn_pauorcon= (Button)findViewById(R.id.btn_pauorcon);
        btn_stop= (Button)findViewById(R.id.btn_stop);
        //为"播放"按钮添加单击事件监听器
        btn_play.setOnClickListener(new OnClickListener(){
            @Override
            public void onClick(View v){
                mp.reset();                             //重置 MediaPlayer 对象
                try {
                    String path = Environment.getExternalStorageDirectory().getPath();
                    mp.setDataSource (path + File.separator +" Movies " + File.separator+"Falling.3gp");    //设置要播放的视频
                    mp.setDisplay(video_view.getHolder());
                                                        //将视频画面输出到视图
                    mp.prepare();                       //预加载视频
                    mp.start();                         //开始播放
```

```java
                btn_pauorcon.setText("暂停");
                btn_pauorcon.setEnabled(true);      //设置"暂停"按钮可用
            } catch(Exception e){
                e.printStackTrace();
            }
        }
    });
    //为"停止"按钮添加单击事件监听器
    btn_stop.setOnClickListener(new OnClickListener(){
        @Override
        public void onClick(View v){
            if(mp.isPlaying()){
                mp.stop();                          //停止播放
                btn_pauorcon.setEnabled(false);     //设置"暂停"按钮不可用
            }
        }
    });
    //为"暂停"按钮添加单击事件监听器
    btn_pauorcon.setOnClickListener(new OnClickListener(){
        @Override
        public void onClick(View v){
            if(mp.isPlaying()){
                mp.pause();                         //暂停视频的播放
                ((Button)v).setText("继续");
            } else {
                mp.start();                         //继续视频的播放
                ((Button)v).setText("暂停");
            }
        }
    });
    //为MediaPlayer对象添加完成事件监听器
    mp.setOnCompletionListener(new OnCompletionListener(){
        @Override
        public void onCompletion(MediaPlayer mp){
            Toast.makeText(MainActivity.this,"视频播放完毕!", Toast.LENGTH
                _SHORT).show();
        }
    });
}
@Override
protected void onDestroy(){
    if(mp.isPlaying()){
        mp.stop();                                  //停止播放视频
    }
```

```
            mp.release();                              //释放资源
            super.onDestroy();
        }
    }
```

5.4 传 感 器

Android设备一般都有内置的测量运动、方向和各种环境条件的传感器。这些传感器具有提供高精度和准确度的原始数据的能力,可用于监视设备在三维方向的移动、位置或监视设备周围环境的变化。

例如,有关天气的应用程序可能要使用设备的温度传感器和湿度传感器来计算并报告露点,有关旅行的应用程序可能要使用地磁场传感器和加速度传感器来报告罗盘方位。因此,游戏中可以从重力传感器中读取轨迹,以便推断出复杂的用户手势和意图,如倾斜、振动、旋转或摆动等。

5.4.1 传感器介绍

Android平台支持3种宽泛类别的传感器。

(1) 运动传感器。

运动传感器沿着三轴方向来测量加速度和扭力。这种类型的传感器包括加速度传感器、重力传感器、陀螺仪和选择矢量传感器。

(2) 环境传感器。

这些传感器测量各种环境参数,如周围空气的温度和压力、照度和湿度等。这种类型的传感器包括气压计、光度计和温度计等。

(3) 位置传感器。

这些传感器用于测量设备的物理位置。这种类型的传感器包括方向传感器和磁力计等。能够访问这些设备上有效的传感器,并能通过使用Android传感器框架来获取原始的传感器数据。该传感器框架提供了几个类和接口来帮助用户执行各种传感器相关的任务。

使用传感器可以做以下事情:
- 判断设备上有哪些传感器可用。
- 判断个别传感器的能力,如它们的最大范围、制造商、电力需求和辨识率。
- 获取原始传感器数据,并定义获取传感器数据的最小比率。
- 注册和解除注册用于监听传感器变化的事件监听器。

很少有Android设备支持所有类型的传感器。例如,大多数手持设备和平板设备都有一个加速仪和一个磁力仪,但是很少有气压计和温度计。一个设备上也能够有多个同一给定类型的传感器。例如,一个设备能够有两个重力传感器,每个有不同测量范围。Android平台所支持的传感器如表5-6所示。

表 5-6　Android 平台支持的传感器类型

传 感 器	类 型	介　　绍	常用场景
TYPE_ACCELEROMETER	硬件	以 m/s^2 为单位测量应用于设备三轴（X、Y、Z）的加速力，包括重力	运动检测（振动、倾斜等）
TYPE_AMBIENT_TEMPERATURE	硬件	以摄氏度（℃）为单位测量周围温度	监测空气温度
TYPE_GRAVITY	软件或硬件	以 m/s^2 为单位测量应用于设备三轴（X、Y、Z）的重力	运动检测（振动、倾斜等）
TYPE_GYROSCOPE	硬件	以弧度/秒（rad/s）为单位，测量设备围绕 3 个物理轴（X、Y、Z）的旋转率	旋转检测（旋转、翻转等）
TYPE_LIGHT	软件	以 lx 为单位，测量周围的亮度等级（照度）	控制屏幕的亮度
TYPE_LINEAR_ACCELERATION	软件或硬件	以 m/s^2 为单位测量应用于设备 3 个物理轴（X、Y、Z）的加速力，重力除外	检测一个单独的物理轴的加速度
TYPE_MAGNETIC_FIELD	硬件	以 μT 为单位，测量设备周围 3 个物理轴（X、Y、Z）的磁场	创建一个罗盘
TYPE_ORIENTATION	软件	测量设备围绕 3 个物理轴（X、Y、Z）的旋转角度。在 API Level 3 以后，能够通过使用重力传感器和磁场传感器与 getRotationMatrix 方法相结合来获取倾斜矩阵和旋转矩阵	判断设备的位置
TYPE_PRESSURE	硬件	以 hPa 或 mBar 为单位来测量周围空气的压力	检测空气压力的变化
TYPE_PROXIMITY	硬件	以 cm 为单位，测量一个对象相对于设备屏幕的距离。这个传感器通常用于判断手持设备是否被举到了一个人的耳朵附近	通话期间的电话位置
TYPE_RELATIVE_HUMIDITY	硬件	以百分比（%）为单位测量周围的相对湿度	监测露点以及绝对和相对的湿度
TYPE_ROTATION_VECTOR	软件或硬件	通过提供设备旋转矢量的 3 个要素来测量设备的方向	运动监测和旋转监测

续表

传 感 器	类 型	介 绍	常用场景
TYPE_TEMPERATURE	硬件	以摄氏度(℃)为单位来测量设备的温度。这个传感器在各种不同设备中被实现，在 API Level 14 中被用于替换 TYPE_AMBIENT_TEMPERATURE 传感器	监测温度

5.4.2 传感器框架

Android 传感器框架提供对多种类型的传感器的访问,其中某些传感器是基于硬件的,有些传感器是基于软件的。基于硬件的传感器是内置于手持或平板设备中的物理组件,通过直接测量特定的环境属性来获取数据,例如加速度、磁场强度或角度的变化等。基于软件的传感器不是物理设备,而是从一个或多个有时被叫做虚拟传感器或合成传感器的基于硬件的传感器来获取数据。线性加速度传感器和重力传感器是基于硬件的传感器的实例。通过使用 Android 框架,用户能够访问这些传感器,并获取原始的传感器数据。传感器框架是 android.hardware 包的一部分,并且包括以下类和接口。

(1) SensorManager:创建一个传感器服务的实例,提供了各种用于访问和监听传感器的方法,它还提供了几个传感器常量,用于报告传感器的精度、设置数据获取的速率以及校准传感器等。

(2) Sensor:创建一个特殊传感器的实例,提供了判断传感器能力的各种方法。

(3) SensorEvent:创建一个传感器事件对象,提供了相关传感器事件的信息。一个传感器事件对象包含以下信息:

- 原始传感器数据。
- 产生事件的传感器的类型。
- 数据的精度。
- 事件的时间戳。

(4) SensorEventListener:这个接口创建两个回调方法,这两个方法在传感器值或精度发生变化时接收通知(传感器事件)。

传感器中 3 个参数 X、Y、Z(float 类型的值,取值范围为 $-10\sim10$)的含义如下。

(1) 手机屏幕向左侧:X 轴朝向天空,垂直放置,这时 Y 轴与 Z 轴没有重力分量,因为 X 轴朝向天空,所以它的重力分量最大。这时 X 轴、Y 轴、Z 轴的重力分量的值为 (10,0,0)。

(2) 手机屏幕向右侧:X 轴朝向地面,垂直放置,这时 Y 轴与 Z 轴没有重力分量,因为 X 轴朝向地面,所以它的重力分量最小。这时候 X 轴、Y 轴、Z 轴的重力分量的值为 (−10,0,0)。

(3) 手机屏幕垂直竖立放置:Y 轴朝向天空,垂直放置,这时候 X 轴与 Z 轴没有重力分量,因为 Y 轴朝向天空,所以它的重力分量最大。这时候 X 轴、Y 轴、Z 轴的重力分量的值为(0,10,0)。

(4) 手机屏幕垂直竖立放置：Y 轴朝向地面,垂直放置,这时 X 轴与 Z 轴没有重力分量,因为 Y 轴朝向地面,所以它的重力分量最小。这时 X 轴、Y 轴、Z 轴的重力分量的值为(0,−10,0)。

(5) 手机屏幕向上：Z 轴朝向天空,水平放置,这时候 X 轴与 Y 轴没有重力分量,因为 Z 轴朝向天空,所以它的重力分量最大。这时候 X 轴、Y 轴、Z 轴的重力分量的值为(0,0,10)。

(6) 手机屏幕向下：Z 轴朝向地面,水平放置,这时候 X 轴与 Y 轴没有重力分量,因为 Z 轴朝向地面,所以它的重力分量最小。这时候 X 轴、Y 轴、Z 轴的重力分量的值为(0,0,−10)。

★**注意**：如果在模拟器上查看传感器数值变化,要安装相应的插件才能实现,但最好用真机调试。

使用 Android 传感器必须调用 registerListener(SensorEventListener listener, Sensor sensor, int rateUs)方法注册(参数 listener 为自定义的监听器事件,参数 sensor 为传感器类型的对象,参数 rateUs 为可选数据变化的刷新频率)。

举例：输出加速度传感器坐标值

步骤 1：新建项目,在主活动中建立传感器监听对象类,并在 onCreate()方法中进行注册。代码如下：

```java
public class MainActivity extends Activity {
    private static final String TAG="sensor";        //设置 LOG 标签
    private SensorManager sm;                        //声明 SensorManager 对象
    @Override
    public void onCreate(Bundle savedInstanceState){
        super.onCreate(savedInstanceState);
        setContentView(R.layout.activity_main);
        //创建一个 SensorManager 来获取系统的传感器服务
        sm=(SensorManager)getSystemService(Context.SENSOR_SERVICE);
        int sensorType=Sensor.TYPE_ACCELEROMETER;    //选取加速度感应器
        sm.registerListener(myAccelerometerListener, sm.getDefaultSensor
        (sensorType), SensorManager.SENSOR_DELAY_NORMAL);
    }
    final SensorEventListener myAccelerometerListener=new SensorEventListener(){
        //重写 onSensorChanged 方法,当数据变化的时候被触发调用
        @Override
        public void onSensorChanged(SensorEvent event){
            if(event.sensor.getType()==Sensor.TYPE_ACCELEROMETER){
                Log.i(TAG, "onSensorChanged");
                float X_lateral=event.values[0];      //传感器 X 方向角度值
                float Y_longitudinal=event.values[1]; //传感器 Y 方向角度值
```

```
                float Z_vertical=event.values[2];    //传感器Z方向角度值
                Log.i(TAG, "\n heading "+X_lateral);
                Log.i(TAG, "\n pitch "+Y_longitudinal);
                Log.i(TAG, "\n roll "+Z_vertical);
            }
        }
        //重写 onAccuracyChanged 方法,当获得数据的精度发生变化的时候被调用
        @Override
        public void onAccuracyChanged(Sensor sensor, int accuracy){
            Log.i(TAG, "onAccuracyChanged");
        }
    };
}
```

步骤2:重写活动的 onResume()方法和 onPause()方法。代码如下:

```
@Override
protected void onResume(){
    super.onResume();
    sm.registerListener(myAccelerometerListener, sm.getDefaultSensor(sensorType),
    SensorManager.SENSOR_DELAY_NORMAL);
}
@Override
protected void onPause(){
    sm.unregisterListener(myAccelerometerListener);
    super.onPause();
}
```

★**注意**:即使活动不可见的时候,感应器依然会继续工作,所以一定要在 onPause 方法中关闭触发器,否则会非常耗费电量。

5.4.3 基础实例:战机飞行

本例实现利用重力传感器控制游戏中飞机的移动。
具体实现步骤如下。

步骤1:新建项目 ProjectPlaneFly,复制图片素材(如图 5-3 所示)到 res/drawable-mdpi 文件夹。

步骤2:建立内部类 GameSurfaceView(继承自 android. view. SurfaceView 类),添加视图类的构造函数并实现 Callback 接口(android. view. SurfaceHolder. Callback);实现 Runable 接口并重写其 run()方法。代码如下:

图 5-3 战机飞行图片素材

```
public class MyView extends SurfaceView implements Callback, Runnable {
    private MainActivity activity;
    private SurfaceHolder holder;
```

```java
        private Canvas canvas;
        private Paint paint;
        private Bitmap bg, plane;                           //声明背景与飞机图像
        private int x, y;
        private int speedX, speedY;                         //飞机移动速度
        float dataX, dataY;                                 //获取传感器数据
        private boolean isGame;
        public MyView(Context context){
            super(context);
            activity=(MainActivity)context;
            holder=getHolder();
            holder.addCallback(this);
            paint=new Paint();
            paint.setAntiAlias(true);
            bg=BitmapFactory.decodeResource(getResources(), R.drawable.bg);
            plane=BitmapFactory.decodeResource(getResources(), R.drawable.plane);
        }
        @Override
        public void surfaceChanged(SurfaceHolder holder, int format, int width,
                int height){
        }
        @Override
        public void surfaceCreated(SurfaceHolder holder){
            isGame=true;
            x=getWidth()/2-plane.getWidth()/2;              //飞机默认位置的X坐标
            y=getHeight()/2-plane.getHeight()/2;            //飞机默认位置的Y坐标
            new Thread(this).start();                       //启动线程
        }
        @Override
        public void surfaceDestroyed(SurfaceHolder holder){
            isGame=false;                                   //游戏结束
            activity.finish();                              //关闭主活动
        }
        /**
         * 根据传感器数据设置飞机移动速度
         */
        private void setSpeed(){
            if(dataX<-0.5 && dataX>=-1.5)
                speedX=5;
            else if(dataX<-1.5 && dataX>=-3)
                speedX=10;
            if(dataX>0.5 && dataX<=1.5)
                speedX=-5;
            else if(dataX>1.5 && dataX<=3)
```

```
            speedX=-10;
        else if(dataX>=-0.5 && dataX<=0.5)
            speedX=0;
        if(dataY>=9.2)
            speedY=5;
        else if(dataY<8)
            speedY=-5;
        else if(dataY>=8 && dataY<9.2)
            speedY=0;
    }
    /**
     * 飞机移动
     */
    private void move(){
        if(x>=speedX && x<=getWidth()-plane.getWidth()-speedX)
            x+=speedX;
        if(y>=speedY && y<=getHeight()-plane.getHeight()-speedY)
            y+=speedY;
    }
    /**
     * 绘制游戏画面
     */
    private void render(){
        canvas=holder.lockCanvas();
        canvas.drawBitmap(bg, 0, 0, paint);           //绘制背景
        canvas.drawBitmap(plane, x, y, paint);        //绘制飞机
        holder.unlockCanvasAndPost(canvas);
    }
    @Override
    public void run(){
        while(isGame){
            setSpeed();
            move();
            render();
            try {
                Thread.sleep(100);
            } catch(InterruptedException e){
                e.printStackTrace();
            }
        }
    }
}
```

步骤3：在主活动的 onCreate()方法中，设置屏幕属性并调用 MyView 视图。代码

如下：

```java
public class MainActivity extends Activity {
    private SensorManager manager;
    private Sensor sensor;
    private MyView gameView;
    @Override
    protected void onCreate(Bundle savedInstanceState){
        super.onCreate(savedInstanceState);
        setContentView(R.layout.activity_main);
        manager= (SensorManager)getSystemService(Context.SENSOR_SERVICE);
        sensor=manager.getDefaultSensor(Sensor.TYPE_ACCELEROMETER);
        gameView=new MyView(this);
        setContentView(gameView);
        manager.registerListener(sensorListener, sensor,
        SensorManager.SENSOR_DELAY_GAME);
    }
    final SensorEventListener sensorListener=new SensorEventListener(){
        @Override
        public void onSensorChanged(SensorEvent event){
            gameView.dataX=event.values[0];      //传递传感器 X 方向的角度值
            gameView.dataX=event.values[1];      //传递传感器 Y 方向的角度值
        }
        @Override
        public void onAccuracyChanged(Sensor sensor, int accuracy){
            //TODO: Auto-generated method stub
        }
    };
    @Override
    protected void onResume(){
        super.onResume();
        manager.registerListener(sensorListener,
        sensor, 0);
    }
    @Override
    protected void onPause(){
        super.onPause();
        manager.unregisterListener(sensorListener);
    }
}
```

真机测试，运行效果如图 5-4 所示。

图 5-4　战机飞行实例运行界面

5.5 综合实例一：控制相机拍照

5.5.1 功能描述

本例实现拍照预览功能，用自定义对话框显示及命名图像并保存在 SD 卡中。

5.5.2 关键技术

本例实现的关键是对相机参数的设置。关键代码如下：

```
Camera.Parameters parameters=camera.getParameters();   //获取相机参数对象
parameters.setPictureSize(640, 480);                    //设置预览画面的尺寸
parameters.setPictureFormat(ImageFormat.JPEG);          //指定图片格式
parameters.set("jpeg-quality", 80);                     //设置图片的质量
camera.setParameters(parameters);                       //重新设置相机参数
camera.startPreview();                                  //开始预览
camera.autoFocus(null);                                 //设置自动对焦
```

5.5.3 实现过程

步骤 1：在 res/values 目录下的 strings.xml 文件中添加字符串资源定义。代码如下：

```
<string name="preview">预览</string>
<string name="takephoto">拍照</string>
```

步骤 2：修改 res/layout 目录下的 XML 布局文件，添加一个 SurfaceView 组件（用于显示相机预览画面）和两个按钮组件，分别用于控制预览和拍照。代码如下：

```
<?xml version="1.0" encoding="utf-8"?>
<LinearLayout xmlns:android="http://schemas.android.com/apk/res/android"
    android:orientation="vertical"
    android:layout_width="fill_parent"
    android:layout_height="fill_parent">
<LinearLayout
    android:orientation="horizontal"
    android:layout_width="fill_parent"
    android:layout_height="wrap_content">
<TextView
    android:layout_width="wrap_content"
    android:layout_height="wrap_content"
    android:layout_marginRight="8dp"
    android:text="相片名称:"/>
```

```xml
<EditText
    android:id="@+id/phone_name"
    android:layout_width="fill_parent"
    android:layout_height="wrap_content"/>
</LinearLayout>
<ImageView
    android:id="@+id/show"
    android:layout_width="320dp"
    android:layout_height="240dp"
    android:scaleType="fitCenter"
    android:layout_marginTop="10dp"/>
</LinearLayout>
```

步骤3：在 res/layout 目录下新建一个布局文件 save_photo.xml，用于对话框布局。代码如下：

```xml
<?xml version="1.0" encoding="utf-8"?>
<LinearLayout xmlns:android="http://schemas.android.com/apk/res/android"
    android:layout_width="fill_parent"
    android:layout_height="fill_parent"
    android:orientation="vertical">
    <TextView
        android:layout_width="wrap_content"
        android:layout_height="wrap_content"
        android:layout_marginRight="8dp"
        android:text="相片名称:"/>
    <EditText
        android:id="@+id/phone_name"
        android:layout_width="fill_parent"
        android:layout_height="wrap_content"/>
    <ImageView
        android:id="@+id/show"
        android:layout_width="320dp"
        android:layout_height="240dp"
        android:layout_marginTop="10dp"
        android:scaleType="fitCenter"/>
</LinearLayout>
```

步骤4：在 AndroidManifest.xml 文件中添加访问 SD 卡和控制相机的权限。代码如下：

```xml
<!--授予程序可以向SD卡中保存文件的权限 -->
<uses-permission android:name="android.permission.MOUNT_UNMOUNT_FILESYSTEMS"/>
<uses-permission android:name="android.permission.WRITE_EXTERNAL_STORAGE"/>
<!--授予程序使用摄像头的权限 -->
```

```xml
<uses-permission android:name="android.permission.CAMERA"/>
<uses-feature android:name="android.hardware.camera"/>
<uses-feature android:name="android.hardware.camera.autofocus"/>
```

步骤 5：在主活动文件中完成代码如下：

```java
public class MainActivity extends Activity {
    private Camera camera;                                      //声明相机对象
    private boolean isPreview=false;                            //标识是否为预览模式
    @Override
    public void onCreate(Bundle savedInstanceState){
        super.onCreate(savedInstanceState);
        requestWindowFeature(Window.FEATURE_NO_TITLE);     //设置全屏显示
        setContentView(R.layout.main);
        //判断是否安装 SD 卡
        if(!android.os.Environment.getExternalStorageState().equals(android.
        os.Environment.MEDIA_MOUNTED)){
            Toast.makeText(this, "请安装 SD 卡!", Toast.LENGTH_SHORT).show();
        }
        SurfaceView sv=(SurfaceView)findViewById(R.id.sv_preview);
        final SurfaceHolder sh=sv.getHolder();
        //设置该 SurfaceHolder 不维护缓冲(已过期)
        //sh.setType(SurfaceHolder.SURFACE_TYPE_PUSH_BUFFERS);
        Button preview=(Button)findViewById(R.id.preview);
                                                            //获取"预览"按钮
        preview.setOnClickListener(new View.OnClickListener(){
            @Override
            public void onClick(View v){
                //如果相机为非预览模式，则打开相机
                if(!isPreview){
                    camera=Camera.open();                   //打开相机
                }
                try {
                    camera.setPreviewDisplay(sh);    //设置显示预览的 SurfaceView
                    //获取相机参数
                    Camera.Parameters parameters=camera.getParameters();
                    parameters.setPictureSize(640, 480);    //设置预览画面的尺寸
                    parameters.setPictureFormat(ImageFormat.JPEG);
                                                            //指定图片格式
                    parameters.set("jpeg-quality", 80);     //设置图片的质量
                    camera.setParameters(parameters);       //重新设置相机参数
                    camera.startPreview();                  //开始预览
                    camera.autoFocus(null);                 //设置自动对焦
                } catch(IOException e){
                    e.printStackTrace();
```

```java
                }
            }
        });
        Button takePhoto=(Button)findViewById(R.id.takephoto);
                                                                //获取"拍照"按钮
        takePhoto.setOnClickListener(new View.OnClickListener(){
            @Override
            public void onClick(View v){
                if(camera!=null){
                    camera.takePicture(null, null, jpeg);    //进行拍照
                }
            }
        });
    }
    //实现拍照的回调接口
    final PictureCallback jpeg=new PictureCallback(){
        @Override
        public void onPictureTaken(byte[] data, Camera camera){
            //根据拍照所得的数据创建位图
            final Bitmap bm=BitmapFactory.decodeByteArray(data, 0, data.length);
            //加载 layout/save.xml 文件对应的布局资源
            View saveView=getLayoutInflater().inflate(R.layout.save_photo, null);
            final EditText photoName = (EditText) saveView.findViewById (R. id.
            phone_name);
            //获取对话框上的 ImageView 组件
            ImageView show=(ImageView)saveView.findViewById(R.id.show);
            show.setImageBitmap(bm);                //显示刚刚拍得的照片
            camera.stopPreview();                   //停止预览
            isPreview=false;
            //使用对话框显示 saveDialog 组件
            new AlertDialog.Builder(MainActivity.this).setView(saveView).
            setPositiveButton("保存", new DialogInterface.OnClickListener(){
                @Override
                public void onClick(DialogInterface dialog, int which){
                    File file=new File("/sdcard/pictures/"+photoName.getText
                    ().toString()+".jpg");              //创建文件对象
                    try{
                        file.createNewFile();           //创建一个新文件
                        //创建一个文件输出流对象
                        FileOutputStream fileOS=new FileOutputStream(file);
                        //将图片内容压缩为 JPEG 格式输出到输出流对象中
                        bm.compress(Bitmap.CompressFormat.JPEG, 100, fileOS);
                        fileOS.flush();           //将缓冲区中的数据全部写出到输出流中
                        fileOS.close();           //关闭文件输出流对象
```

```java
                    isPreview=true;
                    resetCamera();
                } catch(IOException e){
                    e.printStackTrace();
                }
            }
        }).setNegativeButton("取消", new DialogInterface.OnClickListener(){
            public void onClick(DialogInterface dialog, int which){
                isPreview=true;
                resetCamera();                              //重新预览
            }
        }).show();
    }
};
//重新预览函数
private void resetCamera(){
    if(isPreview){
        camera.startPreview();
    }
}
//停止预览并释放资源
@Override
protected void onPause(){
    if(camera!=null){
        camera.stopPreview();                              //停止预览
        camera.release();                                  //释放资源
    }
    super.onPause();
}
```

5.6 综合实例二：游戏导航摇杆

5.6.1 功能描述

本例实现在屏幕上绘制一个 360°的平滑导航摇杆，玩家操作大圆内的小圆（小圆的最大活动范围是外边的大圆）来操作游戏。

5.6.2 关键技术

本例实现的关键是要在小圆作圆周运动时得到小圆的坐标，这里根据角度转弧度，再通过三角函数的定理得到小圆坐标位置。封装一个得到小圆坐标的方法。代码如下：

```
public void setSmallCircleXY(float centerX, float centerY, float R, double rad){
    smallCenterX=(float)(R * Math.cos(rad))+centerX;    //获取圆周运动的 X 坐标
    smallCenterY=(float)(R * Math.sin(rad))+centerY;    //获取圆周运动的 Y 坐标
}
```

封装一个得到玩家触点相对于大圆的角度的方法。代码如下:

```
public double getRad(float px1, float py1, float px2, float py2){
    float x=px2 -px1;                                   //得到两点 X 的距离
    float y=py1 -py2;                                   //得到两点 Y 的距离
    //算出斜边长
    float Hypotenuse= (float)Math.sqrt(Math.pow(x, 2)+Math.pow(y, 2));
    //得到这个角度的余弦值(通过三角函数中的定理:邻边/斜边=角度余弦值)
    float cosAngle=x/Hypotenuse;
    //通过反余弦定理获取其角度的弧度
    float rad= (float)Math.acos(cosAngle);
    //当触屏的位置 Y 坐标<摇杆的 Y 坐标,则取反值-0~-180
    if(py2<py1){
        rad=-rad;
    }
    return rad;
}
```

5.6.3 实现过程

步骤 1:新建项目 MyCtrol,设置主程序文件名为 MySurfaceView(继承 SurfaceView 类),实现 onDraw()方法,绘制导航摇杆图形。代码如下:

```
//定义两个圆形的中心点坐标与半径
private float smallCenterX=120, smallCenterY=120, smallCenterR=20;
private float BigCenterX=120, BigCenterY=120, BigCenterR=40;
...                                                     //此处省略部分代码
public void myDraw(){
    ...                                                 //此处省略部分代码
    //绘制大圆
    paint.setAlpha(0x77);
    canvas.drawCircle(BigCenterX, BigCenterY, BigCenterR, paint);
    //绘制小圆
    canvas.drawCircle(smallCenterX, smallCenterY, smallCenterR, paint);
    ...                                                 //此处省略部分代码
}
```

步骤 2:封装一个得到玩家触点相对于大圆的角度的方法。代码如下:

```
public double getRad(float px1, float py1, float px2, float py2){
    float x=px2 -px1;
```

```
        float y=py1 -py2;
        float Hypotenuse= (float)Math.sqrt(Math.pow(x, 2)+Math.pow(y, 2));
        float cosAngle=x/Hypotenuse;
        float rad=(float)Math.acos(cosAngle);
        if(py2<py1){
            rad=-rad;
        }
        return rad;
    }
```

步骤 3：完成触屏监听函数。代码如下：

```
@Override
public boolean onTouchEvent(MotionEvent event){
    //当用户手指抬起时,应该恢复小圆到初始位置
    if(event.getAction()==MotionEvent.ACTION_UP){
        smallCenterX=BigCenterX;
        smallCenterY=BigCenterY;
    else {
        int pointX=(int)event.getX();
        int pointY=(int)event.getY();
        //判断用户点击的位置是否在大圆内
        if(Math.sqrt(Math.pow((BigCenterX - (int)event.getX()), 2)+Math.pow
            ((BigCenterY -(int)event.getY()), 2))<=BigCenterR){
            //让小圆跟随用户触点位置移动
            smallCenterX=pointX;
            smallCenterY=pointY;
        } else {
            setSmallCircleXY ( BigCenterX, BigCenterY, BigCenterR, getRad
            (BigCenterX, BigCenterY, pointX, pointY));
        }
    }
    return true;
}
```

步骤 4：在主程序文件中设置全屏、去除标题栏、保持亮度及显示视图。

运行效果如图 5-5 所示。

★**提示**：如果需要使用摇杆控制游戏中的元素移动,那么首先将整个 360°分成 4 等分(四方行走)或 8 等分(八方行走),对应主角(即游戏中的元素)的 4 方向或者 8 方向;然后通过封装的两点之间得到弧度的函数获取摇杆弧度,将其转换成角度,再将摇杆的角度与之前的 360°分成的 4 等分或 8 等分范围比对及处理即可。

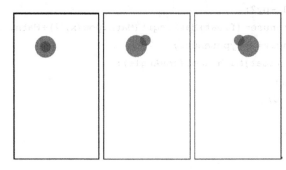

图 5-5　平滑导航摇杆实例运行界面

5.7　综合实例三：多点触屏缩放

5.7.1　功能描述

本例实现一个多触点游戏场景缩放功能，使玩家能用两只手指缩放游戏地图。

5.7.2　关键技术

本例实现的关键是触屏监听器的使用，玩家手指抬起默认还原为第一次触屏标识位，并且保存本次的缩放比例。这里采用的方法是得到第一次触屏时线段的长度，再得到第二次触屏时线段的长度，然后计算出本次的缩放比例。代码如下：

```
if(isFirst){
    //得到第一次触屏时线段的长度
    oldLineDistance=(float)Math.sqrt(Math.pow(event.getX(1)-event.getX(0),2)
    +Math.pow(event.getY(1)-event.getY(0),2));
    isFirst=false;
} else {
    //得到非第一次触屏时线段的长度
    float newLineDistance=(float)Math.sqrt(Math.pow(event.getX(1)-event.getX
    (0),2)+Math.pow(event.getY(1)-event.getY(0),2));
    //获取本次的缩放比例
    rate=oldRate * newLineDistance/oldLineDistance;
}
```

★注意：多触点是 API 5 以后支持的功能，所以 Android 模拟器要选择 SDK 5 或以上的版本，否则运行项目时会报错。

5.7.3　实现过程

步骤 1：新建项目 MyMoreContacts，设置主程序文件名为 MySurfaceView（继承自

SurfaceView 类),复制图片素材(如图 5-6 所示)到资源文件夹。

步骤 2:实现 onDraw()方法,绘制画布与游戏场景图片。代码如下:

图 5-6　多点触摸图片素材

```
public void myDraw(){
    try {
        canvas=sfh.lockCanvas();
        if(canvas !=null){
            canvas.drawColor(Color.WHITE);
            canvas.save();
            //缩放画布(以图片中心点为中心进行缩放,X、Y 轴缩放比例相同)
            canvas.scale(rate, rate, screenW/2, screenH/2);
            //绘制位图 icon
            canvas.drawBitmap (bmpIcon, screenW/2 - bmpIcon.getWidth( )/2,
            screenH/2 -bmpIcon.getHeight()/2, paint);
            canvas.restore();
            //为便于观察,这里绘制两个触点时形成的线段
            canvas.drawLine(x1, y1, x2, y2, paint);
        }
    } catch(Exception e){
    } finally {
        if(canvas !=null)
            sfh.unlockCanvasAndPost(canvas);
    }
}
```

步骤 3:完成触屏监听事件,使用了两个常用的函数获取触屏点的 X、Y 坐标:MotionEvent.getX(int pointerIndex)、MotionEvent.getY(int pointerIndex);除此之外,常用的方法还有获取触屏点的压力值函数 float getPressure(int pointerIndex)。代码如下:

```
public boolean onTouchEvent(MotionEvent event){
//用户手指抬起默认还原为第一次触屏标识位,并且保存本次的缩放比例
    if(event.getAction()==MotionEvent.ACTION_UP){
        isFirst=true;
        oldRate=rate;
    } else {
        x1=(int)event.getX(0);
        y1=(int)event.getY(0);
        x2=(int)event.getX(1);
        y2=(int)event.getY(1);
        if(event.getPointerCount()==2){
            if(isFirst){
                //得到第一次触屏时线段的长度
```

```
            oldLineDistance=(float)Math.sqrt(Math.pow(event.getX(1)-event.
            getX(0),2)+Math.pow(event.getY(1)-event.getY(0),2));
            isFirst=false;
        } else {
            //得到非第一次触屏时线段的长度
            float newLineDistance=(float)Math.sqrt(Math.pow(event.getX(1)-
            event.getX(0),2)+Math.pow(event.getY(1)-event.getY(0),2));
            //获取本次的缩放比例
            rate=oldRate * newLineDistance/oldLineDistance;
        }
    }
    return true;
}
```

真机测试,运行效果如图 5-7 所示。

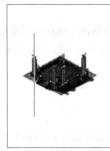

图 5-7　多点触摸实例运行界面

5.8　本章小结

本章讲解了在 Android 中如何控制相机、播放音频与视频,重点说明了两种播放方式的区别。另外,讲解了 Android 传感器的框架和使用方法。这些知识在游戏开发中都是必不可少的,好的游戏配音与开场视频是提高游戏可玩性和吸引力的重要环节,而传感器的运用会大大丰富游戏的操控方式,是游戏发展的必然趋势。

5.9　思考与练习

(1) 尝试开发一个手机照片管理程序,可以显示、删除手机中的所有照片。
(2) 尝试开发一个音乐播放器,可以用列表方式播放手机 SD 中的音频文件,并有跳转、选曲等控制功能。
(3) 尝试设计并开发一款重力小球游戏。可以通过重力感应控制小球在场景中的运动,添加背景音乐,当小球与屏幕边缘碰撞后会发出各种音效。

第 6 章
Android 数据存储与网络编程

学习目标：
- 掌握使用 SharedPreferences 对象存储数据的方法。
- 掌握 openFileOutput 和 openFileInput 的使用。
- 掌握 SQLite 数据库编程及应用。
- 掌握 Socket 网络编程方法。
- 掌握 HttpURLConnection 网络编程方法。
- 掌握 HttpClient 网络编程方法。
- 掌握 WebView 组件的使用。

本章导读：

在 Android 系统中提供了多种存储技术，通过这些存储技术可以将数据存储在各种存储介质上。本章重点讲解在 Android 游戏开发中经常用到的 3 种数据存储技术：SharedPreferences、Files 和 SQLite 数据库，使用 Socket、HttpURLConnection 及 HttpClient 访问网络的方法，以及 WebView 组件的使用。这些技术主要用于游戏状态以及各种数据的保存、网络游戏的开发及在线技术支持等。

6.1 游戏数据存储

在 Android 中提供了以下 4 种数据存储方式。

（1）SharedPreferences：适用于简单的数据的保存，属于配置性质的保存，不适合数据比较大的情况，默认存放在手机内存里。

（2）FileInputStream/FileOutputStream：流文件存储可以保存较大的数据，比较适合游戏数据的保存和使用，而且通过此方式不仅能把数据存储在手机内存中，也能将数据保存到手机的 SD 卡中。

（3）SQLite：适合游戏的保存和使用，不仅可以保存较大的数据，而且可以将自己的数据存放在文件系统或者数据库(SQLite)中，也能将数据保存到 SD 卡中。

（4）ContentProvider：不推荐用于游戏中的数据保存，虽然此方式能存储较大的数据，还支持多个程序之间的数据进行交换，但在游戏中基本上无法访问外部应用的数据。

6.1.1 SharedPreferences

SharedPreferences 是一种轻量级的数据存储方式，不支持多线程，如同一个小小的 Cookie。可以用键值对的方式把简单数据类型（boolean、int、float、long 和 string）存储在应用程序的私有目录下（DDMS 中的 File Explorer 中的/data/data/shares_prefs）自定义的 XML 文件中。

用 getsharedPreferences 方法获得 SharedPreferences（String name,int mode）对象，其中第 2 个参数用于指定文件的建立模式，从而可以设置数据文件的访问权限，通常是一个常量，如表 6-1 所示。

表 6-1 数据文件的访问权限常量

名 称	描 述
MODE_PRIVATE	新内容覆盖原内容
MODE_APPEND	新内容追加到原内容后
MODE_WORLD_READABLE	允许其他应用程序读取
MODE_WORLD_WRITEABLE	允许其他应用程序写入，会覆盖原数据

首先调用 SharedPreferences 类的 edit()方法获得 SharedPreferences.Editor 对象，然后调用增加值方法 putXXX()存入或用 getXXX() 方法读取数据,最后使用 commit() 方法提交,就可以对 SharedPreferences 中的数据进行操作。

举例：保存用户名

步骤 1：完成 res/layout 目录下的 XML 布局文件。代码如下：

```xml
<?xml version="1.0" encoding="utf-8"?>
<LinearLayout xmlns:android="http://schemas.android.com/apk/res/android"
    android:layout_width="fill_parent"
    android:layout_height="fill_parent"
    android:orientation="vertical">
    <LinearLayout
        android:layout_width="fill_parent"
        android:layout_height="wrap_content"
        android:orientation="horizontal">
        <TextView
            android:layout_width="wrap_content"
            android:layout_height="wrap_content"
            android:text="用户名:"/>
        <EditText
            android:id="@+id/login_user_et"
            android:layout_width="150dip"
            android:layout_height="wrap_content"
            android:digits="abcdefghigklmnopqrstuvwxyzABCDEFGHIJKLMNOPQRSTUVWXYZ"/>
    </LinearLayout>
```

```xml
<LinearLayout
    android:layout_width="fill_parent"
    android:layout_height="wrap_content"
    android:orientation="horizontal">
    <TextView
        android:layout_width="wrap_content"
        android:layout_height="wrap_content"
        android:text="密    码:"/>
    <EditText
        android:id="@+id/login_pswd_et"
        android:layout_width="150dip"
        android:layout_height="wrap_content"
        android:password="true"/>
</LinearLayout>
<LinearLayout
    android:layout_width="fill_parent"
    android:layout_height="wrap_content"
    android:orientation="horizontal">
    <TextView
        android:layout_width="wrap_content"
        android:layout_height="wrap_content"
        android:text="记住密码: "/>
    <CheckBox
        android:id="@+id/login_checkbox"
        android:layout_width="wrap_content"
        android:layout_height="wrap_content"/>
</LinearLayout>
<Button
    android:id="@+id/login_btn"
    android:layout_width="200dip"
    android:layout_height="wrap_content"
    android:text="登录"/>
</LinearLayout>
```

步骤2：定义相关变量。代码如下：

```
public static final String BMI_PREF="BMI_PREF";              //偏好设置名称
public static final String REMEMBER_USERID_KEY="remember";   //记住用户名
public static final String USERID_KEY="userid";              //用户名标记
private static final String DEFAULT_USERNAME="yctuzhang";    //默认用户名
private SharedPreferences mSettings=null;
private EditText userName=null;
private EditText passWord=null;
private CheckBox cb=null;
private Button submitBtn=null;
```

步骤 3：在主活动中获取组件并添加事件处理。代码如下：

```java
public void onCreate(Bundle savedInstanceState){
    super.onCreate(savedInstanceState);
    setContentView(R.layout.main);
    userName=(EditText)findViewById(R.id.login_user_et);
    passWord=(EditText)findViewById(R.id.login_pswd_et);
    cb=(CheckBox)findViewById(R.id.login_checkbox);
    submitBtn=(Button)findViewById(R.id.login_btn);
    mSettings=getSharedPreferences(BMI_PREF, Context.MODE_PRIVATE);
    cb.setChecked(getRemember());                           //记住用户名
    userName.setText(getUserName());                        //设置用户名
    submitBtn.setOnClickListener(new View.OnClickListener(){
        @Override
        public void onClick(View v){
            if(cb.isChecked()){                             //是否保存用户名
                saveRemember(true);
                saveUserName(userName.getText().toString());
            } else {
                saveRemember(false);
                saveUserName("");
            }
        }
    });
}
//保存用户名
private void saveUserName(String userid){
    Editor editor=mSettings.edit();                         //获取编辑器
    editor.putString(USERID_KEY, userid);
    editor.commit();                                        //保存数据
    //editor.clear();                                       //清除数据
}
//设置是否保存的用户名
private void saveRemember(boolean remember){
    Editor editor=mSettings.edit();                         //获取编辑器
    editor.putBoolean(REMEMBER_USERID_KEY, remember);
    editor.commit();
}
//获取保存的用户名
private String getUserName(){
    return mSettings.getString(USERID_KEY, DEFAULT_USERNAME);
}
//获取是否保存的用户名
private boolean getRemember(){
```

```
        return mSettings.getBoolean(REMEMBER_USERID_KEY, true);
    }
}
```

SharedPreferences 将数据文件写在手机内存私有的目录中。在模拟器中测试程序可以通过 ADT 的 DDMS 透视图来查看数据文件的位置。输入数据并退出后,当再次进入该程序的时候,上次写的数据还在。在 data 目录下能找到一个名为 BMI_PREF.xml 的文件,如图 6-1 所示。

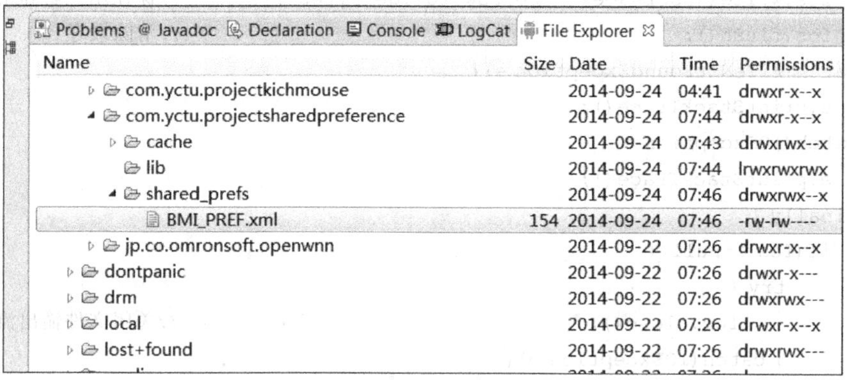

图 6-1 模拟器中的 shared_prefs 目录

导出文件,打开之后的格式以及内容如下:

```
<?xml version='1.0' encoding='utf-8' standalone='yes' ?>
<map>
    <string name="userid">abc</string>
    <boolean name="remember" value="true"/>
</map>
```

从上面的代码可以看出,数据都被保存到 XML 文件中,当开启这个应用的时候,会自动地去 data 目录下找到相应的 XML 文件并且把相应的数据显示出来。

★注意:SharedPreferences 只能保存简单类型的数据,例如 string、int 等。如果需要存取比较复杂的数据类型(类或者图像),则需要对这些数据进行编码,通常将其转换成 Base64 编码,然后将转换后的数据以字符串的形式保存在 XML 文件中。

6.1.2 使用 Files 对象存储数据

用 Files 对象存储数据主要有两种方式:一是 Java 语言的 I/O 流体系,即使用 FileOutputStream 类提供的 openFileOutput() 方法和 FileInputStream 类提供的 openFileInput() 方法访问磁盘上的内容文件;二是使用 Environment 类提供的 getExternalStorageDirectory() 方法对 SD 卡进行数据读写。

举例:内部文件访问

步骤 1:在布局文件中添加两个 TextView(文本框)、两个 EditText(编辑框)和一个

Button(按钮)组件。

步骤 2：输入数据关键代码，如下所示：

```
String username=usernameET.getText().toString();      //获得用户名
String password=passwordET.getText().toString();      //获得密码
FileOutputStream fos=null;
try {
    fos=openFileOutput("login", MODE_PRIVATE);        //获得文件输出流
    fos.write((username+" "+password).getBytes());    //保存用户名和密码
    fos.flush();                                      //清除缓存
} catch(FileNotFoundException e){
    e.printStackTrace();
} catch(IOException e){
    e.printStackTrace();
} finally {
    if(fos !=null){
        try {
            fos.close();                              //关闭文件输出流
        } catch(IOException e){
            e.printStackTrace();
        }
    }
}
```

步骤 3：读取数据关键代码，如下所示：

```
FileInputStream fis=null;
byte[] buffer=null;
try {
    fis=openFileInput("login");                       //获得文件输入流
    buffer=new byte[fis.available()];                 //定义保存数据的数组
    fis.read(buffer);                                 //从输入流中读取数据
} catch(FileNotFoundException e){
    e.printStackTrace();
} catch(IOException e){
    e.printStackTrace();
} finally {
    if(fis !=null){
        try {
            fis.close();                              //关闭文件输入流
        } catch(IOException e){
            e.printStackTrace();
        }
    }
}
```

```
TextView usernameTV=(TextView)findViewById(R.id.username);
TextView passwordTV=(TextView)findViewById(R.id.password);
String data=new String(buffer);                       //获得数组中保存的数据
String username=data.split(" ")[0];                   //获得username
String password=data.split(" ")[1];                   //获得password
usernameTV.setText("用户名:"+username);               //显示用户名
passwordTV.setText("密  码:"+password);               //显示密码
```

每个 Android 设备都支持共享的外部存储,这可以是手机内存等不可移除的存储介质,也可以是 SD 卡等可以移除的存储介质。保存的外部存储的文件都是全局可读的,而且在用户使用 USB 连接计算机后,可以修改这些文件。

举例：对 SD 卡进行操作

(1) 创建文件。

步骤 1：在主活动文件中,重写 onCreate()方法。代码如下：

```
File root=Environment.getExternalStorageDirectory();
if(root.exists()&& root.canWrite()){
    File file=new File(root, "pic.png");
    try {
        if(file.createNewFile()){
            Toast.makeText (MainActivity. this, file. getName () +"创建成功!",
            Toast.LENGTH_SHORT).show();
        } else {
            Toast.makeText (MainActivity. this, file. getName () +"创建失败!",
            Toast.LENGTH_SHORT).show();
        }
    } catch(IOException e){
        e.printStackTrace();
    }
} else {
    Toast.makeText (MainActivity.this, "没有 SD 卡或不可写!", Toast.LENGTH_
    SHORT).show();
}
```

步骤 2：修改 AndroidManifest.xml 配置文件,增加外部存储写入权限。代码如下：

```
<uses-permission android:name="android.permission.WRITE_EXTERNAL_STORAGE"/>
```

(2) 遍历 SD 卡。

步骤 1：在布局文件中添加一个 ListView 文本框组件。代码如下：

```
<ListView
    android:id="@+id/sd_list"
    android:layout_width="fill_parent"
    android:layout_height="wrap_content"
    android:dividerHeight="3dp"
```

/>

步骤 2：在主活动文件中，重写 onCreate 方法。代码如下：

```
ListView lv=(ListView)findViewById(R.id.sd_list);        //获得列表组件
File rootPath=Environment.getExternalStorageDirectory(); //获得SD卡根路径
List<String>items=new ArrayList<String>();               //创建列表
for(File file : rootPath.listFiles()){                   //遍历SD卡获得名称
    items.add(file.getName());
}
ArrayAdapter<String>adapter=new ArrayAdapter<String>(this, android.R.layout.simple_list_item_1, items);
lv.setAdapter(adapter);                                  //设置列表适配器
```

在遍历 SD 卡根目录时，可以判断是否为文件夹。修改代码如下：

```
for(File file : rootPath.listFiles()){
    if(file.isDirectory()){                              //判断是否为文件夹
        items.add(file.getName()+"是文件夹");
    }else{
        items.add(file.getName()+"是文件");
    }
}
```

(3) 将图片复制到 SD 卡上。

步骤 1：复制一张图片到 res/drawable-mdpi 文件夹。

步骤 2：在主活动文件中，重写 onCreate()方法。代码如下：

```
File path=Environment.getExternalStorageDirectory();     //获取SD卡根路径
File file=new File(path, "game.png");                    //创建文件对象
InputStream fis=getResources().openRawResource(R.drawable.game);
                                                         //打开输入流
FileOutputStream fos=null;
try {
    fos=new FileOutputStream(file);                      //建立输出流
    byte buffer[]=new byte[fis.available()];             //定义保存数据的数组
    fis.read(buffer);                                    //读取数据
    fos.write(buffer);                                   //写入数据
} catch(Exception e){
    e.printStackTrace();
}
```

步骤 3：修改 AndroidManifest.xml 配置文件，增加外部存储写入权限。代码如下：

```
<uses-permission android:name="android.permission.WRITE_EXTERNAL_STORAGE"/>
```

6.1.3 SQLite 数据库应用

Android 平台集成了一个嵌入式关系型数据库 SQLite，支持 NULL、INTEGER（整型）、REAL（浮点型）、TEXT（字符串文本）和 BLOB（二进制对象）数据类型，最大的特点是可以把各种类型的数据保存到任何字段中，而不用关心字段声明的数据类型是什么。例如，可以在整型字段中存放字符串，或者在布尔型字段中存放浮点数，或者在字符型字段中存放日期型值。

★注意：定义为 INTEGER PRIMARY KEY 的字段只能存储 64 位整数，当向这种字段保存除整数以外的数据时，会报告 datatype missmatch 的错误。另外，在编写 CREATE TABLE 语句时，可以省略跟在字段名称后面的数据类型信息。

Android 提供了 SQLiteOpenHelper 类，用于实现对数据库的管理，提供了 onCreate（SQLiteDatabase db）和 onUpgrade（SQLiteDatabase db, int oldVersion, int newVersion）两个重要的方法，前者用于初次使用软件时生成数据库表，后者用于升级软件时更新数据库表结构。

当调用 SQLiteOpenHelper 的 getWritableDatabase() 或者 getReadableDatabase() 方法获取用于操作数据库的 SQLiteDatabase 实例时，如果数据库不存在，Android 系统会自动生成一个数据库。接着调用 onCreate() 方法，onCreate() 方法在初次生成数据库时才会被调用，在 onCreate() 方法里可以生成数据库表结构及添加一些在应用中会使用到的初始化数据。onUpgrade() 方法在数据库的版本发生变化时会被调用，一般在软件升级时才需改变版本号，并且在 onUpgrade() 方法中实现表结构的更新。当软件的版本升级次数比较多时，在 onUpgrade 方法中可以根据原版号和目标版本号进行判断，然后作出相应的表结构及数据更新。

★注意：getWritableDatabase 和 getReadableDatabase 的区别是：当数据库写满时，调用前者会报错，调用后者不会，所以如果不是更新数据库的话，最好调用后者来获得数据库连接。

Android 提供的 SQLiteDatabase 类封装了一些操作数据库的 API，使用该类可以完成对数据进行添加（Create）、查询（Retrieve）、更新（Update）和删除（Delete）操作，这些操作简称为 CRUD。SQLiteDatabase 类具有 execSQL() 和 rawQuery() 方法，execSQL() 方法可以执行 insert、delete、update 和 CREATE TABLE 之类有更改行为的 SQL 语句，rawQuery 方法用于执行 select 语句。execSQL() 方法的使用代码如下：

```
SQLiteDatabase db=databaseHelper.getWritableDatabase();
db.execSQL("insert into person(name, age)values('zhanghui', 24)");
db.close();
```

上面的 SQL 语句会往 person 表中添加一条记录，在实际应用中，语句中的 zhanghui 这些参数值会由用户从输入界面提供，如果把用户输入的内容原样组拼到上面的 insert 语句中，当用户输入的内容含有单引号时，组拼出来的 SQL 语句就会存在语法错误。要解决这个问题需要对单引号进行转义，也就是把单引号转换成两个单引号。有些时候用户还会输入像"&"这些特殊 SQL 符号,为保证组拼好的 SQL 语句语法正确,必须对 SQL 语句中的这些特殊 SQL 符号都进行转义,显然,对每条 SQL 语句都做这样的处理工作是比较烦琐的。SQLiteDatabase 类提供了一个重载后的 execSQL(String sql,Object[] bindArgs)方法,使用这个方法可以解决前面提到的问题,因为这个方法支持使用占位符参数(?)。代码如下:

```
SQLiteDatabase db=databaseHelper.getWritableDatabase();
db.execSQL("insert into person (name, age) values (?, ?)", new Object []{"zhanghui", 4});
db.close();
```

execSQL(String sql,Object[] bindArgs)方法的第 1 个参数为 SQL 语句,第 2 个参数为 SQL 语句中占位符参数的值,参数值在数组中的顺序要和占位符的位置对应。SQLiteDatabase 的 rawQuery()用于执行 select 语句,代码如下:

```
SQLiteDatabase db=databaseHelper.getReadableDatabase();
Cursor cursor=db.rawQuery("select * from person", null);
while(cursor.moveToNext()){
    int personid=cursor.getInt(0);           //获取第一列的值,第一列的索引从 0 开始
    String name=cursor.getString(1);         //获取第二列的值
    int age=cursor.getInt(2);                //获取第三列的值
}
cursor.close();
db.close();
```

rawQuery()方法的第一个参数为 select 语句;第二个参数为 select 语句中占位符参数的值,如果 select 语句中没有使用占位符,该参数可以设置为 null。带占位符参数的 select 语句使用例子如下:

```
Cursor cursor=db.rawQuery("select * from person where name like ? and age=?", new String[]{"%辉%", "4"});
```

Cursor 是结果集游标,用于对结果集进行随机访问,与 JDBC 中的 ResultSet 作用很相似。它们同时返回一个 Cursor 对象,代表数据集的游标,有点类似于 JavaSE 中的 ResultSet。Cursor 对象的常用方法如表 6-2 所示。

表 6-2　Cursor 对象的常用方法

名　　称	描　　述
move(int offset)	以当前位置为参考,移动到指定行
moveToFirst()	移动到第一行
moveToLast()	移动到最后一行
moveToPosition(int position)	移动到指定行
moveToPrevious()	移动到前一行
moveToNext()	移动到下一行

名　　　称	描　　　述
isFirst()	是否指向第一条
isLast()	是否指向最后一条
isBeforeFirst()	是否指向第一条之前
isAfterLast()	是否指向最后一条之后
isNull(int columnIndex)	指定列是否为空(列基数为0)
isClosed()	游标是否已关闭
getCount()	总数据项数
getPosition()	返回当前游标所指向的行数
getColumnIndex(String columnName)	返回某列名对应的列索引值
getString(int columnIndex)	返回当前行指定列的值

SQLiteDatabase还专门提供了对应于添加、删除、更新、查询的操作方法：insert()、delete()、update()和query()。

Insert()方法用于添加数据，各个字段的数据使用ContentValues进行存放。ContentValues类似于MAP，它提供了存取数据对应的put(String key, Xxx value)和getAsXxx(String key)方法，key为字段名称，value为字段值，Xxx指的是各种常用的数据类型，例如String、Integer等。代码如下：

```
SQLiteDatabase db=databaseHelper.getWritableDatabase();
ContentValues values=new ContentValues();
values.put("name", "小辉");
values.put("age", 35);
long rowid=db.insert("person", null, values);    //返回新添记录的行号，与主键 id无关
```

无论第3个参数是否包含数据，执行Insert方法必然会添加一条记录，如果第3个参数为空，会添加一条除主键之外其他字段值为null的记录。

Insert()方法内部实际上通过构造insert SQL语句完成数据的添加，Insert()方法的第2个参数用于指定空值字段的名称。如果第3个参数values为null或者元素个数为0，由于Insert()方法要求必须添加一条除了主键之外其他字段值为null的记录，为了满足SQL语法的需要，insert语句必须给定一个字段名，例如insert into person(name) values(null)，否则insert语句就成了insert into person() values()。这显然不满足标准SQL的语法。对于字段名，建议使用主键之外的字段，如果使用了INTEGER类型的主键字段，执行类似insert into person(personid) values(null)的insert语句后，该主键字段值也不会为null。如果第3个参数values不为null并且元素的个数大于0，可以把第2个参数设置为null。

delete()方法的例子如下：

```
SQLiteDatabase db=databaseHelper.getWritableDatabase();
db.delete("person", "personid<?", new String[]{"2"});
db.close();
```

update()方法的例子如下:

```
SQLiteDatabase db=databaseHelper.getWritableDatabase();
ContentValues values=new ContentValues();
values.put("name",小张);                    //key 为字段名, value 为值
db.update("person", values, "personid=?", new String[]{"1"});
db.close();
```

query()方法实际上是把 select 语句拆分成了若干个组成部分,然后作为方法的输入参数:

```
SQLiteDatabase db=databaseHelper.getWritableDatabase();
Cursor cursor=db.query("person", new String[]{"personid, name, age"}, "name like ?", new String[]{"%zh%"}, null, ull, "personid desc", "1, 2");
while(cursor.moveToNext()){
    int personid=cursor.getInt(0);          //获取第一列的值,第一列的索引从 0 开始
    String name=cursor.getString(1);        //获取第二列的值
    int age=cursor.getInt(2);               //获取第三列的值
}
cursor.close();
db.close();
```

上面的代码用于从 person 表中查找 name 字段含有 zh 的记录,匹配的记录按 personid 降序排序,对排序后的结果略过第一条记录,只获取两条记录。

query(table, columns, selection, selectionArgs, groupBy, having, orderBy, limit) 方法各参数的含义如表 6-3 所示。

表 6-3 query 方法的参数

名称	描述
table	表名,相当于 select 语句 from 关键字后面的部分。如果是多表联合查询,可以用逗号将两个表名分开
columns	要查询出来的列名。相当于 select 语句 select 关键字后面的部分
selection	查询条件子句,相当于 select 语句 where 关键字后面的部分,在条件子句允许使用占位符"?"
selectionArgs	对应于 selection 语句中占位符的值,值在数组中的位置与占位符在语句中的位置必须一致,否则就会有异常
groupBy	分组的列名,相当于 select 语句 group by 关键字后面的部分
having	分组条件,相当于 select 语句 having 关键字后面的部分
orderBy	排序的列名,相当于 select 语句 order by 关键字后面的部分,如 personid desc, age asc
limit	分页参数,指定偏移量和获取的记录数,相当于 select 语句 limit 关键字后面的部分

注意:selection、groupBy、having、orderBy、limit 这几个参数中不包括 WHERE、GROUP BY、HAVING、ORDER BY、LIMIT 等 SQL 关键字。

当完成了对数据库的操作后,要调用 SQLiteDatabase 的 close()方法释放数据库连接,否则容易出现 SQLiteException。

在实际开发中,为了更好地管理和维护数据库,会封装一个继承自 SQLiteOpenHelper 类的数据库操作类,然后以这个类为基础,再封装业务逻辑。

使用 SQLiteDatabase 的 beginTransaction()方法可以开启一个事务,程序执行到 endTransaction()方法时会检查事务的标志是否为成功,如果程序执行到 endTransaction()之前调用了 setTransactionSuccessful()方法设置事务的标志为成功则提交事务,如果没有调用 setTransactionSuccessful()方法则回滚事务。代码如下:

```
SQLiteDatabase db=....;
db.beginTransaction();                        //开始事务
try {
    db.execSQL("insert into person(name, age)values(?, ?)", new Object[]{"小辉", 4});
        db.execSQL("update person set name=?where personid=?", new Object[]{"abc", 1});
    //调用此方法会在执行到 endTransaction()时提交当前事务,如果不调用此方法会回滚
      事务
        db.setTransactionSuccessful();
} finally {
    db.endTransaction();                      //由事务的标志决定是提交事务还是回滚事务
}
db.close();
```

举例:玩家列表

步骤 1:新建项目 ProjectPlayerList(Android SDK 22.2.1,Target SDK 4.2),复制图片素材到 res/ drawable-mdpi 文件夹(如图 6-2 所示)。

图 6-2　玩家列表图标素材

步骤 2:建立 DBHelper 类(继承自 SQLiteOpenHelper 类),作为维护和管理数据库的基类。代码如下:

```
public class DBHelper extends SQLiteOpenHelper {
    private static final String DATABASE_NAME="mydata";
    private static final int DATABASE_VERSION=1;
    private static final String TABLE_NAME="users";
    public DBHelper(Context context){
        super(context, DATABASE_NAME, null, DATABASE_VERSION);
    }
    @Override
```

```java
    public void onCreate(SQLiteDatabase db){
        String sql=" create table " + TABLE_NAME +" (_id integer primary key autoincrement, username varchar(50), password varchar(50))";
        db.execSQL(sql);
    }
    @Override
    public void onUpgrade(SQLiteDatabase db, int oldVersion, int newVersion){
        db.execSQL("drop table if exists "+TABLE_NAME);
        onCreate(db);
    }
```

步骤3：DBManager 建立在 DBHelper 之上，封装了常用的业务方法。代码如下：

```java
public class DBManager {
    private SQLiteDatabase db;
    private DBHelper dbhelper;
    public DBManager(Context context){
        dbhelper=new DBHelper(context);
        db=dbhelper.getWritableDatabase();
    }
    /**
     * 插入数据
     * @param user
     * @return
     */
    public boolean insert(User user){
        try {
            ContentValues values=new ContentValues();
            values.put("username", user.getUsername());
            values.put("password", user.getPassword());
            db.insert("users", null, values);
            return true;
        } catch(Exception e){
            e.printStackTrace();
            return false;
        }
    }
    /**
     * 添加数据 SQL 语句
     * @param user
     * @return
     */
    public boolean add(User user){
        try {
            String sql="insert into users(username,password)values(?,?)";
```

```java
                Object[] bindArgs={ user.getUsername(), user.getPassword()};
                db.execSQL(sql, bindArgs);
                return true;
            } catch(Exception e){
                return false;
            }
        }
    }
    /**
     * 更新数据
     * @param id
     */
    public void update(int id){
        String sql="update users set password=? where _id=?";
        db.execSQL(sql, new Object[] { "computer", id });
    }
    /**
     * 查询数据
     * @param id
     */
    public User query(int id){
        User user=new User();
        String col[]=new String[] { "username", "password" };
        Cursor cursor=db.query("users", col, "_id="+id, null, null, null, null);
        if(cursor.getCount()>0){
            cursor.moveToFirst();
            user.setUsername(cursor.getString(0));
            user.setPassword(cursor.getString(1));
            return user;
        }
        cursor.close();
        return null;
    }
    /**
     * 分页
     * @param startResult    偏移量，默认从 0 开始
     * @param maxResult      每页显示的条数
     * @return
     */
    public List<User> findOnePage(int startResult, int maxResult){
        List<User> users=new ArrayList<User>();
        Cursor cursor=db.rawQuery("select * from users limit ?, ?", new String
            [] { String.valueOf(startResult), String.valueOf(maxResult) });
        while(cursor.moveToNext()){
            //int id=cursor.getInt(0);
```

```
            String username=cursor.getString(1);
            String password=cursor.getString(2);
            users.add(new User(username, password));
        }
        return users;
    }
    /**
     * 获取记录总页数
     * @return
     */
    public int getCount(){
        Cursor cursor=db.query("users", new String[] { "count(*)" }, null, null,
        null, null, null);
        if(cursor.moveToNext()){
            return cursor.getInt(0);
        }
        return 0;
    }
```

步骤 4：创建分页函数监听接口。代码如下：

```
public interface OnPageChangeListener {
    /**
     * 点击分页按钮时触发此操作
     * @param curPage          当前页
     * @param numPerPage       每页显示个数
     */
    public void pageChanged(int curPage, int numPerPage);
}
```

步骤 5：编写分页控件。代码如下：

```
public class PageControl extends LinearLayout implements OnClickListener {
    private ImageButton firstImg;
    private ImageButton preImg;
    private ImageButton nextImg;
    private ImageButton endImg;
    private TextView totalPageText;
    private TextView curPageText;
    private int numPerPage=10;
    private int curPage=1;
    private int count=0;
    private OnPageChangeListener pageChangeListener;
    public PageControl(Context context){
        super(context);
        initPageComposite(context);
```

```java
    }
    public PageControl(Context context, AttributeSet attrs){
        super(context, attrs);
        initPageComposite(context);
    }
    public PageControl(Context context, AttributeSet attrs, int defStyle){
        super(context, attrs, defStyle);
        initPageComposite(context);
    }
    private void initPageComposite(Context context){
        this.setPadding(5, 5, 5, 5);
        firstImg=new ImageButton(context);
        firstImg.setId(1);
        firstImg.setImageResource(R.drawable.firstpage);
        firstImg.setPadding(0, 0, 0, 0);
        LayoutParams layoutParam=new LayoutParams(LayoutParams.WRAP_CONTENT,
        LayoutParams.WRAP_CONTENT);
        layoutParam.setMargins(0, 0, 5, 0);
        firstImg.setLayoutParams(layoutParam);
        firstImg.setOnClickListener(this);
        this.addView(firstImg);
        preImg=new ImageButton(context);
        preImg.setId(2);
        preImg.setImageResource(R.drawable.prepage);
        preImg.setPadding(0, 0, 0, 0);
        layoutParam=new LayoutParams(LayoutParams.WRAP_CONTENT, LayoutParams.
        WRAP_CONTENT);
        layoutParam.setMargins(0, 0, 5, 0);
        preImg.setLayoutParams(layoutParam);
        preImg.setOnClickListener(this);
        this.addView(preImg);
        nextImg=new ImageButton(context);
        nextImg.setId(3);
        nextImg.setImageResource(R.drawable.nextpage);
        nextImg.setPadding(0, 0, 0, 0);
        layoutParam=new LayoutParams(LayoutParams.WRAP_CONTENT, LayoutParams.
        WRAP_CONTENT);
        layoutParam.setMargins(0, 0, 5, 0);
        nextImg.setLayoutParams(layoutParam);
        nextImg.setOnClickListener(this);
        this.addView(nextImg);
        endImg=new ImageButton(context);
        endImg.setId(4);
        endImg.setImageResource(R.drawable.lastpage);
```

```java
        endImg.setPadding(0, 0, 0, 0);
        layoutParam=new LayoutParams(LayoutParams.WRAP_CONTENT, LayoutParams.
WRAP_CONTENT);
        layoutParam.setMargins(0, 0, 5, 0);
        endImg.setLayoutParams(layoutParam);
        endImg.setOnClickListener(this);
        this.addView(endImg);
        totalPageText=new TextView(context);
        layoutParam=new LayoutParams(LayoutParams.WRAP_CONTENT, LayoutParams.
MATCH_PARENT);
        layoutParam.setMargins(5, 0, 5, 0);
        totalPageText.setLayoutParams(layoutParam);
        totalPageText.setText("总页数");
        this.addView(totalPageText);
        curPageText=new TextView(context);
        layoutParam=new LayoutParams(LayoutParams.WRAP_CONTENT, LayoutParams.
MATCH_PARENT);
        layoutParam.setMargins(5, 0, 5, 0);
        curPageText.setLayoutParams(layoutParam);
        curPageText.setText("当前页");
        this.addView(curPageText);
    }
    /**
    * 初始化分页组件的显示状态
    * @param newCount
    */
    public void initPageShow(int newCount){
        count=newCount;
        int totalPage=count %numPerPage==0 ? count/numPerPage : count
                /numPerPage+1;
        curPage=1;
        firstImg.setEnabled(false);
        preImg.setEnabled(false);
        if(totalPage<=1){
            endImg.setEnabled(false);
            nextImg.setEnabled(false);
        } else {
            endImg.setEnabled(true);
            nextImg.setEnabled(true);
        }
        totalPageText.setText("总页数 "+totalPage);
        curPageText.setText("当前页 "+curPage);
    }
    /**
```

* 分页按钮被点击时更新状态，该方法要在 initPageShow 后调用
 */
```java
@Override
public void onClick(View view){
    if(pageChangeListener==null){
        return;
    }
    int totalPage=count %numPerPage==0 ? count/numPerPage : count
        /numPerPage+1;
    switch(view.getId()){
    case 1:
        curPage=1;
        firstImg.setEnabled(false);
        preImg.setEnabled(false);
        if(totalPage>1){
            nextImg.setEnabled(true);
            endImg.setEnabled(true);
        }
        break;
    case 2:
        curPage--;
        if(curPage==1){
            firstImg.setEnabled(false);
            preImg.setEnabled(false);
        }
        if(totalPage>1){
            nextImg.setEnabled(true);
            endImg.setEnabled(true);
        }
        break;
    case 3:
        curPage++;
        if(curPage==totalPage){
            nextImg.setEnabled(false);
            endImg.setEnabled(false);
        }
        firstImg.setEnabled(true);
        preImg.setEnabled(true);
        break;
    case 4:
        curPage=totalPage;
        nextImg.setEnabled(false);
        endImg.setEnabled(false);
        firstImg.setEnabled(true);
```

```
                preImg.setEnabled(true);
                break;
            default:
                break;
        }
        totalPageText.setText("总页数 "+totalPage);
        curPageText.setText("当前页 "+curPage);
        pageChangeListener.pageChanged(curPage, numPerPage);
    }
    public OnPageChangeListener getPageChangeListener(){
        return pageChangeListener;
    }
    /**
     * 设置分页监听事件
     * @param pageChangeListener
     */
    public void setPageChangeListener(OnPageChangeListener pageChangeListener){
        this.pageChangeListener=pageChangeListener;
    }
}
```

步骤 6：建立 User 实体类，实现 Serializable 接口。代码如下：

```
public class User implements Serializable {
    private static final long serialVersionUID=1L;
    private String username;
    private String password;
    public User(){}
    public User(String username, String password){
        this.username=username;
        this.password=password;
    }
    public String getUsername(){
        return username;
    }
    public void setUsername(String username){
        this.username=username;
    }
    public String getPassword(){
        return password;
    }
    public void setPassword(String password){
        this.password=password;
    }
}
```

步骤 7：完成 XML 布局文件，添加 ListView 组件和自定义的分页组件。代码如下：

```xml
<LinearLayout xmlns:android="http://schemas.android.com/apk/res/android"
    xmlns:tools="http://schemas.android.com/tools"
    android:layout_width="match_parent"
    android:layout_height="match_parent"
    android:orientation="vertical">
    <ListView
        android:id="@+id/wordList"
        android:layout_width="match_parent"
        android:layout_height="wrap_content"
        android:dividerHeight="1dp"
        android:headerDividersEnabled="true"
        android:footerDividersEnabled="true">
    </ListView>
    <com.yctu.pagedemo.PageControl
        android:id="@+id/wordListPageControl"
        android:layout_width="match_parent"
        android:layout_height="0dp"
        android:layout_weight="1"/>
</LinearLayout>
```

步骤 8：完成列表项布局文件。代码如下：

```xml
<LinearLayout xmlns:android="http://schemas.android.com/apk/res/android"
    android:layout_width="fill_parent"
    android:layout_height="fill_parent"
    android:orientation="horizontal"
    android:paddingLeft="5dp"
    android:paddingRight="5dp">
    <!--头像-->
    <ImageView
        android:id="@+id/player"
        android:layout_width="30dp"
        android:layout_height="30dp"
        />
    <!--名称-->
    <TextView
        android:id="@+id/username"
        android:layout_width="130dp"
        android:layout_height="30dp"
        android:paddingLeft="10dp"/>
    <!--密码-->
    <TextView
        android:id="@+id/password"
```

```
            android:layout_width="150dp"
            android:layout_height="30dp"/>
</LinearLayout>
```

步骤9：在主活动文件中实现自定义的OnPageChangeListener接口，在列表组件中显示数据并能用分页组件浏览数据。代码如下：

```java
public class MainActivity extends Activity implements OnPageChangeListener {
    private DBManager dbManager;
    private ListView wordListView;
    private SimpleAdapter adapter2;
    private PageControl pageControl;
    @Override
    protected void onCreate(Bundle savedInstanceState){
        super.onCreate(savedInstanceState);
        setContentView(R.layout.activity_main);
        wordListView=(ListView)findViewById(R.id.wordList);
        dbManager=new DBManager(this);
        List<User>users=dbManager.findOnePage(1, 10);
        List<HashMap<String, Object>> data=new ArrayList<HashMap<String, Object>>();
        for(User user : users){
            HashMap<String, Object>item=new HashMap<String, Object>();
            item.put("player", R.drawable.player);
            item.put("username", user.getUsername());
            item.put("password", user.getPassword());
            data.add(item);
        }
        adapter=new SimpleAdapter(this, data, R.layout.list_item, new String[]
        { "player", "username", "password" }, new int[] {
                    R.id.player, R.id.username, R.id.password });
        wordListView.setAdapter(adapter);
        //初始化分页组件
        pageControl=(PageControl)findViewById(R.id.wordListPageControl);
        pageControl.setPageChangeListener(this);
        pageControl.initPageShow(dbManager.getCount());
        wordListView.setOnItemClickListener(new OnItemClickListener(){
            @Override
            public void onItemClick(AdapterView<?> arg0, View v, int position,
            long arg3){
                @SuppressWarnings("unchecked")
                HashMap<String, Object> data = (HashMap<String, Object>)
                wordListView.getItemAtPosition(position);
```

```
                String username=data.get("username").toString();
                Toast.makeText(getApplicationContext(), username, Toast.LENGTH
                    _SHORT).show();
            }
        });
    }

    /**
     * 点击分页按钮时触发该方法的执行
     * @param curPage          当前页
     * @param numPerPage       每页显示记录数
     */
    @Override
    public void pageChanged(int curPage, int numPerPage){
        List<User> users = dbManager.findOnePage((curPage - 1) * numPerPage,
            numPerPage);
        List<HashMap<String, Object>> data = new ArrayList<HashMap<String,
            Object>>();
        for(User user : users){
            HashMap<String, Object> item=new HashMap<String, Object>();
            item.put("player", R.drawable.player);
            item.put("username", user.getUsername());
            item.put("password", user.getPassword());
            data.add(item);
        }
        adapter=new SimpleAdapter(this, data, R.layout.list_item, new String[]
            { "player", "username", "password" }, new int[] {R.id.player, R.id.
            username, R.id.password });
        wordListView.setAdapter(adapter);
    }
}
```

可以在数据表中添加示例数据。代码如下：

```
dbManager=new DBManager(this);
for(int i=0; i<50; i++){
    dbManager.insert(new User("zhang"+i, "abc"+i));
}
```

运行效果如图 6-3 所示。

图 6-3　玩家列表实例运行界面

6.2　基于 Socket 的网络编程

Socket(套接字)是一种通信机制,在游戏开发中被广泛应用,可以实现单机或跨网络通信。其创建需要明确地区分 Client(客户端)和 Server(服务器端),支持多个客户端连接到同一个服务器。

举例：用 Socket 实现信息交换

步骤 1：新建普通 Java 项目并创建类 MyServer.java 作为服务器端。代码如下：

```java
public class MyServer {
    public static void main(String[] args){
        ServerSocket serverSocket=null;              //声明 ServerSocket 对象
        Socket socket=null;                          //声明 Socket 对象
        DataInputStream din=null;                    //声明输入流对象
        DataOutputStream dout=null;                  //声明输出流对象
        try{
            //实例化 ServerSocket 对象(监听 8888 端口)
            serverSocket=new ServerSocket(8888);
            System.out.println("监听 8888 端口!");
        }
        catch(Exception e){
            e.printStackTrace();
        }
        while(true){
```

```
try{
    socket=serverSocket.accept();           //等待客户端连接
    din=new DataInputStream(socket.getInputStream());
                                            //得到输入流
    dout=new DataOutputStream(socket.getOutputStream());
                                            //得到输出流
    String msg=din.readUTF();               //读一个字符串
    System.out.println("ip: "+socket.getInetAddress());
                                            //打印客户端 IP
    System.out.println("msg: "+msg);        //打印客户端发来的消息
    System.out.println("===================");
    int info=new Random().nextInt(1000);    //生成一个随机数
    dout.writeUTF("Hello Client!| 验证码:"+info);
                                            //向客户端发送消息
}
catch(Exception e){
    e.printStackTrace();                    //打印异常信息
}
finally{
    try{
        if(dout !=null){
            dout.close();                   //关闭输出流
        }
        if(din !=null){
            din.close();                    //关闭输入流
        }
        if(socket !=null){
            socket.close();                 //关闭 Socket 连接
        }
    }
    catch(Exception e){
        e.printStackTrace();                //打印异常信息
    }
}
```

步骤 2：新建 Android 项目作为客户端，修改 res/layout 目录下的 XML 布局文件。代码如下：

```
<LinearLayout xmlns:android="http://schemas.android.com/apk/res/android"
    xmlns:tools="http://schemas.android.com/tools"
    android:layout_width="fill_parent"
    android:layout_height="fill_parent"
```

```xml
    android:orientation="vertical">
    <EditText
        android:id="@+id/txt_msg"
        android:layout_width="fill_parent"
        android:layout_height="wrap_content"
        android:text="Hello Server!"/>
    <Button
        android:id="@+id/btn_send"
        android:layout_width="wrap_content"
        android:layout_height="wrap_content"
        android:text="发送信息到服务器"/>
    <TextView
        android:id="@+id/txt_back"
        android:layout_width="wrap_content"
        android:layout_height="wrap_content"
        android:text="服务器发来的消息:"/>
</LinearLayout>
```

步骤 3：在 AndroidManifest.xml 文件中添加权限。代码如下：

```xml
<uses-permission android:name="android.permission.INTERNET"/>
```

步骤 4：在主活动中完成代码，如下所示：

```java
public class MainActivity extends Activity {
    private EditText txt_msg;              //声明输入文本内容的编辑框组件
    private Button btn_send;               //声明发送按钮组件
    private TextView txt_back;             //声明一个显示结果的文本框组件
    private Socket socket=null;
    private DataOutputStream dout=null;
    private DataInputStream din=null;
    private String backMsg;                //声明一个代表返回内容的字符串
    private Handler handler;
    @Override
    protected void onCreate(Bundle savedInstanceState){
        super.onCreate(savedInstanceState);
        setContentView(R.layout.activity_main);
        txt_msg=(EditText)findViewById(R.id.txt_msg);
        btn_send=(Button)findViewById(R.id.btn_send);
        txt_back=(TextView)findViewById(R.id.txt_back);
        btn_send.setOnClickListener(new View.OnClickListener(){
            @Override
            public void onClick(View v){
                new Thread(new Runnable(){//创建新线程
                    @Override
                    public void run(){
```

```
                    send();
                    Message m=handler.obtainMessage();
                                                    //获取 Handler 消息对象
                    handler.sendMessage(m);         //向 Handler 对象发送消息
                }
            }).start();                             //启动线程
        }
    });
    handler=new Handler(){                          //实例化 Handler 对象(用于刷新主线程)
        @Override
        public void handleMessage(Message msg){
                                                    //重写 handleMessage 方法
            if(backMsg !=null){
                txt_back.setText(backMsg);          //显示接收到的消息
            }
            super.handleMessage(msg);
        }
    };
}
/**
 * 与服务器端连接并处理信息的接收与发送
 */
private void send(){
    try {
        socket=new Socket("10.0.2.2", 8888);                //连接服务器端
        dout=new DataOutputStream(socket.getOutputStream());//实例化输出流对象
        din=new DataInputStream(socket.getInputStream());   //实例化输入流对象
        dout.writeUTF(txt_msg.getText().toString());        //向服务器发送消息
        backMsg=din.readUTF();                              //从输入流读取数据
    } catch(Exception e){
        e.printStackTrace();
    } finally {
        try {
            if(dout !=null){
                dout.close();
            }
            if(din !=null){
                din.close();
            }
            if(socket !=null){
                socket.close();
            }
        } catch(Exception e){
            e.printStackTrace();
```

```
        }
      }
    }
}
```

运行服务器端与客户端,效果如图 6-4 所示。

图 6-4　Socket 实例运行效果

★注意：(1)测试 IP 地址要使用 10.0.2.2(模拟器不能配置代理)；(2)在 Android 4.0 以后的开发中,连接服务器端的操作不能在主线程进行。

6.3　基于 HTTP 的网络编程

在 Android 中基于 HTTP 进行网络通信主要有两种方法：一是使用 HttpURLConnection 类实现,二是使用 HttpClient 类实现。

6.3.1　使用 HttpURLConnection 类访问网络

在 java.net 包中提供了 HttpURLConnection 类,用于发送 HTTP 请求(GET 和 POST)以及获取 HTTP 响应。该类是抽象类,不能直接实例化获取,需要使用 URL 的 openConnection()方法来获得,这时并没有真正地执行连接操作,只是创建了一个新的实例。

举例：使用 HttpURLConnection 类的 GET 请求方式

步骤 1：新建一个 JSP 文件 index.jsp 作为服务器端。代码如下：

```
<%@page contentType="text/html; charset=gb2312" language="java"%>
<%
  String content="";
  if(request.getParameter("content")!=null){
      content=request.getParameter("content");
      content=new String(content.getBytes("iso-8859-1"),"gb2312");
  }
%>
<%="发布一条信息,内容如下："%>
<%=content%>
```

将 index.jsp 文件放到 Tomcat 安装路径下的 webapps/news 目录下,然后启动 Tomcat。

步骤 2：新建 Android 项目作为客户端,修改 res/layout 目录下的 XML 布局文件。

代码如下:

```xml
<?xml version="1.0" encoding="utf-8"?>
<LinearLayout xmlns:android="http://schemas.android.com/apk/res/android"
    android:layout_width="fill_parent"
    android:layout_height="fill_parent"
    android:gravity="center_horizontal"
    android:orientation="vertical">
    <EditText
        android:id="@+id/content"
        android:layout_width="match_parent"
        android:layout_height="wrap_content"/>
    <Button
        android:id="@+id/btn_send"
        android:layout_width="wrap_content"
        android:layout_height="wrap_content"
        android:text="发送"/>
    <ScrollView
        android:layout_width="match_parent"
        android:layout_height="wrap_content">
        <TextView
            android:id="@+id/result"
            android:layout_width="match_parent"
            android:layout_height="wrap_content"/>
    </ScrollView>
</LinearLayout>
```

步骤 3：在 AndroidManifest.xml 文件中添加权限。代码如下:

```xml
<uses-permission android:name="android.permission.INTERNET"/>
```

步骤 4：在主活动中完成代码，如下所示:

```java
public class MainActivity extends Activity {
    private EditText content;                      //声明输入文本内容的编辑框组件
    private Button btn_send;                       //声明发送按钮组件
    private Handler handler;
    private String result="";                      //声明一个代表显示内容的字符串
    private TextView resultTV;                     //声明一个显示结果的文本框组件
    @Override
    protected void onCreate(Bundle savedInstanceState){
        super.onCreate(savedInstanceState);
        setContentView(R.layout.activity_main);
        content=(EditText)findViewById(R.id.content);
        resultTV=(TextView)findViewById(R.id.result);
        btn_send=(Button)findViewById(R.id.btn_send);
```

```java
        //为按钮添加单击事件监听器
        btn_send.setOnClickListener(new OnClickListener(){
            @Override
            public void onClick(View v){
                if("".equals(content.getText().toString())){
                    Toast.makeText(MainActivity.this, "发表内容不能为空!",
                    Toast.LENGTH_SHORT).show();      //显示消息提示
                    return;
                }
                //创建一个新线程,用于发送并读取微博信息
                new Thread(new Runnable(){
                    public void run(){
                        send();                     //发送文本内容到Web服务器
                        Message m=handler.obtainMessage();
                                                    //获取一个Message
                        handler.sendMessage(m);     //发送消息
                    }
                }).start();                         //开启线程
            }
        });
        //创建一个Handler对象
        handler=new Handler(){
            @Override
            public void handleMessage(Message msg){
                if(result!=null){
                    resultTV.setText(result);
                    content.setText("");            //清空文本框
                }
                super.handleMessage(msg);
            }
        };
    }
    /**
     * 与服务器端连接并处理信息的接收与发送
     */
    public void send(){
    String target="";
    String sendinfo="";
    try {
        //对发送的消息内容重新编码
        sendinfo=URLEncoder.encode(content.getText().toString().trim(), "gb2312");
    } catch(Exception e){
        e.printStackTrace();
    }
```

```
        //要访问的 URL 地址
        target="http://10.0.2.2:8080/news/index.jsp?content=" + sendinfo;
        URL url;
        try {
            url=new URL(target);
            //创建一个 HTTP 连接
            HttpURLConnection urlConn = (HttpURLConnection)url.openConnection
            ();InputStreamReader in=new InputStreamReader(
                    urlConn.getInputStream(), "gb2312");    //获得读取的内容
            BufferedReader buffer=new BufferedReader(in);   //获取输入流对象
            String inputLine=null;              //声明存储读取输入流中内容的字符串变量
            //通过循环逐行读取输入流中的内容
            while((inputLine=buffer.readLine())!=null){
                result+=inputLine+"\n";
            }
            in.close();                         //关闭字符输入流对象
            urlConn.disconnect();               //断开连接
        } catch(Exception e){
            e.printStackTrace();
        }
    }
}
```

运行效果如图 6-5 所示。

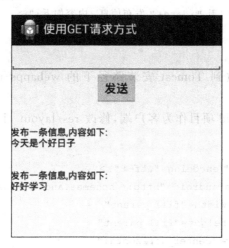

图 6-5　HttpURLConnection 类 GET 请求方式实例运行效果

采用 GET 方式发送请求只适合发送大小在 1024B 以内的数据,当要发送的数据比较大时,就需要采用 POST 方式来发送请求。发送 POST 请求时常用的方法如表 6-4 所示。

表 6-4 发送 POST 请求时常用的方法

方 法	描 述
setDoInput(boolean newValue)	设置是否向连接中写入数据
setDoOutput((boolean newValue))	设置是否从连接中读取数据
setUseCaches((boolean newValue))	设置是否缓存数据
setInstanceFollowRedirects(boolean followRedirects)	设置是否自动执行 HTTP 重定向
setRequestProperty(String field,String newValue)	设置一般请求属性

在 Android 中，使用 HttpURLConnection 类发送请求时，默认采用的是 GET 方式，如果要用 POST 方式，需要通过 setRequestMethod()方法指定。

举例：使用 HttpURLConnection 类的 POST 请求方式

步骤 1：新建一个 JSP 文件 index.jsp 作为服务器端。代码如下：

```
<%@page contentType="text/html; charset=gb2312" language="java"%>
<%
  String username=request.getParameter("username");
  String content=request.getParameter("content");
  if(username!=null && content!=null){
    username=new String(username.getBytes("iso-8859-1"),"gb2312");
    content=new String(content.getBytes("iso-8859-1"),"gb2312");
    String date=new java.text.SimpleDateFormat("yyyy-MM-dd HH:mm:ss").format
    (new java.util.Date());
%>
<%="[ "+username+" ]于 "+date+" 发布信息，内容如下："%>
<%=content%>
<%}%>
```

将 index.jsp 文件放到 Tomcat 安装路径下的 webapps/news 目录下，然后启动 Tomcat。

步骤 2：新建 Android 项目作为客户端，修改 res/layout 目录下的 XML 布局文件。代码如下：

```
<?xml version="1.0" encoding="utf-8"?>
<LinearLayout xmlns:android="http://schemas.android.com/apk/res/android"
    android:layout_width="fill_parent"
    android:layout_height="fill_parent"
    android:gravity="center_horizontal"
    android:orientation="vertical">
    <EditText
        android:id="@+id/username"
        android:layout_width="match_parent"
        android:layout_height="wrap_content"
        android:hint="输入姓名">
    </EditText>
```

```xml
<EditText
    android:id="@+id/content"
    android:layout_width="match_parent"
    android:layout_height="wrap_content"
    android:hint="输入内容"
    android:inputType="textMultiLine"/>
<Button
    android:id="@+id/btn_send"
    android:layout_width="wrap_content"
    android:layout_height="wrap_content"
    android:text="发送"/>
<ScrollView
    android:layout_width="match_parent"
    android:layout_height="wrap_content">
    <TextView
        android:id="@+id/result"
        android:layout_width="match_parent"
        android:layout_height="wrap_content"/>
</ScrollView>
</LinearLayout>
```

步骤3：在 AndroidManifest.xml 文件中添加权限。代码如下：

```xml
<uses-permission android:name="android.permission.INTERNET"/>
```

步骤4：在主活动中完成代码如下：

```java
public class MainActivity extends Activity {
    private EditText username;              //声明一个输入姓名的编辑框组件
    private EditText content;               //声明一个输入文本内容的编辑框组件
    private Button btn_send;                //声明一个发送按钮组件
    private Handler handler;                //声明一个Handler对象
    private String result="";               //声明一个显示内容的字符串
    private TextView resultTV;              //声明一个显示结果的文本框组件
    @Override
    protected void onCreate(Bundle savedInstanceState){
        super.onCreate(savedInstanceState);
        setContentView(R.layout.activity_main);
        content=(EditText)findViewById(R.id.content);
        resultTV=(TextView)findViewById(R.id.result);
        username=(EditText)findViewById(R.id.username);
        btn_send=(Button)findViewById(R.id.btn_send);
        //为按钮添加单击事件监听器
        btn_send.setOnClickListener(new OnClickListener(){
            @Override
            public void onClick(View v){
```

```java
            if("".equals(username.getText().toString())
                    || "".equals(content.getText().toString())){
                Toast.makeText(MainActivity.this,"请将内容输入完整!",
                Toast.LENGTH_SHORT).show();
                return;
            }
            //创建一个新线程,用于从网络上获取文件
            new Thread(new Runnable(){
                public void run(){
                    send();
                    Message m=handler.obtainMessage();      //获取一个Message
                    handler.sendMessage(m);                  //发送消息
                }
            }).start();                                      //开启线程
        }
    });
    handler=new Handler(){
        @Override
        public void handleMessage(Message msg){
            if(result !=null){
                resultTV.setText(result);                   //显示获得的结果
                content.setText("");                         //清空内容编辑框
                username.setText("");                        //清空昵称编辑框
            }
            super.handleMessage(msg);
        }
    };
}
/**
 * 与服务器端连接并处理信息的接收与发送
 */
public void send(){
    String target="http://10.0.2.2:8080/news/index.jsp";
                                                            //要提交的目标地址
    URL url;
    try {
        url=new URL(target);
        HttpURLConnection urlConn = (HttpURLConnection) url.openConnection
        ();urlConn.setRequestMethod("POST");                 //指定使用POST请求方式
        urlConn.setDoInput(true);                            //向连接中写入数据
        urlConn.setDoOutput(true);                           //从连接中读取数据
        urlConn.setUseCaches(false);                         //禁止缓存
        urlConn.setInstanceFollowRedirects(true);            //自动执行HTTP重定向
        urlConn.setRequestProperty("Content-Type", "application/x-www-
```

```
            form-urlencoded");                          //设置内容类型
            DataOutputStream out=new DataOutputStream(urlConn.getOutputStream
            ());                                        //获取输出流
            //连接要提交的数据
            String param="username="+URLEncoder.encode(username.getText().
            toString(),"gb2312")+"&content="+URLEncoder.encode(content.
            getText().toString(),"gb2312");
            out.writeBytes(param);                      //将要传递的数据写入数据输出流
            out.flush();                                //输出缓存
            out.close();                                //关闭数据输出流
            //判断是否响应成功
            if(urlConn.getResponseCode()==HttpURLConnection.HTTP_OK){
                InputStreamReader in = new InputStreamReader ( urlConn.
                getInputStream(),"gb2312");             //获得读取的内容
                BufferedReader buffer=new BufferedReader(in);
                                                        //获取输入流对象
                String inputLine=null;
                while((inputLine=buffer.readLine())!=null){
                    result+=inputLine+"\n";
                }
                in.close();                             //关闭字符输入流
            }
            urlConn.disconnect();                       //断开连接
        } catch(MalformedURLException e){
            e.printStackTrace();
        } catch(IOException e){
            e.printStackTrace();
        }
    }
}
```

运行效果如图 6-6 所示。

图 6-6　HttpURLConnection 类 POST 请求方式实例运行效果

6.3.2 使用 HttpClient 类访问网络

在 Android 中集成的 HttpClient 类对 Java 中访问网络的方法进行了封装,输入输出流操作被统一封装成 HttpGet、HttpPost 和 HttpResponse 类。其中,HttpGet 类用于发送 GET 请求,HttpPost 类用于发送 POST 请求,HttpResponse 类用于处理响应对象。

举例:使用 HttpClient 类的 GET 请求方式

步骤 1:新建一个 JSP 文件 index.jsp 作为服务器端。代码如下:

```jsp
<%@page contentType="text/html; charset=gb2312" language="java"%>
<%
  String param=request.getParameter("param");
  if("".equals(param)||param!=null){
      if("get".equals(param)){
         out.println("服务器端已收到 GET 请求!");
      }
  }
%>
```

将 index.jsp 文件放到 Tomcat 安装路径下的 webapps/news 目录下,然后启动 Tomcat。

步骤 2:新建 Android 项目作为客户端,修改 res/layout 目录下的 XML 布局文件,在线性布局中添加一个用于发送请求的按钮组件和一个用于接收消息的文本框组件。代码如下:

```xml
<Button
    android:id="@+id/btn_send"
    android:layout_width="wrap_content"
    android:layout_height="wrap_content"
    android:text="发送"/>
<TextView
    android:id="@+id/result"
    android:layout_width="match_parent"
    android:layout_height="wrap_content"/>
```

步骤 3:在 AndroidManifest.xml 文件中添加权限。代码如下:

```xml
<uses-permission android:name="android.permission.INTERNET"/>
```

步骤 4:在主活动中完成代码,如下所示:

```java
public class MainActivity extends Activity {
    private Button btn_send;              //声明发送按钮组件
    private TextView resultTV;            //声明显示结果的文本框组件
    private Handler handler;
    private String result="";             //声明接收消息的字符串变量
```

```java
@Override
protected void onCreate(Bundle savedInstanceState){
    super.onCreate(savedInstanceState);
    setContentView(R.layout.activity_main);
    resultTV=(TextView)findViewById(R.id.result);
    btn_send=(Button)findViewById(R.id.btn_send);
    btn_send.setOnClickListener(new View.OnClickListener(){
        @Override
        public void onClick(View v){
            new Thread(new Runnable(){
                @Override
                public void run(){
                    send();
                    Message m=handler.obtainMessage();
                    handler.sendMessage(m);
                }
            }).start();
        }
    });
    handler=new Handler(){
        @Override
        public void handleMessage(Message msg){
            if(result !=null){
                resultTV.setText(result);
            }
            super.handleMessage(msg);
        }
    };
}
public void send(){
    String str="http://10.0.2.2:8080/news/index.jsp?param=get";
    HttpClient httpClient=new DefaultHttpClient();
                                        //创建 HttpClient 对象
    HttpGet httpRequest=new HttpGet(str);    //创建 HttpGet 对象
    try {
        //执行 HttpClient 请求
        HttpResponse httpResponse=httpClient.execute(httpRequest);
        if(httpResponse.getStatusLine().getStatusCode()==HttpStatus.SC_OK){
            //获取返回字符串
            result=EntityUtils.toString(httpResponse.getEntity(),"gb2312");
        }
    } catch(Exception e){
        e.printStackTrace();
    }
}
```

            }
    }

同使用 HttpURLConnection 类发送请求一样，对于复杂的请求数据也需要使用 POST 方式发送。如果需要发送请求参数，可以调用 HttpPost 的 setParams()方法来添加参数，也可以调用 setEntity()方法设置请求参数。

举例：使用 HttpClient 类的 POST 请求方式

步骤 1：新建一个 JSP 文件 index.jsp 作为服务器端。代码如下：

```jsp
<%@page contentType="text/html; charset=gb2312" language="java"%>
<%
    String param=request.getParameter("param");
    if("post".equals(param)){
      String username=request.getParameter("username");
      String content=request.getParameter("content");
      if(username!=null && content!=null){
        username=new String(username.getBytes("iso-8859-1"), "gb2312");
        content=new String(content.getBytes("iso-8859-1"), "gb2312");
        String date= new java.text.SimpleDateFormat("yyyy-MM-dd HH:mm:ss").
        format(new java.util.Date());
        out.println("[ "+username+" ]于 "+date+" 发表信息，内容如下：");
        out.println(content);
      }
    }
%>
```

将 index.jsp 文件放到 Tomcat 安装路径下的 webapps/news 目录下，然后启动 Tomcat。

步骤 2：新建 Android 项目作为客户端，修改 XML 布局文件（与 6.3.1 节实例中介绍的使用 HttpURLConnection 类的 POST 请求方式的布局相同）。

步骤 3：在 AndroidManifest.xml 文件中添加权限。代码如下：

```xml
<uses-permission android:name="android.permission.INTERNET"/>
```

步骤 4：在主活动中完成代码，如下所示：

```java
public class MainActivity extends Activity {
    …                                          //此处省略变量声明部分的代码
    @Override
    protected void onCreate(Bundle savedInstanceState){
        super.onCreate(savedInstanceState);
        setContentView(R.layout.activity_main);
        …                                      //此处省略组件获取部分的代码
        btn_send.setOnClickListener(new OnClickListener(){
            @Override
```

```java
public void onClick(View v){
    if("".equals(username.getText().toString())
        || "".equals(content.getText().toString())){
        Toast.makeText(MainActivity.this,"请将内容输入完整!",
            Toast.LENGTH_SHORT).show();
        return;
    }
    //创建一个新线程,用于从网络上获取文件
    new Thread(new Runnable(){
        public void run(){
            send();
            Message m=handler.obtainMessage();
                                            //获取一个 Message
            handler.sendMessage(m);         //发送消息
        }
    }).start();                             //开启线程
}
});
handler=new Handler(){
    @Override
    public void handleMessage(Message msg){
        if(result !=null){
            resultTV.setText(result);       //显示获得的结果
            content.setText("");            //清空内容编辑框
            username.setText("");           //清空昵称编辑框
        }
        super.handleMessage(msg);
    }
};
}
public void send(){
    String target="http://10.0.2.2:8080/news/index.jsp";
                                            //目标地址
    HttpClient httpclient=new DefaultHttpClient();  //创建 HttpClient 对象
    HttpPost httpRequest=new HttpPost(target);      //创建 HttpPost 对象
    //将要传递的参数保存到 List 集合中
    List<NameValuePair>params=new ArrayList<NameValuePair>();
    params.add(new BasicNameValuePair("param","post"));
                                            //标记参数
    params.add(new BasicNameValuePair("username",username.getText().
        toString()));
    params.add(new BasicNameValuePair("content",content.getText().
        toString()));
    try{
```

```
                httpRequest.setEntity(new UrlEncodedFormEntity(params,"gb2312"));
                HttpResponse httpResponse=httpclient.execute(httpRequest);
                                                            //执行请求
                if(httpResponse.getStatusLine().getStatusCode()==HttpStatus.SC_OK){
                    result+=EntityUtils.toString(httpResponse.getEntity());
                                                            //获取返回字符串
                }else{
                    result="请求失败!";
                }
            }catch(Exception e){
                e.printStackTrace();
            }
        }
    }
```

6.4 用 WebView 组件显示网页

在 Android 中，要用内置的浏览器（WebKit 开源引擎）显示网页，需要通过 WebView 组件实现。WebView 组件常用的方法如表 6-5 所示。

表 6-5　WebView 组件的常用方法

方　　法	描　　述
loadUrl(String url)	加载指定的 URL 网页
loadUrl（String url，Map＜String，String＞additionalHttpHeaders）	加载指定的 URL 并携带 HTTP header 数据
loadData(String data，String mimeType，String encoding)	将指定的字符串数据加载到浏览器
loadDataWithBaseURL（String baseUrl，String data，String mimeType，String encoding，String historyUrl）	从指定的 URL 中加载数据
capturePicture()	创建当前屏幕快照
goBack()	执行后退页面操作
goForward()	执行前进页面操作
stopLoading()	停止加载当前页面
reload()	刷新当前页面

举例：使用 WebView 组件显示网页。

步骤 1：修改 res/layout 目录下的 XML 布局文件，添加一个 WebView 组件。代码如下：

```
<WebView
    android:id="@+id/myWebView"
    android:layout_width="fill_parent"
```

```
android:layout_height="fill_parent"/>
```

步骤 2：在主活动文件的 onCreate()方法中获取 WebView 组件并指定 URL 地址。代码如下：

```
WebView myWebView= (WebView)findViewById(R.id.myWebView);
myWebView.getSettings().setSupportZoom(true);                    //支持网页缩放
myWebView.getSettings().setSupportMultipleWindows(true);         //支持多窗口
myWebView.getSettings().setBuiltInZoomControls(true);            //显示缩放按钮
myWebView.loadUrl("http://ist.yctu.edu.cn");                     //加载网页
```

步骤 3：在 AndroidManifest.xml 文件中指定允许访问网络资源的权限。
运行效果如图 6-7 所示。

图 6-7 使用 WebView 组件显示网页的运行效果

举例：使用 WebView 组件加载本地文件

步骤 1：新建一个网页文件 index.html 和 images 文件夹（存放一张图片 pic.png），复制到 android 项目根目录的 assets 文件夹中。网页代码如下：

```
<html>
    <head></head>
    <style>
        body{width:480px;}
        .pic{text-align:center;}
        .content{padding:0 15px;}
    </style>
    <body>
        <div class="content">游戏中的小房子</div>
        <div class="pic"><img src="file:///android_asset/images/pic.png"/></div>
```

```
</body>
</html>
```

步骤 2：修改 res/layout 目录下的 XML 布局文件，添加一个 WebView 组件。

步骤 3：在主活动文件的 onCreate()方法中获取 WebView 组件并指定 URL 地址。代码如下：

```
myWebView=(WebView)findViewById(R.id.myWebView);
myWebView.getSettings().setSupportZoom(true);                  //支持网页缩放
myWebView.getSettings().setBuiltInZoomControls(true);          //显示缩放按钮
myWebView.getSettings().setDefaultTextEncodingName("gb2312");  //设置文本编码
myWebView.loadUrl("file:///android_asset/index.html");         //加载网页
```

步骤 4：在 AndroidManifest.xml 文件中指定允许访问网络资源的权限。

运行效果如图 6-8 所示。

在 Android 游戏开发中，可以使用 WebView 组件加载 HTML 代码的方式来显示游戏的帮助或升级提示等信息。WebView 组件提供了 loadData()和 loadDataWithBaseURL()两种加载 HTML 代码的方法，使用 loadData()方法会产生乱码，而 loadDataWithBaseURL()方法不会产生乱码。loadDataWithBaseURL()方法的参数如表 6-6 所示。

图 6-8 使用 WebView 组件加载本地文件的运行效果

表 6-6 loadDataWithBaseURL()方法的参数

方法	描述
baseUrl	指定当前页面的 URL(为 null 则为空白页)
data	指定要显示的字符串数据
mimeType	指定要显示内容的 MIME 类型(为 null 则使用默认的 text/html)
Encoding	指定数据的编码方式
historyUrl	指定当前页的历史 URL(为 null 则为空白页)

举例：使用 WebView 组件加载 HTML 代码

步骤 1：在 Android 项目根目录的 assets 文件夹新建一个 images 文件夹，复制一张图片 pic.png 到 images 文件夹中。

步骤 2：修改 res/layout 目录下的 XML 布局文件，添加一个 WebView 组件。

步骤 3：在主活动文件的 onCreate()方法中获取 WebView 组件并创建一个显示 HTML 代码的字符串构建器，最后应用 loadDataWithBaseURL()方法加载。代码如下：

```
String data="<html><head></head>游戏中的小房子<img src=\"file:///android_
asset/images/pic.png\"/></div></body></html>";
myWebView.loadDataWithBaseURL(null, data, "text/html", "utf-8", null);
                                                                    //加载数据
```

步骤 4：在 AndroidManifest.xml 文件中指定允许访问网络资源的权限。

★**注意**：如果要让 WebView 组件支持 JavaScript，可以使用 WebSettings 对象的 setJavaScriptEnabled()方法。但是对于通过 window.alert()方法弹出的对话框并不可用，还需要使用 WebView 组件的 setWebChromeClient()方法。

6.5 本章小结

本章主要学习了 Android 中的数据存储技术，常见的存储技术有 SharedPreferences、Files 和 SQLite 数据库 4 种。其中 SharedPreferences 适合存储简单的数据，如整数、布尔值等；Files 适合存储私有的数据及 SD 卡数据；SQLite 数据库适合存储复杂的数据，它是一种轻便的数据库。本章还讲解了简单的网络开发知识，包括 Socket、HttpURLConnection 和 HttpClient。最后，本章讲解了 WebView 组件加载网页和 HTML 代码的方法。

6.6 思考与练习

（1）应用 SharedPreferences 实现对用户输入数据的保存。
（2）应用 SQLite 技术开发一个信息管理应用，能对数据进行增加、删除、修改和查询。
（3）应用 Socket 实现与服务器之间的消息传递与接收。
（4）应用 HttpClient 类的 POST 方式实现一个登录访问页面的实例。
（5）应用 WebView 组件实现获取指定城市的天气预报。

第 7 章

游戏中的数学与物理学

学习目标：
- 熟悉游戏开发中常用的数学知识。
- 熟悉游戏开发必备的物理学知识。
- 掌握 3 种碰撞检测方法。
- 掌握模拟游戏中粒子的模拟方法。

本章导读：

在游戏开发中，除了要对开发语言、设计模式等知识熟练掌握之外，还必须对数学和物理学方面的知识有所了解。本章主要介绍游戏中常用的数学知识、物理学知识、碰撞检测以及粒子系统，主要用于提升游戏的视觉效果和真实感。

7.1 游戏中常用的数学知识

在游戏编程中，常用的坐标系统就是二维坐标系和三维坐标系。在 2D 平面上用一个二元组表示 (x, y)，在 3D 空间中用一个三元组表示 (x, y, z)。对于几何学上的坐标系（如图 7-1 所示），在手机上由于没有负坐标，采用了二维坐标系（如图 7-2 所示）。

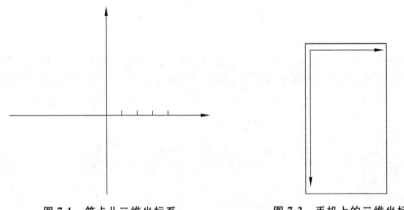

图 7-1　笛卡儿二维坐标系　　　　图 7-2　手机上的二维坐标系

三维坐标系是所谓的 XYZ 坐标系,在手机上无法真正表示三维,因此一般是用二维来模拟三维。此时,正方形一般用平行四边形来表示。

1. 距离的计算

在游戏编程中经常要计算两个物体之间的距离,用于碰撞检测和搜索路径等。2D场景中的距离计算方法如下:设点 $P_1(x_1, y_1)$ 和 $P_2(x_2, y_2)$ 分别为线上的点。两点之间距离 d 的计算方法如式(7-1)所示:

$$d = \sqrt{(x_2-x_1)^2 + (y_2-y_1)^2} \tag{7-1}$$

★提示:为了节省运算开销,通常编程计算时不进行开方操作,而是直接对平方值进行比较得出结果,这样简化了计算。虽然这种方法不是很准确,但在大多数情况下还是可以很好地满足游戏编程的要求,在运算频度高的情况下对性能有一定的优化作用。

3D 场景中的距离计算方法如式(7-2)所示:

$$d = \sqrt{(x_2-x_1)^2 + (y_2-y_1)^2 + (z_2-z_1)^2} \tag{7-2}$$

两点的中点坐标计算方法如下:设有点 $P_1(x_1, y_1)$ 和 $P_2(x_2, y_2)$,两点的中点 P_3 的坐标计算如式(7-3)所示:

$$p_3 = \left(\frac{x_1+x_2}{2}, \frac{y_1+y_2}{2}\right) \tag{7-3}$$

2. 抛物线

抛物线总是轴对称的。有两个因素决定了抛物线的形状,第一个是顶点,即抛物线与对称轴的交点;第二个是对称轴。抛物线有两种形状,一种是对称轴垂直,另一种是对称轴水平。

对称轴垂直的抛物线方程如式(7-4)所示:

$$y = a(x-h)^2 + k \tag{7-4}$$

其中,顶点是 (h, k),对称轴为 $x=h$。

对称轴水平的抛物线方程如式(7-5)所示:

$$x = a(y-k)^2 + h \tag{7-5}$$

其中,顶点是 (h, k),对称轴为 $y=k$。

常数 a 代表了抛物线的开口方向和开口大小。如果 a 是正数,对于 $y=a(x-h)^2+k$ 的抛物线来说开口向上,对于 $x=a(y-k)^2+h$ 的抛物线来说开口向右。如果 a 是负数,对于 $y=a(x-h)^2+k$ 的抛物线来说开口向下,对于 $x=a(y-k)^2+h$ 的抛物线来说开口向左。a 的绝对值越大,开口越小。

3. 圆和球

圆是所有到定点长度等于定长的点的集合,这个定长称为半径,定点称为圆心。

圆的方程如式(7-6)所示:

$$(x-h)^2 + (y-k)^2 = r^2 \tag{7-6}$$

其中,圆心是 (h, k),半径是 r。

球是一个圆绕着圆心旋转所得到的几何体。

球体的方程如式(7-7)所示：

$$(x-h)^2 + (y-k)^2 + (z-l)^2 = r^2 \tag{7-7}$$

其中，圆心是(h, k, l)，半径是r。

使用方程时应注意圆心坐标的正负号。

4. 三角函数

1) 角度和弧度

每个角都由相交于一点的两条射线组成，把其中一条射线称为始边，另一条称为终边。而角的始边总是沿着X轴的正方向。从X轴正方向开始，沿逆时针方向旋转所得的角称为正角，沿顺时针方向旋转所得的角称为负角，注意，该旋转也决定了终边的位置。一个周角是360°，也可以表示成2π，这是角度和弧度之间进行转换的基础。

角度转换成弧度的公式如式(7-8)所示：

$$角度 \times \frac{\pi}{180°} = 弧度 \tag{7-8}$$

弧度转换成角度的公式如式(7-9)所示：

$$弧度 \times \frac{180°}{\pi} = 角度 \tag{7-9}$$

2) 三角函数

所有的三角函数都是在直角三角形中定义的，如式(7-10)所示：

$$正弦：\sin \alpha = \frac{b}{c}$$

$$余弦：\cos \alpha = \frac{a}{c}$$

$$正切：\tan \alpha = \frac{b}{a}$$

$$余切：\cot \alpha = \frac{1}{\tan \alpha} = \frac{a}{b}$$

$$正割：\csc \alpha = \frac{1}{\sin \alpha} = \frac{c}{b}$$

$$余割：\sec \alpha = \frac{1}{\cos \alpha} = \frac{c}{a} \tag{7-10}$$

常用角度的三角函数值如表7-1所示。

表7-1 常用角度的三角函数值

α（角度）	α（弧度）	$\sin\alpha$	$\cos\alpha$	$\tan\alpha$
0	0	0	1	0
30	$\pi/6$	0.5	$\sqrt{3}/2$(0.866)	$\sqrt{3}/3$(0.5774)
45	$\pi/4$	$\sqrt{2}/2$(0.7071)	$\sqrt{2}/2$(0.7071)	1
60	$\pi/3$	$\sqrt{3}/2$(0.866)	0.5	$\sqrt{3}$

续表

α（角度）	α（弧度）	sinα	cosα	tanα
90	π/2	1	0	不可算
120	2π/3	√3/2(0.866)	−0.5	−√3
180	π	0	−1	0
270	3π/2	−1	0	不可算
360	0	0	1	0

在真正进入游戏主循环之前，可以建立一个三角函数的查找表，这样在游戏中需要用到三角函数值的时候就不用重新计算，只需进行查表工作就可以，运行速度大为加快。角的正弦值在第一、第二象限是正值，角的余弦值在第一、第四象限是正值，角的正切值在第一、第三象限是正值。对于所有的反三角函数，如果传入的参数为正，那么返回值都是正，即意味着该角位于第一象限；如果传入的参数为负，那么反正弦函数 asin() 和反正切函数 atan() 的返回角度位于第四象限，而反余弦函数 acos() 的返回角度位于第二象限。

3) 正弦函数

正弦函数如式(7-12)所示：

$$y = A\sin(Bx) + C \tag{7-11}$$

其中，振幅是 A，周期是 $\dfrac{360°}{|B|}$，偏移 X 轴 C。A 越大，振幅越高；B 越大，周期越小。

5．向量

在游戏中使用一个量的时候，一定要区分它是标量还是向量，两者最大的差别就是是否具有方向。游戏开发中用到向量的地方主要有以下几种情况。

1) 计算投影和夹角

很多时候游戏中需要的向量操作就是将向量投影到某个特定的平面上，或者对两个向量进行计算求得两个向量之间的夹角。

2) 判断方向

在游戏中经常将两个向量相乘，这两个向量分别代表两个游戏角色的朝向。对这两个向量进行点乘，如果结果为正，说明二者之间的夹角小于 90°，大致位于同一个方向；如果结果为负，说明两个游戏角色面朝不同的方向。

3) 参与复杂计算

在有些游戏编程中，需要进行大规模的矩阵运算，而向量在矩阵运算中起着重要的作用。

6．在碰撞检测中的应用

可以在游戏中利用圆或球的边界进行碰撞检测，也可以利用其他图形。圆和球都可以方便地进行数学计算，在检测的速度上也优于其他图形。虽然其精确度不高，但是可以作为外围检测。如果两个圆的圆心距离小于两圆的半径和,即发生碰撞。设两圆的方程分别为$(x-h_1)^2+(y-k_1)^2=r_1^2$和$(x-h_2)^2+(y-k_2)^2=r_2^2$。如果$\sqrt{(h_2-h_1)^2+(k_2-k_1)^2} \leqslant (r_1+r_2)$,则两圆发生碰撞。

利用圆的边界进行碰撞检测是一种较快的方法,但是极有可能产生错误的碰撞检测结果,所以避免错误的方法是寻找一种更适合的图形来检测,只要这个图形可以用数学公式表示出来即可。也可以使用多重圆形进行多重检测,先检测外面的圆,如果发生碰撞则检测内部的圆,减小错误判断的几率。但多重检测会消耗更多的CPU时间。

7.2 游戏中常用的物理学知识

物理学研究的是客观世界的各种规律,而游戏中恰恰需要对客观世界进行模拟,这就必须用到物理学方面的知识。具体来说,游戏开发中主要运用的是物理学中与运动有关的内容。

1. 速度v、加速度a、位移s和时间t

如果是匀速直线运动,则$s=vt$;如果是匀加速直线运动,则$a=(v_t-v_0)/t$,$v_t^2-v_0^2=2as$,$s=v_0t+at^2/2$。如果还有重力因素,那么许多物体在空中运行的轨迹是抛物线,典型的如"愤怒的小鸟",其受力如图7-3所示。

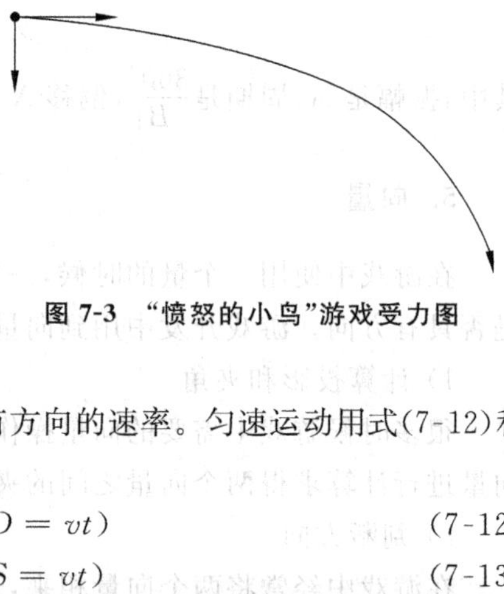

2. 一维空间运动

1) 速度和速率

物体只要移动就会有速率,速率用来表示物体运动的快慢。相应地,如果一个物体有速率,那么它就会有速度,速度是速率的向量形式,即速度是有方向的速率。匀速运动用式(7-12)和式(7-13)来表示:

图7-3 "愤怒的小鸟"游戏受力图

$$位移 = 速度 \times 时间(D = vt) \qquad (7\text{-}12)$$

$$路程 = 速率 \times 时间(S = vt) \qquad (7\text{-}13)$$

平均速度如式(7-14)所示:

$$\bar{v} = \frac{\Delta x}{t} = \frac{x_f - x_i}{t} \qquad (7\text{-}14)$$

其中,Δx表示位移,t表示时间间隔,v_f表示末速度,v_i表示初速度。

2) 加速度

加速度用来衡量速率的变化快慢程度,如式(7-15)所示:

$$a = \frac{\Delta v}{\Delta t} = \frac{v_f - v_i}{t_f - t_i} \qquad (7\text{-}15)$$

其中,t_f表示末速度对应的时间,t_i表示初速度对应的时间。

如果a的方向和v一致,那么表示加速运动;如果相反,表示减速运动。

3. 二维和三维空间运动

1）使用向量

二维和三维空间运动与一维空间运动最大的区别在于它们还有其他方向。在一维空间里，使用正负号来表示位移、速度和加速度这些向量。但在二维和三维空间中，就必须结合向量来描述运动。

在二维空间中，利用末位置减去初位置来表示位移。2D 和 3D 的平均速度如式(7-16)所示：

$$\bar{v} = \frac{\Delta r}{t} = \frac{r_f - r_i}{t} \tag{7-16}$$

其中，Δr 表示位移向量，t 为时间段。

2D 和 3D 中的移动方程如式(7-17)所示：

$$v_f = v_i + at$$
$$\Delta r = \frac{1}{2}(v_f + v_i)t$$
$$\Delta r = v_i t + \frac{1}{2}at^2 \tag{7-17}$$

这些方程对于任意矢量 a、v_f、v_i 和 Δr 以及标量时间都适用。在计算时，要将极坐标表示的向量转换成用直角坐标表示。

2）抛物运动

二维空间计算抛物线的方法是将向量分解成竖直分量和水平分量，然后进行计算。由于分量之间是完全独立的，所以可以分别对其进行计算。如果只看竖直方向，那么速度、位移、加速度就回到了一维空间的竖直运动，先运动到最高点，然后下落。速度在最高点时由正值变成 0，接着下落变成负值。加速度一直都是 -9.8m/s^2。除非有外力施加于运动物体之上，比如弹簧、空气阻力等。如果只看水平方向，那么就是匀速直线运动，排除空气阻力等其他因素。

4. 牛顿定律

1）力

当准备用程序来模拟物体的运动时，首先对要移动的物体进行受力分析。物体受力的总和决定了它的运动模式。

（1）重力。

物体运动的竖直分量加速度为 $-g$ 或 -9.8m/s^2。可以利用重力加速度和物体的质量求出物体所受重力。重力是一个向量，方向指向地心。重力 $w = mg$，其中 m 是物体质量，g 是重力加速度。如果在游戏编程中，在不同星球之间切换，要考虑各个星球的不同重力加速度。重力的单位是牛顿，记为 N，$1\text{N} = 1\text{kg} \cdot \text{m/s}^2$。

（2）支持力。

支持力作用在物体表面，抵消重力并使它不向下落。支持力是正交的，总是垂直于物体表面。对于斜面上的物体，支持力减小了物体竖直方向的加速度。

(3) 摩擦力。

摩擦力分为两种：静摩擦力和滑动摩擦力。静摩擦力可以使物体保持稳定的状态，而滑动摩擦力则可以使物体减速。如果物体所受的其他力总和小于静摩擦力，那么物体保持静止。一旦其他的力大于静摩擦力，物体开始运动，静摩擦力就变成滑动摩擦力。两种摩擦力都取决于它们的接触面。接触面越光滑，摩擦力越小。

要计算摩擦力，就必须知道摩擦系数。静摩擦力 $F_S = -\mu_S N$，其中 N 为支持力。滑动摩擦力 $F_K = -\mu_K N$，其中 N 为支持力。

2) 牛顿定律

牛顿第一定律：当物体所受合力为 0 时，它将保持原有的运动状态不变。

牛顿第二定律：$F=ma$。F 是合力，m 是物体质量，a 是物体的加速度。①一个物体受到的合力越大，速度改变得越快。②如果两个物体受到的合力相同，那么质量小的物体速度改变得更快。

牛顿第三定律：对于每个力，都有一个与之方向相反、大小相同的反作用力。

5．能量

1) 功和动能

功等于力乘以力方向上的位移，即 $W = F\Delta x$，其中 Δx 为位移，F 为位移方向上的力。功的单位是 N·m，也就是焦耳(J)。动能是物体由于运动而具有的能量。物体移动得越快，动能越大。

动能的定义是质量的一半乘以速率的平方。计算方法如式(7-18)所示：

$$KE = \frac{1}{2}mv^2 \tag{7-18}$$

其中，m 为质量，v 为速率。动能是一个标量。单位也是焦耳。

2) 势能

势能包括重力势能和弹性势能。重力势能是根据物体距离地面的高度来衡量的能量。$GPE = mgy$，其中，m 为质量，g 为重力加速度，y 为高度。弹性势能是物体由于发生弹性形变而具有的能量。弹性势能的大小与物体弹性形变的大小和弹簧的劲度系数有关。

6．动量和碰撞

1) 与静止物体的碰撞

与静止物体的碰撞可以用向量反射来研究，这种运动存在着一种对称性，球射入的角度必然等于它射出的角度，即入射角等于反射角。

(1) 向量的轴平行反射。

如果边是竖直方向的，则 $\boldsymbol{v}_f = [-v_{ix}, v_{iy}]$；如果边是水平方向的，则 $\boldsymbol{v}_f = [v_{ix}, -v_{iy}]$。入射向量为 $\boldsymbol{v}_i = [v_{ix}, v_{iy}]$。

(2) 向量的非轴平行反射。

如图 7-4 所示，球的入射方向为 $\boldsymbol{v}_i$，求射出方向 $\boldsymbol{v}_f$，先给公式 $\boldsymbol{v}_f = 2 \cdot \boldsymbol{P} + \boldsymbol{v}_i$，其中 $\boldsymbol{v}_i$

是初速度,P 为 $-v_i$ 基于分界线法线的发射向量。计算过程如下:

① 写出边界 B 的矩阵 $[\Delta x, \Delta y]$。
② 求出边界的垂直向量 $N = [\Delta y, -\Delta x]$。
③ 将 N 单位化得 $N' = [\Delta y / \|N\|, -\Delta x / \|N\|]$。
④ 求发射向量 P,$P = (-v_i \cdot N') \cdot N'$。
⑤ 计算反射向量 v_f,$v_f = 2 \cdot P + v_i$。

2) 动量和冲量

(1) 动量计算公式如下:

$$P = mv$$

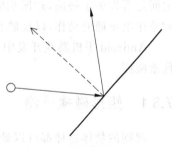

图 7-4　向量的非轴平行反射

其中,m 是物体的质量,v 是物体的速度。如果物体的速度用矩阵表示,那么物体的动量也用矩阵表示。

(2) 冲量计算公式如下:

$$Ft = \Delta p$$

其中,F 是合力,t 是时间,Δp 是动量增量。

3) 碰撞建模

动量定理:$Ft = mv' - mv = P' - P$,其中 F 是对象所受的包括重力在内的所有外力的合力,可以是恒力,也可以是变力。当合外力为变力时,F 是合外力对作用时间的平均值。P 为物体初动量,P' 为物体末动量,t 为合外力的作用时间。

动量定理的变形:

$$m_1 v_{1i} + m_2 v_{2i} = m_1 v_{1f} + m_2 v_{2f}$$

下标 1 表示第一个物体,下标 2 表示第二个物体。每个碰撞的情况都介于弹性碰撞和非弹性碰撞之间。弹性碰撞是一种没有动量损失的碰撞,但现实中的碰撞往往伴随着能量损失,因此可以用还原系数 ε 来表示能量损失的大小,计算方法如式(7-19)所示:

$$v_{1f} - v_{2f} = -\varepsilon(v_{1i} - v_{2i}), \quad 0 < \varepsilon < 1 \tag{7-19}$$

7. 游戏中的"非物理学"现象

虽然在游戏中应用物理学知识会收到很好的效果,但是物理学引擎过于模仿现实也并不总是好事。应该适当允许"非物理学"现象的出现。"非物理学"现象指的是那些游戏中明显违背常理的运动方式,例如让玩家在最高点二次起跳等。

7.3　碰 撞 检 测

在游戏中碰撞检测无时不在,比如在射击游戏中,游戏主角与敌机发生碰撞,游戏主角与敌机子弹发生碰撞等,然后根据检测的结果作出不同的处理。可能有些进行碰撞检测的物体形状很复杂,这些需要进行组合碰撞检测,就是将复杂的物体处理成一个一个的基本形状的组合,然后分别进行不同的检测。

碰撞问题包括碰撞检测和碰撞响应两个方面的内容。碰撞检测用来检测不同对象之间是否发生了碰撞,碰撞响应是在碰撞发生后根据碰撞点和其他参数促使发生碰撞的对象作出正确的动作,以反映真实的动态效果。

Android 手机游戏开发中 3 种最常用的碰撞检测方式分别是矩形碰撞、圆形碰撞和像素碰撞。

7.3.1 矩形碰撞检测

规则的物体碰撞都可以处理成矩形碰撞,实现的原理就是检测两个矩形是否重叠。矩形碰撞就是利用两个矩形之间的位置关系来进行判断,如果一个矩形的像素在另外一个矩形之中或者之上,都可以认为这两个矩形发生了碰撞,两个矩形不发生碰撞的情况只有 4 种(如图 7-5 所示)。

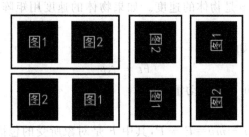

图 7-5　两个矩形不发生碰撞的 4 种情况

举例:矩形碰撞检测

步骤 1:建立内部类 MySurfaceView(继承自 android.view.SurfaceView 类),添加构造函数并实现 Callback 接口(android.view.SurfaceHolder.Callback);实现 Runable 接口并重写其 run()方法。代码如下:

```
public class MySurfaceView extends SurfaceView implements Callback, Runnable{
    private SurfaceHolder sfh;              //用于控制 SurfaceView
    private Paint paint;                    //声明一个画笔
    private Thread th;                      //声明一个线程
    private boolean flag;                   //标识线程消亡
    private Canvas canvas;                  //声明一个画布
    private int x1=10, y1=110, w1=40, h1=40;   //定义第一个矩形的位置和宽高
    private int x2=60, y2=160, w2=40, h2=40;   //定义第二个矩形的位置和宽高
    private boolean isCollsion;             //标识是否发生碰撞
    public MySurfaceView(Context context){
        super(context);
        sfh=this.getHolder();               //实例化 SurfaceHolder 对象
        sfh.addCallback(this);              //为 SurfaceView 添加状态监听
        paint=new Paint();                  //实例化画笔对象
        paint.setColor(Color.BLACK);        //设置画笔颜色为黑色
        setFocusable(true);                 //设置获取焦点
    }
    @Override
```

```java
public void surfaceCreated(SurfaceHolder holder){
    flag=true;
    th=new Thread(this);                    //实例线程
    th.start();                             //启动线程
}
/**
 * 游戏绘图
 */
public void myDraw(){
    try {
        canvas=sfh.lockCanvas();
        if(canvas !=null){
            canvas.drawColor(Color.WHITE);
            //判断是否发生了碰撞
            if(isCollsion){                 //发生碰撞
                paint.setColor(Color.RED);
                paint.setTextSize(20);
                canvas.drawText("发生碰撞!", 0, 30, paint);
            } else{                         //没发生碰撞
                paint.setColor(Color.BLACK);
            }
            //绘制两个矩形
            canvas.drawRect(x1, y1, x1+w1, y1+h1, paint);
            canvas.drawRect(x2, y2, x2+w2, y2+h2, paint);
        }
    }catch(Exception e){
    }finally{
        if(canvas !=null)
            sfh.unlockCanvasAndPost(canvas);
    }
}
/**
 * 触屏事件监听
 */
@Override
public boolean onTouchEvent(MotionEvent event){
    //让矩形1随着触屏位置移动(触屏点设为此矩形的中心点)
    x2=(int)event.getX()-w2/2;
    y2=(int)event.getY()-h2/2;
    if(isCollsionWithRect(x1, y1, w1, h1, x2, y2, w2, h2)){
                                            //当矩形之间发生碰撞
        isCollsion=true;                    //设置标志位为真
    }else{                                  //当矩形之间没有发生碰撞
```

```
            isCollsion=false;                    //设置标志位为假
        }
        return true;
}
/**
 * 矩形碰撞的函数
 * @param x1    第一个矩形的 X 坐标
 * @param y1    第一个矩形的 Y 坐标
 * @param w1    第一个矩形的宽
 * @param h1    第一个矩形的高
 * @param x2    第二个矩形的 X 坐标
 * @param y2    第二个矩形的 Y 坐标
 * @param w2    第二个矩形的宽
 * @param h2    第二个矩形的高
 */
public boolean isCollsionWithRect(int x1, int y1, int w1, int h1, int x2, int y2, int w2, int h2){
    if(x1>=x2 && x1>=x2+w2){              //当矩形 1 位于矩形 2 的右侧
        return false;
    } else if(x1<=x2 && x1+w1<=x2){       //当矩形 1 位于矩形 2 的左侧
        return false;
    } else if(y1>=y2 && y1>=y2+h2){       //当矩形 1 位于矩形 2 的下方
        return false;
    } else if(y1<=y2 && y1+h1<=y2){       //当矩形 1 位于矩形 2 的上方
        return false;
    }
    //所有不会发生碰撞的情况都不满足时,肯定就是碰撞了
    return true;
}
/**
 * 游戏逻辑
 */
private void logic(){

}
@Override
public void run(){
    while(flag){
        long start=System.currentTimeMillis();
        myDraw();
        logic();
        long end=System.currentTimeMillis();
        try {
            if(end-start<50){
```

```
                Thread.sleep(50 - (end - start));
            }
        } catch(InterruptedException e){
            e.printStackTrace();
        }
    }
}
@Override
public void surfaceChanged(SurfaceHolder holder, int format, int width, int height)
{
    //TODO: Auto-generated method stub
}
@Override
public void surfaceDestroyed(SurfaceHolder holder){
    //TODO: Auto-generated method stub
}
}
```

步骤 2：在主活动的 onCreate()方法中,设置屏幕属性并调用 MySurfaceView 视图。代码如下：

```
@Override
protected void onCreate(Bundle savedInstanceState){
    super.onCreate(savedInstanceState);
    this.requestWindowFeature(Window.FEATURE_NO_TITLE);
    this.getWindow().setFlags(WindowManager.LayoutParams.FLAG_FULLSCREEN,
    WindowManager.LayoutParams.FLAG_FULLSCREEN);
    setContentView(new MySurfaceView(this));
}
```

运行效果如图 7-6 所示(拖动第 2 矩形接近第一个矩形)：

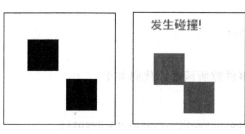

图 7-6　矩形碰撞检测实例效果

7.3.2　圆形碰撞检测

圆形碰撞检测是圆形和圆形的碰撞,要判断两个圆形是否发生重叠,应计算两个圆心之间的距离,看是否小于两个圆的半径之和。当两圆的圆心距小于两圆半径之和时,则发生了碰撞。

举例：圆形碰撞检测

参考矩形碰撞检测的代码,进行相应的修改即可。

步骤1：定义两个圆的半径和坐标。代码如下：

```
private int r1=20, r2=20;
private int x1=50, y1=100, x2=200, y2=150;
```

步骤2：在myDraw()函数中修改绘制矩形为绘制两个圆形。代码如下：

```
canvas.drawCircle(x1, y1, r1, paint);
canvas.drawCircle(x2, y2, r2, paint);
```

步骤3：重写圆形碰撞检测函数。代码如下：

```
/**
 * 圆形碰撞
 * @param x1    圆形 1 的圆心 X 坐标
 * @param y1    圆形 1 的圆心 Y 坐标
 * @param x2    圆形 2 的圆心 X 坐标
 * @param y2    圆形 2 的圆心 Y 坐标
 * @param r1    圆形 1 的半径
 * @param r2    圆形 2 的半径
 */
private boolean isCollsionWithCircle(int x1, int y1, int x2, int y2, int r1, int r2){
        //Math.sqrt:开平方
        //Math.pow(double x, double y):x 的 y 次方
        if(Math.sqrt(Math.pow(x1 - x2, 2)+Math.pow(y1 - y2, 2))<=r1+r2){
            //如果两圆的圆心距小于或等于两圆半径之和则认为发生碰撞
            return true;
        }
        return false;
}
```

步骤4：重写触屏事件监听函数。代码如下：

```
@Override
public boolean onTouchEvent(MotionEvent event){
    x2=(int)event.getX();
    y2=(int)event.getY();
    if(isCollsionWithCircle(x1, y1, x2, y2, r1, r2)){        //碰撞检测
        isCollsion=true;                                      //设置标志位为真
    }else{
        isCollsion=false;                                     //设置标志位为假
```

```
        }
        return true;
    }
}
```

运行效果如图 7-7 所示。

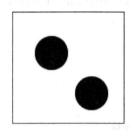

图 7-7　圆形碰撞检测实例效果

7.3.3　像素碰撞检测

首先遍历算出一张位图所有的像素点坐标，然后与另外一张位图上的所有点坐标进行对比，一旦有一个像素点的坐标相同，就立刻取出这两个坐标相同的像素点，通过位运算取出这两个像素点的最高位（透明度）进行对比，如果两个像素点都是非透明像素，则判定这两张位图发生碰撞。

举例：像素碰撞检测
参考矩形碰撞检测的代码，进行相应的修改即可。
步骤 1：定义两个矩形的矩形碰撞数组。代码如下：

```
private Rect clipRect1=new Rect(0, 0, 15, 15);
private Rect clipRect2=new Rect(rectW1 -15, rectH1 -15, rectW1, rectH1);
private Rect[] arrayRect1=new Rect[]{clipRect1, clipRect2};
private Rect clipRect3=new Rect(0, 0, 15, 15);
private Rect clipRect4=new Rect(rectW2 -15, rectH2 -15, rectW2, rectH2);
private Rect[] arrayRect2=new Rect[]{clipRect3, clipRect4};
```

步骤 2：在 myDraw() 函数中绘制两个矩形。代码如下：

```
//绘制两个矩形
canvas.drawRect(rectX1, rectY1, rectX1+rectW1, rectY1+rectH1, paint);
canvas.drawRect(rectX2, rectY2, rectX2+rectW2, rectY2+rectH2, paint);
//绘制碰撞区域为非填充，并设置画笔为白色
paint.setStyle(Style.STROKE);
paint.setColor(Color.WHITE);
//绘制第一个矩形的所有矩形碰撞区域
for(int i=0; i<arrayRect1.length; i++){
    canvas.drawRect(arrayRect1[i].left+this.rectX1, arrayRect1[i].top+this.rectY1,
        arrayRect1[i].right + this.rectX1, arrayRect1[i].bottom + this.rectY1,
```

```
            paint);
        }
        //绘制第二个矩形的所有矩形碰撞区域
        for(int i=0;i<arrayRect2.length; i++){
            canvas.drawRect(arrayRect2[i].left+this.rectX2, arrayRect2[i].top+this.
                rectY2, arrayRect2 [i]. right + this. rectX2, arrayRect2 [i]. bottom + this.
                rectY2, paint);
        }
```

步骤3：重写碰撞检测函数。代码如下：

```
/**
 * 矩形碰撞的函数
 * @param left      表示矩形左上角的 X 坐标
 * @param top       表示矩形左上角的 Y 坐标
 * @param right     表示矩形右下角的 X 坐标
 * @param buttom    表示矩形右下角的 Y 坐标
 */
public boolean isCollsionWithRect(Rect[] rectArray, Rect[] rect2Array){
    Rect rect=null;
    Rect rect2=null;
    for(int i=0; i<rectArray.length; i++){
        //依次取出第一个矩形数组的每个矩形实例
        rect=rectArray[i];
        //获取到第一个矩形数组中每个矩形元素的属性值
        int x1=rect.left+this.rectX1;
        int y1=rect.top+this.rectY1;
        int w1=rect.right - rect.left;
        int h1=rect.bottom - rect.top;
        for(int j=0; j<rect2Array.length; j++){
            //依次取出第二个矩形数组的每个矩形实例
            rect2=rect2Array[i];
            //获取到第二个矩形数组中每个矩形的属性值
            int x2=rect2.left+this.rectX1;
            int y2=rect2.top+this.rectY2;
            int w2=rect2.right - rect2.left;
            int h2=rect2.bottom - rect2.top;
            //进行循环遍历两个矩形碰撞数组所有元素之间的位置关系
            if(x1>=x2 && x1>x2+w2){
            }else if(x1<=x2 && x1+w1<=x2){
            }else if(y1>=y2 && y1>=y2+h2){
            }else if(y1<=y2 && y1+h1<=y2){
            }else {
                //只要有一个矩形碰撞数组与另一个矩形碰撞数组发生碰撞则认为发生碰撞
                return true;
```

                }
            }
        }
        return false;
    }

步骤4：重写触屏事件监听函数。代码如下：

```
@Override
public boolean onTouchEvent(MotionEvent event){
    //TODO: Auto-generated method stub
    //让矩形1随着触屏位置移动(触屏点设为此矩形的中心点)
    rectX1=(int)event.getX()-rectW1/2;
    rectY1=(int)event.getY()-rectH1/2;
    //当矩形之间发生碰撞
    if(isCollsionWithRect(arrayRect1, arrayRect2)){
        isCollsion=true;                    //设置标志位为真
    //当矩形之间没有发生碰撞
    }else{
        isCollsion=false;                   //设置标志位为假
    }
    return true;
}
```

运行效果如图7-8所示。

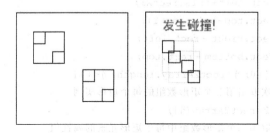

图7-8　像素碰撞检测实例效果

　　设置多个矩形碰撞区域,虽然精确,但会造成代码的效率降低,在游戏开发中不推荐使用,而应尽量使用多矩形、多圆形的检测方式代替像素碰撞检测。

　　对多个矩形进行碰撞检测,首先应该处理一个矩形的碰撞检测。关于矩形的碰撞检测,在之前介绍矩形碰撞时已经封装过方法。修改代码如下：

```
public boolean isCollsionWithRect(Rect rect, Rect rect2){
    int x1=rect.left;                    //x1, y1:矩形1的左上角
    int y1=rect.top;
    int w1=rect.right - rect.left;       //w1:矩形1的宽
    int h1=rect.bottom - rect.top;       //h1:矩形1的高
    int x2=rect2.left;                   //x2, y2:矩形2的左上角
```

```
            int y2=rect2.top;
            int w2=rect2.right - rect2.left;        //w2:矩形 2 的宽
            int h2=rect2.bottom - rect2.top;        //h2:矩形 2 的高
            if(x1>=x2 && x1>=x2+w2){
                return false;
            } else if(x1<=x2 && x1+w1<=x2){
                return false;
            } else if(y1>=y2 && y1>=y2+h2){
                return false;
            } else if(y1<=y2 && y1+h1<=y2){
                return false;
            }
            return true;
        }
```

封装好一个矩形的碰撞检测，对其进行修改，使之支持多矩形碰撞检测。代码如下：

```
        public boolean isCollsionWithRect(Rect[] rectArray, Rect[] rect2Array){
            Rect rect=null;
            Rect rect2=null;
            for(int i=0; i<rectArray.length; i++){
                //依次取出第一个矩形数组的每个矩形实例
                rect=rectArray[i];
                //获取第一个矩形数组中每个矩形元素的属性值
                int x1=rect.left+this.rectX1;
                int y1=rect.top+this.rectY1;
                int w1=rect.right - rect.left;
                int h1=rect.bottom - rect.top;
                for(int j=0; j<rect2Array.length; j++){
                    //依次取出第二个矩形数组的每个矩形实例
                    rect2=rect2Array[j];
                    //获取第二个矩形数组中每个矩形元素的属性值
                    int x2=rect2.left+this.rectX2;
                    int y2=rect2.top+this.rectY2;
                    int w2=rect2.right - rect2.left;
                    int h2=rect2.bottom - rect2.top;
                    //循环遍历两个矩形碰撞数组所有元素之间的位置关系
                    if(x1>=x2 && x1>=x2+w2){
                    } else if(x1<=x2 && x1+w1<=x2){
                    } else if(y1>=y2 && y1>=y2+h2){
                    } else if(y1<=y2 && y1+h1<=y2){
                    } else {
                        //只要有一个矩形碰撞数组与另一个矩形碰撞数组发生碰撞则认为发生碰撞
                        return true;
                    }
```

```
        }
    }
    return false;
}
```

上面的代码就是遍历两个矩形碰撞数组每个矩形之间的位置关系,一旦有一个矩形数组中的矩形与另外一个矩形数组的矩形发生碰撞,就可认为发生了多矩形碰撞。多圆形碰撞与多矩形碰撞是类似的。另外还有一种 Region 碰撞检测的方法。Region 是一个类,这个类比较常用的方法是判断一个点是否在矩形区域内,该方法是使用 Regions 类中的 contains(int x, int y)函数实现的。

7.4 游戏中的粒子系统

粒子系统是构造具有模糊形状物体的计算模型的方法,这一类物体包括一些自然界中常见的现象,如火焰、云、浪花、烟雾等,它们的共性是没有固定的形状,没有规则的几何外形,更重要的一点就是它们的外观会随着时间的推移而发生不确定的变化。

为了模拟粒子的生长和死亡的过程,每个粒子均有一定的生命周期,使其经历出生、成长、衰老和死亡的过程。因此,粒子系统要解决的问题就是粒子的存在和运动遵循的规则及其所受的作用。对粒子的作用可以分为两大类:

(1) 宏观作用。粒子作为一个整体要遵循的规则,这类作用相对简单一些。例如,物体要受重力的作用,那么构成物体的每一个粒子都要受重力的作用。可以认为这类作用的规则就是物体所受作用均匀分布到每一个粒子。但宏观作用需要根据实际情况来判断,在特殊情况下,数值也可以是不均匀的,例如风的效果对于每一个雪花粒子和雨粒子的影响就是不一样的。

(2) 微观作用。对于物体整体而言,这种作用是不存在或没有意义的;但是对于构成物体的单个粒子本身而言,就应该考虑这种作用。例如,火焰粒子表示的就是"炙热的碳粒"。对于某个粒子,如果处于高密度区,它就会向低密度区扩散,例如气体粒子和烟粒子。还有些粒子之间由于相互碰撞而导致的能量交换也属于这种作用。由于粒子的数目通常可以达到几千,多的甚至达到数十万,所以在对它们进行动态变化和渲染的计算中需要采用简化的假设。

举例:喷泉粒子效果

步骤 1:新建类文件 Particle.java 作为单个粒子对象,声明用于计算粒子位置的相关变量,编写初始化成员变量的构造函数。代码如下:

```
public class Particle{
    private int color;                      //粒子颜色
    int r;                                  //粒子半径
    double vertical_v;                      //垂直速度
    double horizontal_v;                    //水平速度
    int startX;                             //开始位置的 X 坐标
```

```java
    int startY;                                      //开始位置的Y坐标
    int x;                                           //当前位置的X坐标
    int y;                                           //当前位置的Y坐标
    double startTime;                                //起始时间
    public Particle(int color, int r, double vertical_v, double horizontal_v, int
    x, int y, double startTime){
        this.color=color;                            //初始化粒子颜色
        this.r=r;                                    //初始化粒子半径
        this.vertical_v=vertical_v;                  //初始化垂直速度
        this.horizontal_v=horizontal_v;              //初始化水平速度
        this.startX=x;                               //初始化开始位置的X坐标
        this.startY=y;                               //初始化开始位置的Y坐标
        this.x=x;                                    //初始化实时X坐标
        this.y=y;                                    //初始化实时Y坐标
        this.startTime=startTime;                    //初始化开始运动的时间
    }
}
```

步骤2：新建类文件ParticleSet.java作为粒子集对象来存放所有粒子，并提供成员方法用于管理和添加粒子对象。代码如下：

```java
public class ParticleSet{
    ArrayList<Particle>particleSet;                  //声明存放Particle对象的集合
    public ParticleSet(){                            //构造函数(实例化粒子集合)
        particleSet=new ArrayList<Particle>();
    }
    /**
     * 向粒子集合中添加指定个数的粒子对象
     * @param count 粒子个数
     * @param startTime 创建时间
     */
    public void add(int count, double startTime){
        for(int i=0; i<count; i++){                  //创建Particle对象
            int tempColor=this.getColor(i);          //获得粒子颜色
            int tempR=1;                             //粒子半径
            double tempv_v=-30+10 * (Math.random()); //产生粒子垂直方向的速度
            double tempv_h=10 - 20 * (Math.random());//产生粒子水平方向的速度
            int tempX=160;                           //粒子的X坐标是固定的
            int tempY=(int)(100 -10 * (Math.random())); //产生粒子的Y坐标
            //创建Particle对象
            Particle particle=new Particle(tempColor, tempR, tempv_v, tempv_h,
            tempX, tempY, startTime);
            particleSet.add(particle);               //添加创建好的Particle对象到列表中
        }
    }
```

```java
/**
 * 获取指定索引的颜色
 * @param i 颜色标识
 * @return 得到的颜色
 */
public int getColor(int i){
    int color=Color.RED;
    switch(i%4){
    case 0:
        color=Color.RED;                    //将颜色设为红色
        break;
    case 1:
        color=Color.GREEN;                  //将颜色设为绿色
        break;
    case 2:
        color=Color.YELLOW;                 //将颜色设为黄色
        break;
    case 3:
        color=Color.GRAY;                   //将颜色设为灰色
        break;
    }
    return color;                           //返回得到的颜色
}
```

步骤 3：新建类文件 ParticleThread.java（继承自 Thread 类），用于对粒子集对象 ParticleSet 进行操作，主要功能是添加粒子并修改粒子的位置。代码如下：

```java
public class ParticleThread extends Thread{
    boolean flag;                           //线程执行标志位
    ParticleView father;                    //ParticleView对象引用
    int sleepSpan=80;                       //线程休眠时间
    double time=0;                          //物理引擎的时间轴
    double span=0.15;                       //每次计算粒子的位移时采用的时间间隔
    public ParticleThread(ParticleView father){
        this.father=father;
        this.flag=true;                     //设置线程执行标志位为true
    }
    public void run(){
        while(flag){
            father.ps.add(5, time);         //每次添加5个粒子
            ArrayList<Particle>tempSet=father.ps.particleSet;
                                            //获取粒子集合
            int count=tempSet.size();       //记录粒子集合的大小
            for(int i=0; i<count; i++){     //遍历粒子集合,修改其轨迹
```

```
            Particle particle=tempSet.get(i);
            double timeSpan=time -particle.startTime;
                                            //开始到现在的时间间隔
//计算粒子的 X 坐标
            int tempx=(int)(particle.startX+particle.horizontal_v*timeSpan);
            int tempy = (int)(particle.startY + 4.9 * timeSpan * timeSpan +
                particle.vertical_v*timeSpan);
            if(tempy>ParticleView.DIE_OUT_LINE){
                                            //如果超过屏幕下边沿
                tempSet.remove(particle);   //从粒子集合移除该 Particle 对象
                count=tempSet.size();       //重新设置粒子个数
            }
            particle.x=tempx;               //修改粒子的 X 坐标
            particle.y=tempy;               //修改粒子的 Y 坐标
        }
        time+=span;                         //将时间延长
        try{
            Thread.sleep(sleepSpan);        //休眠一段时间
        }
        catch(Exception e){
            e.printStackTrace();            //捕获并打印异常
        }
    }
  }
}
```

以上代码没有采用获得系统时间的方法,而是自定义一个时间轴,这样可以方便地确定时间轴行进的快慢程度,而不必依赖系统时间。

步骤 4:新建类文件 ParticleView.java(继承自 SurfaceView 类),声明成员变量,添加构造函数,实现 SurfaceHolder.Callback 接口并编写绘制方法。代码如下:

```
public class ParticleView extends SurfaceView implements SurfaceHolder.
Callback{
    public static final int DIE_OUT_LINE=420;
                        //粒子的 Y 坐标(超过屏幕下边沿时会从粒子集合移除)
    DrawThread dt;                          //后台刷新屏幕线程
    ParticleSet ps;                         //ParticleSet 对象引用
    ParticleThread pt;                      //ParticleThread 对象引用
    String fps="FPS:N/A";                   //声明帧速率字符串
    public ParticleView(Context context){
        super(context);                     //调用父类构造器
        this.getHolder().addCallback(this); //添加 Callback 接口
        dt=new DrawThread(this,getHolder());//创建 DrawThread 对象
        ps=new ParticleSet();               //创建 ParticleSet 对象
```

```java
            pt=new ParticleThread(this);            //创建ParticleThread对象
        }
        //方法:绘制屏幕
        public void doDraw(Canvas canvas){
            canvas.drawColor(Color.WHITE);          //清屏
            //获得ParticleSet对象中的粒子集合对象
            ArrayList<Particle>particleSet=ps.particleSet;
            Paint paint=new Paint();                //创建画笔对象
            for(int i=0;i<particleSet.size();i++){  //遍历粒子集合,绘制每个粒子
                Particle p=particleSet.get(i);
                paint.setColor(p.color);            //设置画笔颜色为粒子颜色
                int tempX=p.x;                      //获得粒子的X坐标
                int tempY=p.y;                      //获得粒子的Y坐标
                int tempRadius=p.r;                 //获得粒子半径
                RectF oval=new RectF(tempX, tempY, tempX+2*tempRadius, tempY+2*
                tempRadius);
                canvas.drawOval(oval, paint);       //绘制椭圆粒子
            }
            paint.setColor(Color.WHITE);            //设置画笔颜色
            paint.setTextSize(18);                  //设置文字大小
            paint.setAntiAlias(true);               //设置抗锯齿
            canvas.drawText(fps, 15, 15, paint);    //画出帧速率字符串
        }
        @Override
        public void surfaceChanged(SurfaceHolder holder, int format, int width,
        int height){
        }
        @Override
        public void surfaceCreated(SurfaceHolder holder){
            if(!dt.isAlive()){                      //判断DrawThread是否启动
                dt.start();
            }
            if(!pt.isAlive()){                      //判断ParticleThread是否启动
                pt.start();
            }
        }
        @Override
        public void surfaceDestroyed(SurfaceHolder holder){
            dt.flag=false;                          //停止线程的执行
            dt=null;                                //将dt指向的对象声明为null
        }
}
```

步骤5:新建类文件ParticleView.java(继承自Thread类),用于刷新屏幕。代码

如下：

```java
public class DrawThread extends Thread{
    ParticleView pv;                                    //声明 ParticleView 对象
    SurfaceHolder surfaceHolder;                        //声明 SurfaceHolder 对象
    boolean flag;                                       //标识线程执行
    int sleepSpan=15;                                   //睡眠时间
    long start=System.nanoTime();                       //记录起始时间(用于计算帧速率)
    int count=0;                                        //记录帧数(用于计算帧速率)
    public DrawThread(ParticleView pv, SurfaceHolder surfaceHolder){
        this.pv=pv;
        this.surfaceHolder=surfaceHolder;
        this.flag=true;                                 //设置线程执行标志位为 true
    }
    public void run(){                                  //线程执行方法
        Canvas canvas=null;                             //声明一个 Canvas 对象
        while(flag){
            try{
                canvas=surfaceHolder.lockCanvas(null);
                                                        //获取 ParticleView 的画布
                synchronized(surfaceHolder){
                    pv.doDraw(canvas);    //调用 ParticleView 的 doDraw 方法进行绘制
                }
            }
            catch(Exception e){
                e.printStackTrace();
            }
            finally{
                if(canvas !=null){                      //如果 canvas 不为空
                 surfaceHolder.unlockCanvasAndPost(canvas);
                }
            }
            this.count++;
            if(count==20){                              //如果计满 20 帧
                count=0;                                //清空计数器
                long tempStamp=System.nanoTime();       //获取当前时间
                long span=tempStamp - start;            //获取时间间隔
                start=tempStamp;                        //为 start 重新赋值
                double fps=Math.round(100000000000.0/span * 20)/100.0;
                                                        //计算帧速率
                //将计算出的帧速率设置到 BallView 的相应字符串对象
                pv.fps="FPS:"+fps;
            }
            try{
                Thread.sleep(sleepSpan);                //线程休眠一段时间
            }
            catch(Exception e){
```

```
            e.printStackTrace();
        }
    }
  }
}
```

步骤 6：在主活动的 onCreate()方法中，设置屏幕属性并调用 MyView 视图。代码如下：

```
@Override
protected void onCreate(Bundle savedInstanceState){
    super.onCreate(savedInstanceState);
    …                                    //此处省略隐藏标题栏及全屏的代码
    ParticleView gameView=new ParticleView(this);
    setContentView(gameView);
}
```

运行效果如图 7-9 所示。

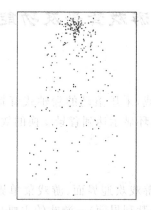

图 7-9 喷泉粒子实例运行效果

7.5 本章小结

本章主要讲述了游戏编程中常用的数学知识和物理学知识，并举例说明了这些知识在实际开发中的应用。另外，重点讲解了矩形、圆形、像素 3 种碰撞检测方法，介绍了游戏开发中模拟流体的粒子特效制作方法，这是游戏开发中的重点内容。

7.6 思考与练习

（1）尝试开发碰碰车游戏。可以通过触屏或重力感应控制小车在场景中跑动，当撞到设置的障碍物时要有较真实的物理效果。

（2）尝试开发粒子特效，模拟五彩烟花喷射的效果。

第 8 章 案例演练——疯狂战机

本游戏是一款操控飞机进行战斗的射击类游戏,游戏中的音效和场景的渲染赋予了游戏紧张、激烈的气氛。操作简单,玩家可以通过控制角色使用不同武器消灭敌人来过关。其中融入了动作类游戏中常用的技术,画面流畅,声音震撼,具有较好的可玩性和丰富的用户体验。

8.1 游戏背景及功能概述

8.1.1 游戏类型

本游戏属于滚屏射击类游戏,采用滚动的卷轴式背景,将一幅图片首尾相接作为背景,在游戏过程中通过不停地循环显示达到背景变换的效果。

8.1.2 功能简介

游戏的运行过程主要包含游戏欢迎界面、游戏菜单界面和游戏运行界面(包括游戏战斗界面、游戏失败界面和游戏胜利界面)。游戏的主要功能如下:

(1) 启动运行游戏,进入欢迎动画界面。
(2) 欢迎动画播放完毕将进入菜单界面,包括"开始游戏"、"继续游戏"、"游戏帮助"、"声音设置"、"退出游戏"5 个按钮。
(3) 单击"游戏帮助"按钮将进入游戏帮助界面。
(4) 单击"声音设置"按钮将打开"声音设置"窗口。
(5) 单击"开始游戏"按钮进入正式游戏界面。在游戏界面,玩家可以通过触屏控制飞机移动,飞机自动发射子弹攻击敌机。
(6) 玩家在已有的生命值和复活次数范围内消灭所有敌机,将通过当前关卡,显示胜利界面。

8.2 游戏的策划及准备工作

本节主要介绍空战游戏的策划、开发前的准备工作以及图片和音效等素材的搜集。

8.2.1 游戏的策划

下面简单介绍本游戏的策划,游戏开发中涉及的方面很多,此处只列出游戏情境、采用的呈现技术、操作方式、音效设计和目标平台几个部分。

1. 游戏情境

本游戏作为一个简单的射击类游戏案例,没有过多的情境设计,主要的工作是主战飞机(即玩家战机)生命的设计(生命值)、道具设计(子弹道具及补血的道具)等。

2. 采用的呈现技术

本游戏案例的表现形式采用竖向滚动的卷轴式背景设计,采用 2D 呈现技术,游戏中的场景采用多层贴图,增加了游戏界面的层次感。

3. 操作方式

本游戏案例采用触屏方式控制主战飞机的飞行路线。

4. 音效设计

在游戏的运行过程中,根据不同的界面添加了适当的声音效果,例如背景音乐、发射炮弹时的音效及爆炸音效等。

5. 目标平台

目标平台为 Android 4.0 以及上版本。

8.2.2 Android 平台下游戏的准备工作

准备工作包括搜集图片、声音等素材,并将资源文件放到项目的指定位置。本案例使用的图片素材清单如表 8-1 所示。

表 8-1 游戏图片素材清单

图片名称	缩 略 图	用 途
map.png		滚动背景

续表

图片名称	缩略图	用途
logo.png		闪屏图片
pbullet0.png		玩家战机子弹
pbullet1.png		玩家战机子弹
pbullet2.png		玩家战机子弹
bullet0.png		敌机子弹
bullet1.png		敌机子弹
bullet2.png		敌机子弹
bullet3.png		敌机子弹
font_win.png		胜利图片
enemy0.png		敌机
enemy1.png		敌机

续表

图片名称	缩略图	用途
enemy2.png		敌机
enemy3.png		敌机
enemy4.png		敌方 BOSS
number.png		生命值
pbomb.png		爆炸效果
player.png		玩家战机
property.png		奖励道具
font_lose.png		失败图片

案例使用的声音素材清单如表 8-2 所示。

表 8-2 游戏声音素材清单

声音文件名称	用途
game.mid	游戏背景音乐
explosion.wav	游戏中的爆炸声
menu.wav	菜单按键音

8.3 游戏的架构

8.3.1 游戏中各个类的简介

本案例涉及的实体相关类如下：
（1）主战飞机类 Plane，为主战飞机的封装类。
（2）敌机类 Enemy，为敌机的封装类。
（3）子弹类 Bullet，为子弹类的封装类，游戏中的所有子弹均为该类的对象。
（4）道具类 Property，为补血道具类，当主战飞机与该类对象碰撞时，为主战飞机补血。

本案例涉及的游戏界面相关类如下：
（1）游戏显示类 PlaneGameActivity，负责调用 GameView 类，启动游戏界面。
（2）游戏主界面类 GameView，为游戏菜单界面的实现类，显示菜单界面。
（3）游戏界面绘制类 GameScreen，负责背景滚动、界面初始化、显示状态信息及界面逻辑处理。
（4）菜单界面类 MenuScreen，用于菜单界面、声音设置界面及帮助界面的绘制。
（5）数据存储类 GameStore，负责游戏状态数据的存储与调用。

本案例涉及的辅助类如下：
（1）工具类 Tools，实现在指定位置绘图及碰撞检测方法。
（2）声音类 GameMusic，用于控制背景音乐及音效的播放。

8.3.2 游戏运行界面

滚屏类游戏是最传统的游戏类型，主要考验玩家的反应能力和手眼配合能力，游戏的情节不是重点，而快速的游戏节奏、激烈火爆的场面、良好的操作才是最重要的。

本游戏主要界面的效果如图 8-1 至图 8-5 所示。

图 8-1　游戏欢迎界面

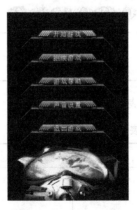
图 8-2　游戏菜单界面

图 8-3　游戏战斗界面

图 8-4　游戏失败界面

图 8-5　游戏胜利界面

8.4 游戏中的实体相关类

8.4.1 主战飞机类 Plane

在游戏主界面的设计中需要用到主战飞机类的对象。在 Plane 类中封装了主战飞机的有关信息以及碰撞检测等方法。代码如下：

```java
public class Player {
    public static final int UP=0;                    //标识飞机向上移动状态
    public static final int DOWN=1;                  //标识飞机向下移动状态
    public static final int LEFT=2;                  //标识飞机向左移动状态
    public static final int RIGHT=3;                 //标识飞机向右移动状态
    public static final int NORMAL=0;                //标识飞机正常状态
    public static final int LEVELUP=1;               //升级
    public static final int BOOM=2;                  //音效
    public static final int RECOVER=3;               //补血
    public static final int SKILL_BOOM=4;            //加弹
    final int MAX_HP=285;                            //飞机最大生命值
    private int playerState;                         //飞机状态
    private int x, y;                                //飞机的X、Y坐标
    private int speed;                               //飞机的速度
    private int lifeNum;                             //生命条数
    private int hp;                                  //血量
    private int width, height;                       //飞机的宽高
    private int direction;
    private int frame;                               //当前的画面帧
    private boolean isMove;
    private int level;                               //飞机的等级(决定子弹的类型)
    private int atk;                                 //子弹的攻击力
    private int propertyNum;                         //炸弹道具的数量
    Bitmap plane;
    GameScreen screen;
    int timeCount;
    public Player(Bitmap plane, GameScreen screen){
        this.screen=screen;
        this.plane=plane;
        width=plane.getWidth()/6;
        height=plane.getHeight();
        init();
    }
    void setData(int x, int y, int lifeNum, int hp, int direction, int level, int atk, int propertyNum)
```

```java
    {
        this.x=x;
        this.y=y;
        this.lifeNum=lifeNum;
        this.hp=hp;
        this.direction=direction;
        this.level=level;
        this.atk=atk;
        this.propertyNum=propertyNum;
    }
    int getDirection()
    {
        return direction;
    }
    //返回炸弹道具总数
    int getPropertyNum()
    {
        return propertyNum;
    }
    //炸弹道具增加方法
    void addPropertyNum()
    {
        if(propertyNum<3)
            propertyNum++;
    }
    //初始化方法
    void init()
    {
        playerState=NORMAL;
        direction=UP;
        frame=2;
        x=GameView.SCREEN_WIDTH/2-width/2;
        y=GameView.SCREEN_HEIGHT-height-40;
        speed=15;
        lifeNum=3;
        hp=MAX_HP;
        isMove=false;
        level=0;
        atk=20;
        propertyNum=1;
    }
    int getState()
    {
        return playerState;
```

```java
    }
    void setState(int state)
    {
        switch(state){
        case LEVELUP:
            if(level<2)
            {
                level=level+1;
                atk+=2*atk;
                playerState=LEVELUP;
                timeCount=0;
            }
            break;
        case RECOVER:
            hp+=100;
            playerState=RECOVER;
            timeCount=0;
            break;
        case SKILL_BOOM:
            if(propertyNum>=1)
            {
                playerState=SKILL_BOOM;
                propertyNum--;
                screen.bullet.createPlayerBoom();
            }
            break;
        case NORMAL:
            playerState=NORMAL;
            break;
        }
    }
    int getAtk(){return atk;}
    int getMaxHp(){return MAX_HP;}
    int getWidth(){return width;}
    int getHeight() {return height;}
    int getX(){return x;}
    int getY(){return y;}
    void setLevelUp(){
        if(level<2)
            level++;
    }
    int getLevel(){return level;}
    int getLifeNum(){return lifeNum;}
    int getHp() {return hp;}
```

```
void setHp(int atk)
{
    if(hp>atk)
        hp-=atk;
    else
    {    hp=0;
        if(lifeNum>1)
        {
            lifeNum-=1;
            playerState=BOOM;
            frame=0;
        }else
            screen.setFont(screen.LOSE);
    }
}
void changeFrame()
{
    switch(playerState){
    case NORMAL:
        switch(direction){
        case UP:
        case DOWN:
            frame=(frame==2?3:2);
            break;
        case LEFT:
            frame=(frame==0?1:0);
            break;
        case RIGHT:
            frame=(frame==42?5:4);
            break;
        }
        break;
    case LEVELUP:
        if(timeCount==10)
            playerState=NORMAL;
        else
            timeCount++;
        switch(direction){
        case UP:
        case DOWN:
            frame=(frame==2?3:2);
            break;
        case LEFT:
            frame=(frame==0?1:0);
```

```
            break;
        case RIGHT:
            frame=(frame==42?5:4);
            break;
        }
        break;
    case RECOVER:
        if(timeCount==20)
            playerState=NORMAL;
        else
            timeCount++;
        switch(direction){
        case UP:
        case DOWN:
            frame=(frame==2?3:2);
            break;
        case LEFT:
            frame=(frame==0?1:0);
            break;
        case RIGHT:
            frame=(frame==42?5:4);
            break;
        }
        break;
    case BOOM:
        if(frame==5)
        {
            playerState=NORMAL;
            hp=MAX_HP;
        }else
            frame++;
        break;
    }
}
void move()
{
    if(isMove)
        switch(playerState){
        case NORMAL:
        case LEVELUP:
        case RECOVER:
            switch(direction){
            case UP:
                if(y>=speed)
```

```
                    y-=speed;
                break;
            case DOWN:
                if(y<=GameView.SCREEN_HEIGHT-height-speed)
                    y+=speed;
                break;
            case LEFT:
                if(x>=speed)
                    x-=speed;
                break;
            case RIGHT:
                if(x<=GameView.SCREEN_WIDTH-width-speed)
                    x+=speed;
                break;
            }
            break;
        default:
            break;
        }
    }
    void changeDirection(int direction)
    {
        isMove=true;
        this.direction=direction;
    }
    void stop()
    {
        isMove=false;
        direction=UP;
    }
    //根据游戏状态绘制游戏界面
    void paint(Canvas canvas, Paint paint)
    {
        switch(playerState){
        case NORMAL:
        Tools.drawImage(plane, x-frame*width, y, x, y, width, height, canvas, paint);
            break;
        case LEVELUP:
            if(timeCount%2==0)
                canvas.drawBitmap(screen.fontImg[2], 0, 90, paint);
        Tools.drawImage(plane, x-frame*width, y, x, y, width, height, canvas, paint);
            break;
```

```
            case RECOVER:
                if(timeCount%2==0)
                canvas.drawBitmap(screen.fontImg[3], 0, 90, paint);
                Tools.drawImage(screen.recoverImg, x-12-timeCount%2 * 60, y-10, x-
                    12, y-10, 60, 60, canvas, paint);
            Tools.drawImage(plane, x-frame * width, y, x, y, width, height, canvas,
                paint);
                break;
            case BOOM:
                Tools.drawImage(screen.enemy.boomImg,
                x-frame * screen.enemy.boomImg.getWidth()/6, y, x, y, screen.enemy.
                boomImg.getWidth()/6, screen.enemy.boomImg.getHeight(), canvas,
                paint);
                break;
        }
    }
    //飞机逻辑控制
    void logic()
    {
        changeFrame();
        move();
    }
}
```

8.4.2 敌机类 Enemy

在游戏战斗场景中，需要用到敌机类 Enemy 的对象，并且按照一定的轨迹运动。在敌机类 Enemy 中同样封装了相关信息及功能，用数组对象池管理敌机。代码如下：

```
public class Enemy {
    //敌机类型
    static final int TYPE_YELLOW=0;
    static final int TYPE_RED=1;
    static final int TYPE_PURPLE=2;
    static final int TYPE_GREEN=3;
    static final int TYPE_BOSS=4;
    //敌机属性下标值
    final int SHOW=1;
    final int UN_SHOW=0;
    final int BOOM=2;
    final int ISVISIABLE=0;              //下标对应的值(0 为不可见, 1 为可见)
    final int TYPE=1;
    final int X=2;
    final int Y=3;
```

```java
final int WIDTH=4;
final int HEIGHT=5;
final int SPEED_X=6;
final int SPEED_Y=7;
final int HP=8;
final int SETP_COUNT=9;
final int SETP_INDEX=10;
final int ATK=11;
final int FRAME=12;
GameScreen gameScreen;
int enemy[][];                                    //敌机对象池数组
Bitmap eImg[];                                    //敌机图片数组
Bitmap boomImg;                                   //爆炸图片
int timeCount;
Random random;
//不同类型敌机的初始位置
int randomX[][]={{50, 120, 175 , 260}, {35, 250}, {65, 215}, {0, 280, 60, 220}, {}};
//紫色敌机飞行路径设定数组
int purpleStep[]={15, 5, 10, 5, 8};
Property property;
public Enemy(GameScreen gameScreen){
    this.gameScreen=gameScreen;
    property=gameScreen.property;
    random=new Random();
    enemy=new int[20][13];
    eImg=new Bitmap[5];
    eImg[0]=BitmapFactory.decodeResource(gameScreen.gameView.
        getResources(), R.drawable.enemy0);
    eImg[1]=BitmapFactory.decodeResource(gameScreen.gameView.
        getResources(), R.drawable.enemy1);
    eImg[2]=BitmapFactory.decodeResource(gameScreen.gameView.
        getResources(), R.drawable.enemy2);
    eImg[3]=BitmapFactory.decodeResource(gameScreen.gameView.
        getResources(), R.drawable.enemy3);
    eImg[4]=BitmapFactory.decodeResource(gameScreen.gameView.
        getResources(), R.drawable.enemy4);
    boomImg=BitmapFactory.decodeResource(gameScreen.gameView.
        getResources(), R.drawable.boomimg);
}
//初始化敌机对象池
void init()
{
    for(int i=0; i<enemy.length; i++){
        for(int j=0; j<enemy[i].length; j++)
```

```java
            enemy[i][ISVISIABLE]=UN_SHOW;  //不可视
        }
    }
//绘制敌机
void paint(Canvas canvas, Paint paint)
{
    for(int i=0; i<enemy.length; i++){
        switch(enemy[i][ISVISIABLE]){
        case SHOW:
            int type=enemy[i][TYPE];
            canvas.drawBitmap(eImg[type], enemy[i][X], enemy[i][Y], paint);
            break;
        case BOOM:
            Tools.drawImage (boomImg, enemy [i] [X] - enemy [i] [FRAME] *
            boomImg.getWidth()/6, enemy[i][Y], enemy[i][X], enemy[i][Y],
            boomImg.getWidth()/6, boomImg.getHeight(), canvas, paint);
            break;
        }
    }
}
//移动方法
void move()
{
    for(int i=0; i<enemy.length; i++){
        if(enemy[i][ISVISIABLE]==SHOW)
        {
            enemy[i][X]+=enemy[i][SPEED_X];
            enemy[i][Y]+=enemy[i][SPEED_Y];
        }
    }
}
//创建敌机
void createEnemy(int type)
{
    switch(type){
    case TYPE_YELLOW:                              //直线飞行
        for(int i=0; i<enemy.length; i++){
            if(enemy[i][ISVISIABLE]==UN_SHOW)
            {
                enemy[i][ISVISIABLE]=SHOW;
                enemy[i][TYPE]=TYPE_YELLOW;
                enemy[i][HP]=18;
                enemy[i][WIDTH]=eImg[type].getWidth();
                enemy[i][HEIGHT]=eImg[type].getHeight();
```

```java
                enemy[i][X]=randomX[TYPE_YELLOW][Math.abs(random.nextInt
                ()%4)];
                enemy[i][Y]=-enemy[i][HEIGHT];
                enemy[i][SPEED_X]=0;
                enemy[i][SPEED_Y]=8;
                enemy[i][ATK]=2*enemy[i][HP];
                enemy[i][FRAME]=0;
                break;
            }
        }
        break;
    case TYPE_GREEN:                                    //交叉飞行
        int count=0;
        int index=Math.abs(random.nextInt()%2);
        for(int i=0; i<enemy.length; i++){
            if(enemy[i][ISVISIABLE]==UN_SHOW)
            {
                enemy[i][ISVISIABLE]=SHOW;
                enemy[i][TYPE]=TYPE_GREEN;
                enemy[i][HP]=20;
                enemy[i][WIDTH]=eImg[type].getWidth();
                enemy[i][HEIGHT]=eImg[type].getHeight();
                if(index==0)
                    enemy[i][X]=randomX[TYPE_GREEN][count];
                else
                    enemy[i][X]=randomX[TYPE_GREEN][count+2];
                enemy[i][Y]=-enemy[i][HEIGHT];
                enemy[i][SPEED_X]=8+count*-16;
                enemy[i][SPEED_Y]=15;
                enemy[i][ATK]=30;
                enemy[i][FRAME]=0;
                count++;
                if(count==2)
                    break;
            }
        }
        break;
    case TYPE_RED:                                      //来回横扫
        index=Math.abs(random.nextInt()%2);
        for(int i=0; i<enemy.length; i++){
            if(enemy[i][ISVISIABLE]==UN_SHOW)
            {
                enemy[i][ISVISIABLE]=SHOW;
                enemy[i][TYPE]=TYPE_RED;
```

```
                enemy[i][HP]=24;
                enemy[i][WIDTH]=eImg[type].getWidth();
                enemy[i][HEIGHT]=eImg[type].getHeight();
                enemy[i][X]=randomX[TYPE_RED][index];
                enemy[i][Y]=-enemy[i][HEIGHT];
                enemy[i][SPEED_X]=14+index*-28;
                enemy[i][SPEED_Y]=6;
                enemy[i][ATK]=2*enemy[i][HP];
                enemy[i][FRAME]=0;
                break;
            }
        }
        break;
    case TYPE_PURPLE:                    //前进后退，左右小范围扫射
        count=0;
        for(int i=0; i<enemy.length; i++){
            if(enemy[i][ISVISIABLE]==UN_SHOW)
            {
                enemy[i][ISVISIABLE]=SHOW;
                enemy[i][TYPE]=TYPE_PURPLE;
                enemy[i][HP]=35;
                enemy[i][WIDTH]=eImg[type].getWidth();
                enemy[i][HEIGHT]=eImg[type].getHeight();
                enemy[i][X]=randomX[TYPE_PURPLE][count];
                enemy[i][Y]=-enemy[i][HEIGHT];
                enemy[i][SPEED_X]=0;
                enemy[i][SPEED_Y]=12;
                enemy[i][SETP_INDEX]=0;
                enemy[i][SETP_COUNT]=0;
                enemy[i][ATK]=2*enemy[i][HP];
                enemy[i][FRAME]=0;
                count++;
                if(count==2)
                    break;
            }
        }
        break;
    case TYPE_BOSS:
        for(int i=0;i<enemy.length;i++)
            enemy[i][ISVISIABLE]=UN_SHOW;
        int i=0;
        enemy[i][ISVISIABLE]=SHOW;
        enemy[i][TYPE]=TYPE_BOSS;
        enemy[i][HP]=1000;
```

```
                enemy[i][WIDTH]=eImg[type].getWidth();
                enemy[i][HEIGHT]=eImg[type].getHeight();
                enemy[i][X]=GameView.SCREEN_WIDTH/2-enemy[i][WIDTH]/2;
                enemy[i][Y]=-enemy[i][HEIGHT];
                enemy[i][SPEED_X]=0;
                enemy[i][SPEED_Y]=12;
                enemy[i][SETP_INDEX]=0;
                enemy[i][SETP_COUNT]=0;
                enemy[i][ATK]=enemy[i][HP]/2;
                enemy[i][FRAME]=0;
                break;
        }
    }
}
//敌机回池
void setVisiable()
{
    for(int i=0; i<enemy.length; i++){
        if(enemy[i][ISVISIABLE]==SHOW)
        {
            if(enemy[i][X]+enemy[i][WIDTH]<=0 || enemy[i][X]>=GameView.
            SCREEN_WIDTH || enemy[i][Y]>=GameView.SCREEN_HEIGHT)
            {
                enemy[i][ISVISIABLE]=UN_SHOW;
            }
        }
    }
}
//碰撞检测
void collidesWith()
{
    for(int i=0; i<enemy.length; i++){
        switch(enemy[i][ISVISIABLE]){
        case SHOW:
            switch(gameScreen.player.getState()){
            case Player.NORMAL:
                if(Tools.collides(enemy[i][X], enemy[i][Y], enemy[i]
                    [WIDTH], enemy[i][HEIGHT], gameScreen.player.getX(),
                    gameScreen.player.getY(), gameScreen.player.getWidth(),
                    gameScreen.player.getHeight()))
                {
                    if(enemy[i][TYPE] !=TYPE_BOSS)
                        enemy[i][ISVISIABLE]=BOOM;
                    else
                        enemy[i][HP] -=gameScreen.player.getAtk() * 2;
```

```java
                    gameScreen.player.setHp(enemy[i][ATK]);
                }
                break;
            default:
                break;
            }
            break;
        case BOOM:
            if(enemy[i][FRAME]==5)
            {
                int ranNum=Math.abs(random.nextInt()%10)+1;
                switch(enemy[i][TYPE]){
                case Enemy.TYPE_GREEN:
                    if(ranNum<=3)
                    property.createProperty(Property.POWER, enemy[i][X],
                    enemy[i][Y]);
                    break;
                case Enemy.TYPE_PURPLE:
                    if(ranNum>=6 && ranNum<=7)
                    property.createProperty(Property.BOOM, enemy[i][X],
                    enemy[i][Y]);
                    break;
                case Enemy.TYPE_RED:
                    if(ranNum>=8 && ranNum<=10)
                    property.createProperty(Property.LEVER_UP, enemy[i][X],
                    enemy[i][Y]);
                    break;
                }
                enemy[i][ISVISIABLE]=UN_SHOW;
            }
            else
                enemy[i][FRAME]++;
            break;
        }
    }
}
//设置敌机飞行路径
void setPath()
{
    for(int i=0;i<enemy.length;i++)
    {
        if(enemy[i][ISVISIABLE]==SHOW)
        {
            switch(enemy[i][TYPE]){
```

```
case TYPE_RED:
    if(enemy[i][X]>=GameView.SCREEN_WIDTH-enemy[i][WIDTH])
        enemy[i][SPEED_X]=-14;
    else if(enemy[i][X]<=8)
        enemy[i][SPEED_X]=14;
    break;
case TYPE_PURPLE:
    enemy[i][SETP_COUNT]++;
    switch(enemy[i][SETP_INDEX]){
    case 0:                        //向下
        if(enemy[i][SETP_COUNT]==purpleStep[0])
        {
            enemy[i][SETP_COUNT]=0;
            enemy[i][SETP_INDEX]=1;
            enemy[i][SPEED_X]=-8;
            enemy[i][SPEED_Y]=0;
        }
        break;
    case 1:                        //向左
        if(enemy[i][SETP_COUNT]==purpleStep[1])
        {
            enemy[i][SETP_COUNT]=0;
            enemy[i][SETP_INDEX]=2;
            enemy[i][SPEED_X]=8;
            enemy[i][SPEED_Y]=0;
        }
        break;
    case 2:                        //向右
        if(enemy[i][SETP_COUNT]==purpleStep[2])
        {
            enemy[i][SETP_COUNT]=0;
            enemy[i][SETP_INDEX]=3;
            enemy[i][SPEED_X]=-8;
            enemy[i][SPEED_Y]=0;
        }
        break;
    case 3:                        //向左
        if(enemy[i][SETP_COUNT]==purpleStep[3])
        {
            enemy[i][SETP_COUNT]=0;
            enemy[i][SETP_INDEX]=4;
            enemy[i][SPEED_X]=0;
            enemy[i][SPEED_Y]=-10;
        }
```

```
            break;
        case 4:                       //向上
            if(enemy[i][SETP_COUNT]==purpleStep[4])
            {
                enemy[i][SETP_COUNT]=0;
                enemy[i][SETP_INDEX]=0;
                enemy[i][SPEED_X]=0;
                enemy[i][SPEED_Y]=12;
            }
            break;
        }
        break;
    case TYPE_BOSS:
        enemy[i][SETP_COUNT]++;
        switch(enemy[i][SETP_INDEX]){
        case 0:                       //向下
            if(enemy[i][SETP_COUNT]==10)
            {
                enemy[i][SETP_COUNT]=0;
                enemy[i][SETP_INDEX]=1;
                enemy[i][SPEED_X]=-8;
                enemy[i][SPEED_Y]=0;
            }
            break;
        case 1:                       //向左
            if(enemy[i][SETP_COUNT]==5)
            {
                enemy[i][SETP_COUNT]=0;
                enemy[i][SETP_INDEX]=2;
                enemy[i][SPEED_X]=8;
                enemy[i][SPEED_Y]=0;
            }
            break;
        case 2:                       //向右
            if(enemy[i][SETP_COUNT]==10)
            {
                enemy[i][SETP_COUNT]=0;
                enemy[i][SETP_INDEX]=3;
                enemy[i][SPEED_X]=-8;
                enemy[i][SPEED_Y]=0;
            }
            break;
        case 3:                       //不动
            if(enemy[i][SETP_COUNT]==5)
```

```
                    {
                        enemy[i][SETP_COUNT]=0;
                        enemy[i][SETP_INDEX]=4;
                        enemy[i][SPEED_X]=0;
                        enemy[i][SPEED_Y]=-10;
                    }
                    break;
                case 4:                              //向上
                    if(enemy[i][SETP_COUNT]==8)
                    {
                        enemy[i][SETP_COUNT]=0;
                        enemy[i][SETP_INDEX]=0;
                        enemy[i][SPEED_X]=0;
                        enemy[i][SPEED_Y]=12;
                    }
                    break;
                }
                break;
            }
        }
    }
}
//敌机逻辑处理
void logic()
{
    timeCount++;
    if(gameScreen.player.getState()!=Player.SKILL_BOOM)
    {
        if(timeCount<400)
        {
            if(timeCount%15==0)
                createEnemy(TYPE_YELLOW);
            if(timeCount%37==0){
                createEnemy(TYPE_GREEN);
            }
            if(timeCount%65==0)
                createEnemy(TYPE_RED);
            if(timeCount%170==0){
                createEnemy(TYPE_PURPLE);
            }
        }else if(timeCount==400)
            createEnemy(TYPE_BOSS);
    }
    move();
```

```
        setPath();
        setVisiable();
        collidesWith();
    }
}
```

8.4.3 子弹类 Bullet

在游戏战斗界面中敌我双方飞机都使用子弹进行攻击,故在程序中封装了子弹类 Bullet。代码如下:

```
public class Bullet {
    final int SHOW=1;                       //子弹显示状态
    final int UN_SHOW=0;                    //子弹未显示状态
    final int ISVISIABLE=0;                 //isvisiable 对应的值(0为不可视,1为可视)
    final int TYPE=1;                       //子弹类型
    final int X=2;                          //X 坐标
    final int Y=3;                          //Y 坐标
    final int SPEED_X=4;                    //X 方向上的速度
    final int SPEED_Y=5;                    //Y 方向上的速度
    final int WIDTH=6;                      //子弹的宽度
    final int HEIGHT=7;                     //子弹的高度
    final int ATK=8;                        //攻击力
    final int FRAME_INDEX=9;                //帧下标
    Bitmap bitmapBoom;                      //爆炸图片
    int playerBullet[][];                   //主战飞机子弹数组(行为子弹数,列为子弹属性)
    int enemyBullet[][];                    //敌机子弹数组(行为子弹数,列为子弹属性)
    Bitmap eImg[];                          //敌机子弹图片数组
    Bitmap pImg[];                          //玩家子弹图片数组
    GameScreen gameScreen;
    Enemy enemy;
    int timeCount;
    Random random;
    int boomPosition[]={20, 20, 219, 60, 150, 170, 40, 350, 160, 295};
                                            //炸弹道具坐标数组
    public Bullet(GameScreen gameScreen, Enemy enemy, Property property){
        this.gameScreen=gameScreen;
        this.enemy=enemy;
        pImg=new Bitmap[3];
        pImg[0]=BitmapFactory.decodeResource(gameScreen.gameView.
            getResources(), R.drawable.pbullet0);
        pImg[1]=BitmapFactory.decodeResource(gameScreen.gameView.
            getResources(), R.drawable.pbullet1);
        pImg[2]=BitmapFactory.decodeResource(gameScreen.gameView.
```

```java
        getResources(), R.drawable.pbullet2);
    eImg=new Bitmap[4];
    eImg[0]=BitmapFactory.decodeResource(gameScreen.gameView.
        getResources(), R.drawable.bullet0);
    eImg[1]=BitmapFactory.decodeResource(gameScreen.gameView.
        getResources(), R.drawable.bullet1);
    eImg[2]=BitmapFactory.decodeResource(gameScreen.gameView.
        getResources(), R.drawable.bullet2);
    eImg[3]=BitmapFactory.decodeResource(gameScreen.gameView.
        getResources(), R.drawable.bullet3);
    bitmapBoom=BitmapFactory.decodeResource(gameScreen.gameView.
        getResources(), R.drawable.pbomb);
    playerBullet=new int[10][10];
    enemyBullet=new int[60][10];
    random=new Random();
}
//初始化
void init()
{
    //玩家子弹的初始化
    for(int i=0; i<playerBullet.length; i++){
        playerBullet[i][0]=UN_SHOW;
    }
    for(int i=0; i<enemyBullet.length; i++){
        enemyBullet[i][ISVISIABLE]=UN_SHOW;
    }
}
//绘制子弹
void paint(Canvas canvas, Paint paint)
{
    //玩家子弹的绘制
    for(int i=0; i<playerBullet.length; i++){
        switch(playerBullet[i][ISVISIABLE])
        {
            case SHOW:
                int type=playerBullet[i][TYPE];
                canvas.drawBitmap(pImg[type], playerBullet[i][X],
                playerBullet[i][Y], paint);
                break;
        }
    }
    if(gameScreen.player.getState()==Player.SKILL_BOOM)
    {
        if(playerBullet[0][FRAME_INDEX]<=3)
```

```java
            {
                for(int i=0; i<5; i++){
                    Tools.drawImage(bitmapBoom, playerBullet[i][X] -
                    playerBullet[i][FRAME_INDEX] * 91, playerBullet[i][Y],
                    playerBullet[i][X], playerBullet[i][Y], playerBullet[i]
                    [WIDTH], playerBullet[i][HEIGHT], canvas, paint);
                }
            }
            else if(playerBullet[0][FRAME_INDEX]%2==0)
                canvas.drawColor(Color.WHITE);
        }
        //敌机子弹的绘制
        for(int i=0; i<enemyBullet.length; i++){
            if(enemyBullet[i][ISVISIABLE]==SHOW)
            {
                int type=enemyBullet[i][TYPE];
                Tools.drawImage(eImg[type], enemyBullet[i][X]-enemyBullet[i]
                [FRAME_INDEX] * enemyBullet[i][WIDTH], enemyBullet[i][Y],
                enemyBullet[i][X], enemyBullet[i][Y], enemyBullet[i][WIDTH],
                enemyBullet[i][HEIGHT], canvas, paint);
            }
        }
    }
    //子弹移动
    void move()
    {
        //玩家子弹移动
        for(int i=0; i<playerBullet.length; i++){
            if(playerBullet[i][ISVISIABLE]==SHOW)
            {
                playerBullet[i][X]+=playerBullet[i][SPEED_X];
                playerBullet[i][Y]+=playerBullet[i][SPEED_Y];
            }
        }
        //敌机子弹移动
        for(int i=0; i<enemyBullet.length; i++){
            if(enemyBullet[i][ISVISIABLE]==SHOW)
            {
                enemyBullet[i][X]+=enemyBullet[i][SPEED_X];
                enemyBullet[i][Y]+=enemyBullet[i][SPEED_Y];
            }
        }
    }
    //创建玩家炸弹道具
```

```java
void createPlayerBoom()
{
    for(int i=0; i<playerBullet.length; i++){
        playerBullet[i][ISVISIABLE]=UN_SHOW;
    }
    for(int i=0; i<enemyBullet.length; i++){
        enemyBullet[i][ISVISIABLE]=UN_SHOW;
    }
    for(int i=0; i<enemy.enemy.length; i++){
        if(enemy.enemy[i][enemy.TYPE] !=Enemy.TYPE_BOSS)
            enemy.enemy[i][enemy.ISVISIABLE]=enemy.UN_SHOW;
        else
            enemy.enemy[i][enemy.HP] -=gameScreen.player.getAtk() * 5;
    }
    for(int i=0; i<boomPosition.length/2; i++){
        playerBullet[i][ISVISIABLE]=Player.SKILL_BOOM;
        playerBullet[i][X]=boomPosition[2 * i];
        playerBullet[i][Y]=boomPosition[2 * i+1];
        playerBullet[i][FRAME_INDEX]=0;
        playerBullet[i][WIDTH]=bitmapBoom.getWidth()/4;
        playerBullet[i][HEIGHT]=bitmapBoom.getHeight();
        playerBullet[i][TYPE]=Player.SKILL_BOOM;
    }
}
//创建玩家子弹
void createPlayerBullet()
{
    //玩家子弹创建
    for(int i=0; i<playerBullet.length; i++){
        if(gameScreen.player.getState()!= Player.BOOM && playerBullet[i]
        [ISVISIABLE]==UN_SHOW)
        {
            playerBullet[i][ISVISIABLE]=SHOW;
            playerBullet[i][TYPE]=gameScreen.player.getLevel();
            int type=playerBullet[i][TYPE];
            if(type==0)
            playerBullet[i][ATK]=2;            //暂定
            playerBullet[i][WIDTH]=pImg[type].getWidth();
            playerBullet[i][HEIGHT]=pImg[type].getHeight();
            playerBullet[i][X]= gameScreen. player. getX () + (gameScreen.
            player.getWidth()/2-playerBullet[i][WIDTH]/2);
            playerBullet[i][Y]=gameScreen.player.getY()-playerBullet[i]
            [HEIGHT]+12;
            playerBullet[i][SPEED_X]=0;
```

```
                    playerBullet[i][SPEED_Y]=-18;
                    break;
            }
        }
    }
    //创建敌机子弹
    void creatEnemyBullet(int enemyType)
    {
        switch(enemyType){
        case Enemy.TYPE_YELLOW:
            for(int i=0; i<enemy.enemy.length; i++){
                if(enemy.enemy[i][enemy.ISVISIABLE]==enemy.SHOW && enemy.
                enemy[i][enemy.TYPE]==Enemy.TYPE_YELLOW)
                {
                for(int j=0; j<enemyBullet.length; j++){
                    if(enemyBullet[j][ISVISIABLE]==UN_SHOW)
                    {
                        enemyBullet[j][ISVISIABLE]=SHOW;
                        enemyBullet[j][ATK]=4;
                        enemyBullet[j][FRAME_INDEX]=0;
                        enemyBullet[j][WIDTH]=eImg[enemyType].getWidth()/2;
                        enemyBullet[j][HEIGHT]=eImg[enemyType].getHeight();
                        enemyBullet[j][TYPE]=Enemy.TYPE_YELLOW;
                        enemyBullet[j][X]=enemy.enemy[i][enemy.X]+enemy.enemy
                        [i][enemy.WIDTH]/2-enemyBullet[j][WIDTH]/2;
                        enemyBullet[j][Y]=enemy.enemy[i][enemy.Y]+enemy.enemy
                        [i][enemy.HEIGHT];
                        enemyBullet[j][SPEED_X]=(enemyBullet[j][X]>gameScreen.
                        player.getX()?-5 : 5);
                        enemyBullet[j][SPEED_Y]=(enemyBullet[j][Y]>gameScreen.
                        player.getY()?-10 : 10);
                        break;
                    }
                }
                }
            }
            break;
        case Enemy.TYPE_RED:
            for(int i=0; i<enemy.enemy.length; i++){
                if(enemy.enemy[i][enemy.ISVISIABLE]==enemy.SHOW && enemy.enemy
                [i][enemy.TYPE]==Enemy.TYPE_RED)
                {
                    for(int j=0; j<enemyBullet.length; j++){
                        if(enemyBullet[j][ISVISIABLE]==UN_SHOW)
```

```java
                        {
                            enemyBullet[j][ISVISIABLE]=SHOW;
                            enemyBullet[j][ATK]=2;
                            enemyBullet[j][FRAME_INDEX]=0;
                            enemyBullet[j][WIDTH]=eImg[enemyType].getWidth()/2;
                            enemyBullet[j][HEIGHT]=eImg[enemyType].getHeight();
                            enemyBullet[j][X]=enemy.enemy[i][enemy.X]+enemy.
                            enemy[i][enemy.WIDTH]/2-enemyBullet[j][WIDTH]/2;
                            enemyBullet[j][Y]=enemy.enemy[i][enemy.Y]+enemy.
                            enemy[i][enemy.HEIGHT];
                            enemyBullet[j][SPEED_X]=0;
                            enemyBullet[j][SPEED_Y]=18;
                            enemyBullet[j][TYPE]=Enemy.TYPE_RED;
                            break;
                        }
                    }
                }
            }
            break;
        case Enemy.TYPE_GREEN:
            for(int i=0; i<enemy.enemy.length; i++){
                if(enemy.enemy[i][enemy.ISVISIABLE]==enemy.SHOW && enemy.enemy
                        [i][enemy.TYPE]==Enemy.TYPE_GREEN)
                {
                    int count=0;
                    for(int j=0; j<enemyBullet.length; j++){
                        if(enemyBullet[j][ISVISIABLE]==UN_SHOW){
                            enemyBullet[j][ISVISIABLE]=SHOW;
                            enemyBullet[j][ATK]=3;
                            enemyBullet[j][FRAME_INDEX]=0;
                            enemyBullet[j][WIDTH]=eImg[enemyType].getWidth()/2;
                            enemyBullet[j][HEIGHT]=eImg[enemyType].getHeight();
                            enemyBullet[j][X]=enemy.enemy[i][enemy.X]+2+count*20;
                            enemyBullet[j][Y]=enemy.enemy[i][enemy.Y]+enemy.
                            enemy[i][enemy.HEIGHT];
                            enemyBullet[j][SPEED_X]=0;
                            enemyBullet[j][SPEED_Y]=20;
                            enemyBullet[j][TYPE]=Enemy.TYPE_GREEN;
                            count++;
                            if(count==2)
                                break;
                        }
```

```
                }
            break;
        case Enemy.TYPE_PURPLE:
            for(int i=0; i<enemy.enemy.length; i++){
                if(enemy.enemy[i][enemy.ISVISIABLE]==enemy.SHOW && enemy.enemy
                [i][enemy.TYPE]==Enemy.TYPE_PURPLE)
                {
                    int count=0;
                    for(int j=0; j<enemyBullet.length; j++){
                        if(enemyBullet[j][ISVISIABLE]==UN_SHOW)
                        {
                            enemyBullet[j][ISVISIABLE]=SHOW;
                            enemyBullet[j][ATK]=5;
                            enemyBullet[j][FRAME_INDEX]=0;
                            enemyBullet[j][WIDTH]=eImg[enemyType].getWidth()/2;
                            enemyBullet[j][HEIGHT]=eImg[enemyType].getHeight();
                            enemyBullet[j][X]=enemy.enemy[i][enemy.X]+3+count * 9;
                            enemyBullet[j][Y]= enemy. enemy [i] [enemy.Y] + enemy.
                            enemy[i][enemy.HEIGHT];
                            enemyBullet[j][SPEED_X]=8 * count-8;
                            enemyBullet[j][SPEED_Y]=16;
                            enemyBullet[j][TYPE]=Enemy.TYPE_PURPLE;
                            count++;
                            if(count==3)
                                break;
                        }
                    }
                }
            }
            break;
        case Enemy.TYPE_BOSS:
            int ran=Math.abs(random.nextInt()%8);
            if(enemy.enemy[0][enemy.ISVISIABLE]==enemy.SHOW && enemy.enemy[0]
            [enemy.TYPE]==Enemy.TYPE_BOSS)
            {
                if(enemy.enemy[0][enemy.Y]>=20)
                if(ran<=4)
                {
                    int count=-5;
                    for(int j=0; j<enemyBullet.length; j++){
                        if(enemyBullet[j][ISVISIABLE]==UN_SHOW)
                        {
                            enemyBullet[j][ISVISIABLE]=SHOW;
                            enemyBullet[j][ATK]=15;
```

```java
                    enemyBullet[j][FRAME_INDEX]=0;
                    enemyBullet[j][WIDTH] = eImg[Enemy.TYPE_RED].
                    getWidth()/2;
                    enemyBullet[j][HEIGHT] = eImg[Enemy.TYPE_RED].
                    getHeight();
                    enemyBullet[j][X]=enemy.enemy[0][enemy.X] + 15 *
                    (count+4)+7;
                    int y=-1*count*count-2*count+3;
                    enemyBullet[j][Y]=enemy.enemy[0][enemy.Y]+90+y;
                    enemyBullet[j][SPEED_X]=(count+1==0?0:(count+1)+
                    (count+1)/Math.abs(count+1) * 5);
                    enemyBullet[j][SPEED_Y]=12;
                    enemyBullet[j][TYPE]=Enemy.TYPE_RED;
                    count++;
                    if(count==4)
                        break;
                }
            }
        }else
        {
            int count=0;
            for(int j=0; j<enemyBullet.length; j++){
                if(enemyBullet[j][ISVISIABLE]==UN_SHOW)
                {
                    enemyBullet[j][ISVISIABLE]=SHOW;
                    enemyBullet[j][ATK]=25;
                    enemyBullet[j][FRAME_INDEX]=0;
                    enemyBullet[j][WIDTH]= eImg[Enemy.TYPE_GREEN].
                    getWidth()/2;
                    enemyBullet[j][HEIGHT]= eImg[Enemy.TYPE_GREEN].
                    getHeight();
                    enemyBullet[j][X]=enemy.enemy[0][enemy.X]+40+count*24;
                    enemyBullet[j][Y]=enemy.enemy[0][enemy.Y]+85;
                    enemyBullet[j][SPEED_X]=0;
                    enemyBullet[j][SPEED_Y]=14;
                    enemyBullet[j][TYPE]=Enemy.TYPE_GREEN;
                    count++;
                    if(count==2)
                        break;
                }
            }
        }
    break;
```

```
        }
    }
    //子弹回池
    void setVisiable()
    {
        //玩家子弹回池
        for(int i=0; i<playerBullet.length; i++){
            if(playerBullet[i][ISVISIABLE]==SHOW && playerBullet[i][Y]<=
            -playerBullet[i][HEIGHT])
                playerBullet[i][ISVISIABLE]=UN_SHOW;
        }
        //敌机子弹回池
        for(int i=0; i<enemyBullet.length; i++){
            if(enemyBullet[i][ISVISIABLE]==SHOW&&(enemyBullet[i][Y]<=
            -enemyBullet[i][HEIGHT]||enemyBullet[i][Y]>=GameView.SCREEN_
            HEIGHT||enemyBullet[i][X]<=-enemyBullet[i][WIDTH]||enemyBullet
            [i][X]>=GameView.SCREEN_WIDTH))
                enemyBullet[i][ISVISIABLE]=UN_SHOW;
        }
    }
    //碰撞检测
    void collidesWith()
    {
        //敌机子弹碰撞检测
        for(int i=0; i<enemyBullet.length; i++){
            if(enemyBullet[i][ISVISIABLE]==SHOW)
            {
                switch(gameScreen.player.getState()){
                case Player.NORMAL:
                case Player.LEVELUP:
                    if(Tools.collides(enemyBullet[i][X], enemyBullet[i][Y],
                    enemyBullet[i][WIDTH], enemyBullet[i][HEIGHT], gameScreen.
                    player.getX(), gameScreen.player.getY(), gameScreen.player.
                    getWidth(), gameScreen.player.getHeight()))
                    {
                        enemyBullet[i][ISVISIABLE]=UN_SHOW;
                        gameScreen.player.setHp(enemyBullet[i][ATK]);
                    }
                    break;
                default:
                    break;
                }
            }
        }
```

```java
//玩家子弹碰撞检测
for(int i=0; i<playerBullet.length; i++){
    if(playerBullet[i][ISVISIABLE]==SHOW)
    {
        for(int j=0; j<enemy.enemy.length; j++){
            if(enemy.enemy[j][enemy.ISVISIABLE]==enemy.SHOW)
            {
                if(Tools.collides(enemy.enemy[j][enemy.X], enemy.enemy
                    [j][enemy.Y], enemy.enemy[j][enemy.WIDTH], enemy.enemy
                    [j][enemy.HEIGHT], playerBullet[i][X], playerBullet[i]
                    [Y], playerBullet[i][WIDTH], playerBullet[i][HEIGHT]))
                {
                    playerBullet[i][ISVISIABLE]=UN_SHOW;
                    enemy.enemy[j][enemy.HP] -=gameScreen.player.getAtk();
                    if(enemy.enemy[j][enemy.HP]<=0)
                    {
                        if(enemy.enemy[j][enemy.TYPE]==Enemy.TYPE_BOSS)
                            gameScreen.gameState=gameScreen.WIN;
                        enemy.enemy[j][enemy.ISVISIABLE]=enemy.BOOM;
                        enemyBullet[j][FRAME_INDEX]=0;
                    }
                }
            }
        }
    }
}
//玩家子弹逻辑
void playerBulletLogic()
{
    for(int i=0; i<playerBullet.length; i++){
        if(playerBullet[i][ISVISIABLE]==Player.SKILL_BOOM)
        {
            playerBullet[i][FRAME_INDEX]++;
            if(playerBullet[i][FRAME_INDEX]==8)
            {
                playerBullet[i][ISVISIABLE]=UN_SHOW;
                gameScreen.player.setState(Player.NORMAL);
                break;
            }
        }
    }
}
//敌机子弹逻辑
```

```
void enemyBulletLogic()
{
    for(int i=0; i<enemyBullet.length; i++){
        if(enemyBullet[i][ISVISIABLE]==SHOW)
        {
            switch(enemyBullet[i][TYPE]){
            case Enemy.TYPE_YELLOW:
                enemyBullet[i][FRAME_INDEX]=(enemyBullet[i][FRAME_INDEX]=
                =1?0:1);
                enemyBullet[i][SPEED_X]= (enemyBullet[i][X] > gameScreen.
                player.getX()?-5:5);
                enemyBullet[i][SPEED_Y]= (enemyBullet[i][Y] > gameScreen.
                player.getY()?-10:10);
                break;
            case Enemy.TYPE_GREEN:
            case Enemy.TYPE_PURPLE:
            case Enemy.TYPE_RED:
                enemyBullet[i][FRAME_INDEX]=(enemyBullet[i][FRAME_INDEX]=
                =1?0:1);
                break;
            }
        }
    }
}
//子弹类逻辑处理
void logic()
{
    timeCount++;
    if(timeCount%4==0)
        createPlayerBullet();
    if(timeCount%15==0){
        createEnemyBullet(Enemy.TYPE_YELLOW);
    }
    if(timeCount%6==0)
        createEnemyBullet(Enemy.TYPE_RED);
    if(timeCount%7==0){
        createEnemyBullet(Enemy.TYPE_GREEN);
    }
    if(timeCount%9==0)
        creatEnemyBullet(Enemy.TYPE_PURPLE);
    if(timeCount%4==0)
        createEnemyBullet(Enemy.TYPE_BOSS);
    enemyBulletLogic();
    playerBulletLogic();
```

```
        move();
        setVisiable();
        collidesWith();
    }
}
```

8.4.4 道具类 Property

该类为补血道具类，当主战飞机与该类对象碰撞时，为主战飞机补血。代码如下：

```java
public class Property {
    static final int BOOM=0;
    static final int POWER=1;
    static final int LEVER_UP=2;
    final int UN_SHOW=0;
    final int SHOW=1;
    final int ISVISIABLE=0;
    final int X=1;
    final int Y=2;
    final int WIDTH=3;
    final int HEIGHT=4;
    final int TYPE=5;
    final int FRAME=6;
    final int SPEED_X=7;
    final int SPEED_Y=8;
    Random random;
    GameScreen gameScreen;
    Bitmap image;
    int property[][];                    //道具对象池
    public Property(GameScreen gameScreen){
        this.gameScreen=gameScreen;
        property=new int[4][9];
        random=new Random();
        image=BitmapFactory.decodeResource(gameScreen.gameView.getResources
        (), R.drawable.property);
    }
    //初始化
    void init()
    {
        for(int i=0; i<property.length; i++){
            property[i][ISVISIABLE]=UN_SHOW;
        }
    }
    /**
```

```
 *  道具创建方法
 *  @param type    道具类型
 *  @param x       道具的初始化 x 坐标
 *  @param y       道具的初始化 y 坐标
 */
void createProperty(int type, int x, int y)
{
    switch(type){
    case BOOM:
        for(int i=0; i<property.length; i++){
            if(property[i][ISVISIABLE]==UN_SHOW)
            {
                property[i][ISVISIABLE]=SHOW;
                property[i][X]=x;
                property[i][Y]=y;
                property[i][SPEED_X]=random.nextInt()%10;
                property[i][SPEED_Y]=random.nextInt()%10;
                property[i][WIDTH]=image.getWidth()/6;
                property[i][HEIGHT]=image.getHeight();
                property[i][TYPE]=type;
                property[i][FRAME]=0;
                break;
            }
        }
        break;
    case POWER:
        for(int i=0; i<property.length; i++){
            if(property[i][ISVISIABLE]==UN_SHOW)
            {
                property[i][ISVISIABLE]=SHOW;
                property[i][X]=x;
                property[i][Y]=y;
                property[i][SPEED_X]=(random.nextInt()>=0?8:-8);
                property[i][SPEED_Y]=(random.nextInt()>=0?10:-10);
                property[i][WIDTH]=image.getWidth()/6;
                property[i][HEIGHT]=image.getHeight();
                property[i][TYPE]=type;
                property[i][FRAME]=0;
                break;
            }
        }
        break;
    case LEVER_UP:
        for(int i=0; i<property.length; i++){
```

```java
                if(property[i][ISVISIABLE]==UN_SHOW)
                {
                    property[i][ISVISIABLE]=SHOW;
                    property[i][X]=x;
                    property[i][Y]=y;
                    property[i][SPEED_X]=random.nextInt()%9;
                    property[i][SPEED_Y]=random.nextInt()%9;
                    property[i][WIDTH]=image.getWidth()/6;
                    property[i][HEIGHT]=image.getHeight();
                    property[i][TYPE]=type;
                    property[i][FRAME]=0;
                    break;
                }
            }
            break;
    }
}

/**
 * 道具绘制
 * @param canvas
 * @param paint
 */
void paint(Canvas canvas, Paint paint)
{
    for(int i=0; i<property.length; i++){
        if(property[i][ISVISIABLE]==SHOW)
            Tools.drawImage(image, property[i][X] - property[i][FRAME] *
            property[i][WIDTH], property[i][Y], property[i][X], property[i]
            [Y], property[i][WIDTH], property[i][HEIGHT], canvas, paint);
    }
}
//道具移动
void move()
{
    for(int i=0; i<property.length; i++){
        if(property[i][ISVISIABLE]==SHOW)
        {
            property[i][X]+=property[i][SPEED_X];
            property[i][Y]+=property[i][SPEED_Y];
        }
    }
}
//道具回池方法
```

```java
void setVisiable()
{
    for(int i=0; i<property.length; i++){
        if(property[i][ISVISIABLE]==SHOW)
        {
            if(property[i][X]<=-property[i][WIDTH] || property[i][X]>=
            GameView.SCREEN_WIDTH || property[i][Y]<=-property[i][HEIGHT] |
            | property[i][Y]>=GameView.SCREEN_HEIGHT)
                property[i][ISVISIABLE]=UN_SHOW;
        }
    }
}
/**
 * 道具碰撞检测
 * @param p  飞机对象
 */
void collidesWith(Player p)
{
    for(int i=0; i<property.length; i++){
        if(property[i][ISVISIABLE]==SHOW && p.getState()==Player.NORMAL)
        {
            if(Tools.collides(property[i][X], property[i][Y], property[i]
            [WIDTH], property[i][HEIGHT], p.getX(), p.getY(), p.getWidth(),
            p.getHeight()))
            {
                switch(property[i][TYPE]){
                case BOOM:
                    gameScreen.player.addPropertyNum();
                    gameScreen.player.setState(Player.SKILL_BOOM);
                    property[i][ISVISIABLE]=UN_SHOW;
                    break;
                case POWER:
                    gameScreen.player.setState(Player.RECOVER);
                    property[i][ISVISIABLE]=UN_SHOW;
                    break;
                case LEVER_UP:
                    gameScreen.player.setState(Player.LEVELUP);
                    property[i][ISVISIABLE]=UN_SHOW;
                    break;
                }
            }
        }
    }
}
```

```
void changeFrame()
{
    for(int i=0; i<property.length; i++){
        if(property[i][ISVISIABLE]==SHOW)
        {
            switch(property[i][TYPE]){
            case BOOM:
                property[i][FRAME]=(property[i][FRAME]==1?0:1);
                break;
            case LEVER_UP:
                property[i][FRAME]=(property[i][FRAME]==4?5:4);
                break;
            case POWER:
                property[i][FRAME]=(property[i][FRAME]==2?3:2);
                break;
            }
        }
    }
}
//修改路径
void setPath()
{
    for(int i=0; i<property.length; i++){
        if(property[i][ISVISIABLE]==SHOW)
        {
            switch(property[i][TYPE]){
            case BOOM:
                property[i][SPEED_X]=random.nextInt()%10;
                property[i][SPEED_Y]=random.nextInt()%10;
                break;
            case LEVER_UP:
                break;
            case POWER:
                break;
            }
        }
    }
}
void logic()
{
    move();
    setVisiable();
    collidesWith(gameScreen.player);
    changeFrame();
```

```
        if(gameScreen.gameView.timeCount%15==0)
            setPath();
    }
}
```

8.5 游戏中的界面相关类

8.5.1 游戏显示类 PlaneGameActivity

该类负责调用 GameView 类，启动游戏界面。代码如下：

```
public class PlaneGameActivity extends Activity {
    GameView gameView;
    public void onCreate(Bundle savedInstanceState){
        super.onCreate(savedInstanceState);
        requestWindowFeature(Window.FEATURE_NO_TITLE);
        getWindow().setFlags(WindowManager.LayoutParams.FLAG_KEEP_SCREEN_ON,
        WindowManager.LayoutParams.FLAG_KEEP_SCREEN_ON);
        getWindow().setFlags(WindowManager.LayoutParams.FLAG_FULLSCREEN,
        WindowManager.LayoutParams.FLAG_FULLSCREEN);
        gameView=new GameView(this);
        setContentView(gameView);
    }
}
```

8.5.2 游戏主界面类 GameView

GameView 类就是玩家进行游戏的架构界面，负责组装 MenuScreen 类、GameScreen 类、GameStore 类，根据游戏的不同状态绘制屏幕及各种逻辑处理。代码如下：

```
public class GameView extends SurfaceView implements Callback, Runnable {
    public static final int GAME_MENU=5;              //游戏中的辅助菜单
    public static final int SPLASH=0;                 //logo 动画
    public static final int GAME=6;                   //游戏界面
    public static final int RECORD=1;                 //记录存档
    public static final int HELP=2;                   //帮助
    public static final int MUSIC_SET=3;              //音乐设置
    public static final int MAIN_MENU=4;              //主菜单
    public static int SCREEN_WIDTH;                   //屏幕宽
    public static int SCREEN_HEIGHT;                  //屏幕高
    public PlaneGameActivity activity;
    private SurfaceHolder holder;
    private Canvas canvas;
```

```java
        private Paint paint;
        public int gameState;                              //当前游戏状态
        Bitmap[] splashImg;                                //logo图片数组
        MenuScreen menuScreen;
        GameScreen gameScreen;
        GameMusic gameMusic;
        GameStore gameStore;
        boolean isGame;                                    //标识游戏运行
        int timeCount;
        public GameView(Context context){
            super(context);
            activity=(PlaneGameActivity)context;
            setFocusable(true);
            holder=getHolder();
            holder.addCallback(this);
            canvas=holder.lockCanvas();
            paint=new Paint();
            paint.setAntiAlias(true);
            gameState=SPLASH;
            splashImg=new Bitmap[2];
            splashImg[0]=BitmapFactory.decodeResource(getResources(), R.drawable.
            logo);
            splashImg[1]=BitmapFactory.decodeResource(getResources(), R.drawable.
            anykey);
            gameMusic=new GameMusic(context);
            gameStore=new GameStore(activity);
        }
        public void surfaceChanged(SurfaceHolder holder, int format, int width, int
        height){
        }
        public void surfaceCreated(SurfaceHolder holder){
            SCREEN_HEIGHT=getHeight();
            SCREEN_WIDTH=getWidth();
            menuScreen=new MenuScreen(this);
            gameScreen=new GameScreen(this);
            isGame=true;
            gameMusic.starMusic(GameMusic.BG_MUSIC);
            new Thread(this).start();
        }
        public void surfaceDestroyed(SurfaceHolder holder){
            isGame=false;
            gameMusic.stopMusic();
            gameMusic.recycle();
            activity.finish();
```

```
}
//绘制闪屏界面
void drawSplash(){
    canvas.drawBitmap(splashImg[0], 0, 0, paint);
    if(timeCount %2==0){
        canvas.drawBitmap(splashImg[1],
                SCREEN_WIDTH/2 - splashImg[1].getWidth()/2,
                SCREEN_HEIGHT - splashImg[1].getHeight()-20, paint);
    }
}
//绘制方法
void paint(){
    canvas=holder.lockCanvas();
    switch(gameState){
    case SPLASH:
        drawSplash();
        break;
    case MAIN_MENU:
    case MUSIC_SET:
    case HELP:
    case RECORD:
        menuScreen.paint(canvas, paint, gameState);
        break;
    case GAME:
        gameScreen.paint(canvas, paint);
        break;
    }
    holder.unlockCanvasAndPost(canvas);
}

public boolean onTouchEvent(MotionEvent event){
    int x=(int)event.getX();
    int y=(int)event.getY();
    switch(gameState){
    case SPLASH:
        gameState=MAIN_MENU;
        break;
    case MAIN_MENU:
    case HELP:
    case RECORD:
    case MUSIC_SET:
        menuScreen.onTouchEvent(x, y);
        break;
    case GAME:
        gameScreen.onTouchEvent(x, y);
```

```
            break;
        }
        return super.onTouchEvent(event);
    }
    void logic(){
        switch(gameState){
        case SPLASH:
            timeCount++;
            if(timeCount==25)
                gameState=MAIN_MENU;
            break;
        case GAME:
            timeCount++;
            gameScreen.logic();
            break;
        case RECORD:
            break;
        }
    }
    public void run(){
        while(isGame){
            logic();
            paint();
            try {
                Thread.sleep(100);
            } catch(InterruptedException e){
                e.printStackTrace();
            }
        }
    }
}
```

8.5.3 游戏界面绘制类 GameScreen

该类负责背景滚动、界面初始化、显示状态信息及界面逻辑处理。代码如下：

```
public class GameScreen {
    final int GAME=0;
    final int LOSE=1;                            //游戏失败
    final int WIN=2;                             //游戏胜利
    final int LEVEL_UP=3;                        //战机升级
    final int RECOVER=4;                         //战机修复
    final int BOSS=5;                            //敌机 boss 出现
    GameView gameView;
```

```java
Bitmap mapImg;                                    //地图图片
Bitmap score[];                                   //游戏的状态栏图片
Bitmap number;
Bitmap plane;
Bitmap fontImg[];                                 //游戏失败等图片
Bitmap recoverImg;                                //战机修复图片
Bitmap bottomBar;
int gameState;
int mapY;
Player player;
Bullet bullet;
Enemy enemy;
Property property;
Rect[] barSelectRect;
public GameScreen(GameView view){
    gameView=view;
    mapImg=BitmapFactory.decodeResource(view.getResources(),
    R.drawable.map);
    plane=BitmapFactory.decodeResource(view.getResources(),
    R.drawable.player);
    score=new Bitmap[2];
    score[0]=BitmapFactory.decodeResource(view.getResources(),
    R.drawable.score);
    score[1]=BitmapFactory.decodeResource(view.getResources(),
    R.drawable.imghp);
    number=BitmapFactory.decodeResource(view.getResources(),
    R.drawable.number);
    fontImg=new Bitmap[5];
    fontImg[0]=BitmapFactory.decodeResource(view.getResources(),
    R.drawable.font_lose);
    fontImg[1]=BitmapFactory.decodeResource(view.getResources(),
    R.drawable.font_win);
    fontImg[2]=BitmapFactory.decodeResource(view.getResources(),
    R.drawable.font_levelup);
    fontImg[3]=BitmapFactory.decodeResource(view.getResources(),
    R.drawable.font_recover);
    fontImg[4]=BitmapFactory.decodeResource(view.getResources(),
    R.drawable.font_boss);
    recoverImg=BitmapFactory.decodeResource(view.getResources(),
    R.drawable.invincible);
    bottomBar=BitmapFactory.decodeResource(view.getResources(),
    R.drawable.bottom_bar);
    player=new Player(plane, this);
    property=new Property(this);
```

```
        enemy=new Enemy(this);
        bullet=new Bullet(this, enemy, property);
        mapY=GameView.SCREEN_HEIGHT-mapImg.getHeight();
        barSelectRect=new Rect[2];
        barSelectRect[0]=new Rect(0, 454, 80, 480);
        barSelectRect[1]=new Rect(240, 454, 320, 480);
        gameState=0;
    }
    //游戏界面的初始化方法
    void init()
    {
        gameState=0;
        mapY=GameView.SCREEN_HEIGHT-mapImg.getHeight();
        enemy.init();
        player.init();
        bullet.init();
        property.init();
        gameView.timeCount=0;
    }
    //设置字体
    void setFont(int type)
    {
        gameState=type;
    }
    //绘制游戏界面
    void paint(Canvas canvas, Paint paint)
    {
        drawMap(canvas, paint);
        player.paint(canvas, paint);
        bullet.paint(canvas, paint);
        enemy.paint(canvas, paint);
        property.paint(canvas, paint);
        drawScore(canvas, paint);
        if(gameState !=GAME)
        {
            canvas.drawBitmap(fontImg[gameState-1], 0, 130, paint);
            if((gameState==LOSE || gameState==WIN)&& gameView.timeCount%2==0)
                canvas.drawBitmap(gameView.splashImg[1],
                GameView.SCREEN_WIDTH/2 - gameView.splashImg[1].getWidth()/2,
                280, paint);
        }
    }
    //绘制地图(背景滚动)
    void drawMap(Canvas canvas, Paint paint)
```

```java
    {
        if(mapY<0)
            canvas.drawBitmap(mapImg, 0, mapY, paint);
        else
        {
            Tools.drawImage(mapImg, 0, mapY-mapImg.getHeight(), 0, 0, GameView.
              SCREEN_WIDTH, mapY, canvas, paint);
            Tools.drawImage (mapImg, 0, mapY, 0, mapY, GameView.SCREEN_WIDTH,
              GameView.SCREEN_HEIGHT-mapY, canvas, paint);
        }
    }
}
//绘制状态栏
void drawScore(Canvas canvas, Paint paint)
{
    canvas.drawBitmap(score[0], 0, 0, paint);
    canvas.drawBitmap(score[1], 2, 4, paint);
    canvas.drawRect(97- (player.getMaxHp()-player.getHp())/3, 4, 97, 11,
      paint);
    for(int i=0; i<player.getPropertyNum(); i++){
        Tools.drawImage(property.image, 110+i*32-property.image.getWidth
          ()/6, 3, 110+i*32, 3, property.image.getWidth()/6, property.image.
          getHeight(), canvas, paint);
    }
    Tools.drawImage(number, 288- (player.getLifeNum()-1) * number.getWidth
      ()/3, 8, 288, 8, number.getWidth()/3, number.getHeight(), canvas, paint);
    canvas.drawBitmap (bottomBar, 0, GameView.SCREEN_HEIGHT - bottomBar.
      getHeight(), paint);
}
/**
 * 触屏时间
 * @param x      触点 X 坐标
 * @param y      触点 Y 坐标
 */
void onTouchEvent(int x, int y){
    if(barSelectRect[0].contains(x, y)){
        gameView.gameState=GameView.MAIN_MENU;
    }else if(barSelectRect[1].contains(x, y))
    {
        gameView.gameStore.saveData(gameView.gameScreen.player);
    }
}
//游戏界面逻辑处理
void logic()
{
```

```
        switch(gameState){
        case GAME:
            mapY+=2;
            if(mapY>=GameView.SCREEN_HEIGHT)
                mapY=GameView.SCREEN_HEIGHT-mapImg.getHeight();
            player.logic();
            bullet.logic();
            enemy.logic();
            property.logic();
            break;
        default:
            break;
        }
    }
}
```

8.5.4 菜单界面类 MenuScreen

该类用于菜单界面、声音设置界面及帮助界面的绘制。代码如下：

```
public class MenuScreen {
    GameView gameView;
    Bitmap[] mainMenuImg;                           //主菜单图片数组
    Bitmap helpImg;                                 //帮助图片
    int menuSelectIndex;                            //菜单选项下标
    Rect[] menuSelectRect;
    Bitmap musicSetImg;                             //声音设置图片
    String music[]={"ON", "OFF"};
    public MenuScreen(GameView view){
        gameView=view;
        mainMenuImg=new Bitmap[4];
        mainMenuImg[0]=BitmapFactory.decodeResource(view.getResources(), R.
        drawable.menu1);
        mainMenuImg[1]=BitmapFactory.decodeResource(view.getResources(), R.
        drawable.menu2);
        mainMenuImg[2]=BitmapFactory.decodeResource(view.getResources(), R.
        drawable.menu3);
        mainMenuImg[3]=BitmapFactory.decodeResource(view.getResources(), R.
        drawable.menu4);
        menuSelectRect=new Rect[5];
        for(int i=0;i<menuSelectRect.length;i++)
            menuSelectRect[i]=new Rect(85, 45+i * 60, 85+148, 45+i * 60+24);
        helpImg=BitmapFactory.decodeResource(view.getResources(), R.drawable.
        help);
```

```java
            musicSetImg= BitmapFactory.decodeResource(view.getResources(), R.
            drawable.set);
    }
    //绘制菜单界面
    void drawMainMenu(Canvas canvas, Paint paint)
    {
        canvas.drawRGB(22, 22, 24);
        canvas.drawBitmap(mainMenuImg[3],
                GameView.SCREEN_WIDTH/2 - mainMenuImg[3].getWidth()/2,
                GameView.SCREEN_HEIGHT-mainMenuImg[3].getHeight(), paint);
        for(int i=0; i<5; i++){
            canvas.drawBitmap(mainMenuImg[0], 0, 45+i*60, paint);
            if(i!=menuSelectIndex)
                Tools.drawImage(mainMenuImg[1], 123, 45+i*60-i*24, 123, 45+i*
                60, 74, 24, canvas, paint);
            else
                Tools.drawImage(mainMenuImg[2], 85, 45+i*60-i*24, 85, 45+i*
                60, 148, 24, canvas, paint);
        }
    }
    //绘制帮助界面
    void drawHelp(Canvas canvas, Paint paint)
    {
        canvas.drawBitmap(helpImg, 0, 0, paint);
    }
    void drawMusicSet(Canvas canvas, Paint paint)
    {
        canvas.drawBitmap(musicSetImg, 0, 185, paint);
        paint.setColor(Color.CYAN);
        if(gameView.gameMusic.isOpen)
            canvas.drawText(music[0], 145, 205, paint);
        else
            canvas.drawText(music[1], 143, 205, paint);
    }
    void paint(Canvas canvas, Paint paint, int gameState)
    {
        switch(gameState){

        case GameView.MAIN_MENU:
            drawMainMenu(canvas, paint);
            break;
        case GameView.MUSIC_SET:
            drawMusicSet(canvas, paint);
            break;
```

```java
        case GameView.HELP:
            drawHelp(canvas, paint);
            break;
        case GameView.RECORD:
            paint.setColor(Color.GRAY);
            canvas.drawRect(50, 70, 270, 100, paint);
            paint.setColor(Color.CYAN);
            canvas.drawText("暂无存档,按任意键返回主菜单", 58, 88, paint);
            break;
        }
    }
    void mainMenuInput(int keyCode)
    {
        switch(keyCode){
        case KeyEvent.KEYCODE_DPAD_UP:
            menuSelectIndex= (menuSelectIndex==0?4:menuSelectIndex-1);
            break;
        case KeyEvent.KEYCODE_DPAD_DOWN:
            menuSelectIndex= (menuSelectIndex==4?0:menuSelectIndex+1);
            break;
        case KeyEvent.KEYCODE_DPAD_CENTER:
            switch(menuSelectIndex){
            case 0:                                    //开始游戏
                gameView.gameState=GameView.GAME;
                gameView.gameScreen.init();
                break;
            case 1:                                    //读档
                boolean  readFlag =  gameView. gameStore. getData ( gameView.
                gameScreen.player);
                if(readFlag)
                    gameView.gameState=GameView.GAME;
                else
                    gameView.gameState=GameView.RECORD;
                break;
            case 2:                                    //帮助
                gameView.gameState=GameView.HELP;
                break;
            case 3:                                    //声音设置
                gameView.gameState=GameView.MUSIC_SET;
                break;
            case 4:                                    //退出游戏
                gameView.isGame=true;
                gameView.activity.finish();
                break;
```

```java
        }
        break;
    }
}
void onTouchEvent(int x , int y)
{
    switch(gameView.gameState){
    case GameView.MAIN_MENU:
        if(menuSelectRect[0].contains(x, y))
        {
            menuSelectIndex=0;
            gameView.gameState=GameView.GAME;
            gameView.gameScreen.init();
            gameView.gameMusic.starMusic(GameMusic.MENU_MUSIC);
        }
        else if(menuSelectRect[1].contains(x, y))
        {
            menuSelectIndex=1;
            gameView.gameMusic.starMusic(GameMusic.MENU_MUSIC);
            boolean readFlag = gameView. gameStore. getData ( gameView.
            gameScreen.player);
            if(readFlag)
                gameView.gameState=GameView.GAME;
            else
                gameView.gameState=GameView.RECORD;
        }
        else if(menuSelectRect[2].contains(x, y))
        {
            menuSelectIndex=2;
            gameView.gameState=GameView.HELP;
            gameView.gameMusic.starMusic(GameMusic.MENU_MUSIC);
        }
        else if(menuSelectRect[3].contains(x, y))
        {
            menuSelectIndex=3;
            gameView.gameState=GameView.MUSIC_SET;
        }
        else if(menuSelectRect[4].contains(x, y))
        {
            menuSelectIndex=4;
            gameView.isGame=true;
            gameView.activity.finish();
        }
        break;
```

```
            case GameView.HELP:
                gameView.gameState=GameView.MAIN_MENU;
                break;
            case GameView.RECORD:
                gameView.gameState=GameView.MAIN_MENU;
                break;
        }
    }
}
```

8.5.5 数据存储类 GameStore

该类负责游戏状态数据的存储与调用。代码如下：

```
public class GameStore {
    SQLiteHelper helper;
    SQLiteDatabase database;
    public GameStore(Context context){
        helper= new SQLiteHelper (context, SQLiteHelper.DATABASE _ NAME, null,
        SQLiteHelper.VERSION);
        database=helper.getWritableDatabase();
    }
    /**
     * 游戏存档
     * @param player  存档数据
     */
    void saveData(Player player)
    {
        ContentValues contentValues=new ContentValues();
        contentValues.put("flag", "true");
        contentValues.put("playerState", player.getState());
        contentValues.put("x", player.getX());
        contentValues.put("y", player.getY());
        contentValues.put("lifeNum", player.getLifeNum());
        contentValues.put("hp", player.getHp());
        contentValues.put("direction", player.getDirection());
        contentValues.put("level", player.getLevel());
        contentValues.put("atk", player.getAtk());
        contentValues.put("propertyNum", player.getPropertyNum());
        database.update("gameStore", contentValues, null, null);
    }

    /**
     * 读取存档
     * @param player  存储数据对象
```

```java
     * @return         是否读取成功
     */
    boolean getData(Player player)
    {
        Cursor cursor = database.rawQuery(" SELECT * FROM gameStore WHERE
        playerState>=?", new String[]{"0"});
        boolean flag=false;
        while(cursor.moveToNext()){
            String msg=cursor.getString(cursor.getColumnIndex("flag"));
            if(msg.equals("true"))
            {
                int playerState = cursor.getInt(cursor.getColumnIndex("
                playerState"));
                int x=cursor.getInt(cursor.getColumnIndex("x"));
                int y=cursor.getInt(cursor.getColumnIndex("y"));
                int lifeNum=cursor.getInt(cursor.getColumnIndex("lifeNum"));
                int hp=cursor.getInt(cursor.getColumnIndex("hp"));
                int direction=cursor.getInt(cursor.getColumnIndex("direction"));
                int level=cursor.getInt(cursor.getColumnIndex("level"));
                int atk=cursor.getInt(cursor.getColumnIndex("atk"));
                int propertyNum=cursor.getInt(cursor.getColumnIndex("propertyNum"));
                player.setState(playerState);
                player.setData(x, y, lifeNum, hp, direction, level, atk, propertyNum);
                return true;
            }
        }
        return flag;
    }
}
class SQLiteHelper extends SQLiteOpenHelper{
    static final String DATABASE_NAME="game.db";
    static final int VERSION=1;
    public SQLiteHelper (Context context, String name, CursorFactory factory,
    int version){
        super(context, name, factory, version);
    }
    @Override
    public void onCreate(SQLiteDatabase db){
        db.execSQL(" CREATE TABLE gameStore (_id INTEGER PRIMARY KEY
        AUTOINCREMENT, flag String, playerState INT, x INT, y INT, lifeNum INT,
        hp INT, direction INT, level INT, atk INT, propertyNum INT)");
        ContentValues contentValues=new ContentValues();
        contentValues.put("flag", "false");
        contentValues.put("playerState", 0);
        contentValues.put("x", 0);
        contentValues.put("y", 0);
```

```java
            contentValues.put("lifeNum", 0);
            contentValues.put("hp", 0);
            contentValues.put("direction", 0);
            contentValues.put("level", 0);
            contentValues.put("atk", 0);
            contentValues.put("propertyNum", 0);
            db.insert("gameStore", null, contentValues);
    }
    @Override
    public void onUpgrade(SQLiteDatabase arg0, int arg1, int arg2){
    }
}
```

8.6 游戏中的辅助类

8.6.1 Tools 类

该类实现在指定位置绘图及碰撞检测方法。代码如下：

```java
public class Tools {
    /**
     * 绘制局部图片
     * @param x         图片的X坐标
     * @param y         图片的Y坐标
     * @param img       图片对象
     * @param viewX     显示在屏幕上的X坐标
     * @param viewY     显示在屏幕上的Y坐标
     * @param width     显示在屏幕上的图片内容宽度
     * @param height    显示在屏幕上的图片内容高度
     */
    static void drawImage(Bitmap img, int x, int y, int viewX, int viewY, int
    width, int height, Canvas canvas, Paint paint)
    {
        canvas.save();
        canvas.clipRect(viewX, viewY, viewX+width, viewY+height);
        canvas.drawBitmap(img, x, y, paint);
        canvas.restore();
    }
    /**
     * 碰撞检测方法
     * @param x    A物体的X坐标
     * @param y    A物体的Y坐标
     * @param w    A物体的宽
     * @param h    A物体的高
     * @param x1   B物体的X坐标
```

```
 * @param y1    B物体的Y坐标
 * @param w1    B物体的宽
 * @param h1    B物体的高
 * @return
 */
static boolean collides(int x, int y, int w, int h, int x1, int y1, int w1, int h1)
{
    if(y1+h1<y || y+h<y1 || x1+w1<x || x+w<x1)
        return false;
    else
        return true;
}
```

8.6.2 GameMusic 类

该类用于控制背景音乐及音效的播放。代码如下：

```
public class GameMusic {
    //音乐类型
    static final int BG_MUSIC=0;
    static final int BOOM_MUSIC=2;
    static final int MENU_MUSIC=1;
    SoundPool soundPool;                    //音效播放器
    MediaPlayer player;                     //背景音乐播放器
    boolean isOpen;                         //音乐开关量
    public GameMusic(Context context){
        isOpen=true;
        player=MediaPlayer.create(context, R.raw.game);
        player.setLooping(true);
        soundPool=new SoundPool(2, AudioManager.STREAM_MUSIC, 5);
        soundPool.load(context, R.raw.menu, 1);
        soundPool.load(context, R.raw.explosion, 1);
    }
    /**
     * 开启音乐
     * @param id    指定id的音乐
     */
    void starMusic(int id)
    {
        if(isOpen)
        {
            switch(id){
            case BG_MUSIC:
                player.start();
```

```
                break;
            case BOOM_MUSIC:
                soundPool.play(BOOM_MUSIC, 1, 1, 0, 0, 1);
                break;
            case MENU_MUSIC:
                soundPool.play(MENU_MUSIC, 1, 1, 0, 0, 1);
                break;
        }
    }
    void stopMusic()
    {
        player.stop();
        isOpen=false;
    }
    void setMusicOpen()
    {
        isOpen=true;
        starMusic(BG_MUSIC);
    }
    void recycle()
    {
        player.release();
        soundPool.release();
    }
}
```

8.7 本章小结

本章通过一款滚屏式飞行射击类游戏的设计与开发,详细介绍了游戏的策划及准备工作、游戏的功能、素材组织以及各个类(实体类、界面类、辅助类)的实现过程,是对前面各章节知识的综合运用。

8.8 思考与练习

(1) 尝试拓展本章游戏案例,添加成绩排行功能。
(2) 尝试拓展本章游戏案例,尽可能多地添加关卡。
(3) 尝试开发一款赛车游戏,实现选车、声音设置、赛车控制等基本功能。

第 9 章

Android 游戏物理引擎

学习目标：
- 了解 Android 常用的 2D 游戏引擎。
- 了解 Android 常用的 3D 游戏引擎。

本章导读：

当游戏需要实现比较复杂的刚体碰撞、滚动或者弹跳时，通过全部自行编程的方式实现非常困难，成本也很高。遇到这种情况时，就可以使用独立的物理引擎来模拟物体的运动，使用物理引擎不仅可以得到更加真实的效果，而且比自行开发要耗时短、效率高。一款好的物理引擎可以非常真实地模拟现实世界，使得游戏更加逼真，提供更好的娱乐体验。本章简要介绍 Android 平台常用的 2D、3D 物理引擎。

9.1 常用 2D 物理引擎

1. Cocos2D

基于 MIT 协议的开源框架，用于构建游戏、应用程序和其他图形界面交互应用。该引擎的主要版本包括 Cocos2D-iPhone、Cocos2D-X、Cocos2D-HTML5 和 JavaScript bindings for Cocos2D-X。同时也拥有非常优秀的编辑器（独立编辑器），例如 SpriteSheet Editors、Particle Editors、Font Editors、Tilemap Editors。

该引擎具有如下优点。

(1) 易用：游戏开发者可以把关注焦点放在游戏设置本身，而不必消耗大量时间学习难懂的 OpenGL ES。此外，Cocos2D 还提供了大量的规范。

(2) 高效：Cocos2D 基于 OpenGL ES 进行图形渲染，从而让移动设备的 GPU 性能发挥到极致。

(3) 灵活：方便扩展，易于集成第三方库。

(4) 免费：基于 MIT 协议的免费开源框架，用户可以放心使用，不用担心商业授权的问题。

(5) 社区支持：关心 Cocos2D 的开发者自发建立了多个社区组织，可以方便地查阅各类技术资料。

另外，2012 年发布的 CocoStudio 工具集是开源游戏引擎 Cocos2D-X 开发团队官方推出的游戏开发工具，目前已经进入稳定版。CocoStudio 总结了他们自己在游戏制作中的经验，为移动游戏开发者和团队量身定做，旨在降低游戏开发的门槛，提高开发效率，同时也为 Cocos2D-X 的进一步发展打下基础。

从全球市场份额数据来看，Cocos2D-X 主要占据除中端市场以外的高端与低端市场，约占 1/4 的市场份额。值得注意的另一个数据是，在中国，Cocos2D-X 则遥遥领先。目前，在中国的 2D 手机游戏开发中，Cocos2D-X 引擎的市场份额超过 70%。

2. AndEngine

AndEngine 基于 libGDX 框架开发，使用 OpenGL ES 进行图形绘制。同时继承了 Box2D 物理引擎，因此能实现一些较为复杂的物理效果。在 Rokon 停止更新以后，AndEngine 成为 Android 最为流行的 2D 游戏引擎。

该引擎具有如下优点。

（1）运行高效，特别是在运算量较大的情况下，使用 C/C++ 本地代码进行开发。

（2）AndEngine 是开源项目，开发者可对源码进行修改，源码在 github 上托管。

（3）拥有 Particle System（粒子系统），能制作雨、雪、流水等效果。另外还有 Streak（动态模糊）、Radial Blur（径向模糊）等效果。

（4）使用 JNI(Java Native Interface)封装了 Box2D 的 C++ 端，使运行效率得到提高。JNI 是本地编程接口，也是 Java 平台的一部分。它使得在 Java 虚拟机(JVM)内部运行的 Java 代码能够与用其他编程语言(如 C、C++ 和汇编语言)编写的应用程序和库进行交互操作。

3. Rokon

Rokon 基于 OpenGL ES 技术开发，是 Android 2D 游戏引擎，物理引擎为 Box2D，因此能够实现一些较为复杂的物理效果。开发文档相当完备，对软件缺陷的修正迅速，被称为 Cocos2D-iPhone 引擎的 Android 版（两者在业务逻辑和编码风格上很相像）。

4. LGame

LGame 有 Android 及 PC(J2SE)两个开发版本。该引擎具有如下优点。

（1）层绘图器 LGraphics 封装有 J2SE 以及 J2ME 提供的全部 Graphics API(PC 版采用 Graphics2D 封装，Android 版采用 Canvas 模拟实现)，能够将 J2SE 或 J2ME 开发经验直接应用于其中，两个版本的主要代码能够相互移植。

（2）Android 版内置有 Admob 接口，可以不必配置 XML 即可直接硬编码 Admob 广告信息。

（3）内置有按照 1∶1 实现的 J2ME 精灵类及相关组件，可以将绝大多数 J2ME 游戏平移到 Android 或 PC 版中。

该引擎除了基本的音效组件、图形组件、物理组件、精灵组件等常用组件以外，还内置有 ioc、xml、http 等常用 Java 组件的封装，jar 包体积较为庞大。

5. JBox2D

JBox2D 是开源的物理引擎 Box2D 的 Java 版本，可以直接用于 Android。由于 JBox2D 的图形渲染使用的是 Processing 库，因此在 Android 平台上使用 JBox2D 时，图形渲染工作只能自行开发。该引擎能够根据开发人员设定的参数，如重力、密度、摩擦系数和弹性系数等，自动地进行 2D 刚体物理运动的全方位模拟。

9.2 常用 3D 物理引擎

1. Unity3D

Unity3D 是成熟的开发引擎，有独立的开发客户端，采用脚本式开发而非编码式开发，可以和 Unreal、Cry 等国际顶级引擎的效果相媲美。它具有跨平台特性，wp7、ios、PC、Mac、XBox360 等终端都可以使用，语言较 C 类语言更容易学习，更贴近 Java(JavaScript、C♯)。

2. JPCT

JPCT 是基于 OpenGL 技术开发的 3D 图形引擎(PC 环境为标准 OpenGL，Android 为 OpenGL ES)，以 Java 语言为基础，拥有功能强大的 Java 3D 解决方案。该引擎与 LGame(为 2D 游戏引擎)相类似，目前拥有 PC(J2SE)以及 Android 两个开发版本。

JPCT 的最大优势之一在于有非常好的向下兼容性。在 PC 环境中，JPCT 甚至可以运行在 JVM 1.1 环境之中，因为 JPCT 内部提供的图形渲染接口完全符合所有的 Java 1.1 规范，连已经消失的 Microsoft VM 乃至更古老的 Netscape 4VM 也不例外。

3. Libgdx

Libgdx 是基于 OpenGL ES 技术开发的 Android 游戏引擎，支持 Android 平台下的 2D 游戏开发。单就性能角度来说，它是一款非常强大的 Android 游戏引擎，其缺点是精灵类等相关组件在使用上不够简化，仅支持 2 次方的图片尺寸(即图片的宽、高都是整数的平方，如 4、9、16 等，但长和宽可以不相等)。

4. Alien3d

Alien3d 基于 OpenGL ES 技术开发，是一款体积非常小的 Android 3D 游戏引擎，核心文件大约只有 40KB，所有相关 jar 文件的总和也不足 150KB。为了压缩体积，根据不同功能采用多 jar 包方式发布。

9.3 本章小结

物理引擎通过给物体赋予真实的物理属性来模拟物体的运动，包括碰撞、移动、旋转等。好的物理引擎不仅能帮助实现碰撞检测、力学公式模拟，而且提供很多机械结构的实现,如滑轮、齿轮、铰链等。更高级的物理引擎不但可以提供刚体的模拟,甚至可以提供软体及流体的模拟。这些都能帮助游戏提升真实感和吸引力。

9.4 思考与练习

(1) 除了本章介绍的物理引擎以外,在 Android 平台中还支持哪些游戏物理引擎?
(2) 了解在 Android 平台中使用 Cocos2D 引擎的环境配置方法。
(3) 通过图书资料或网络了解 Unity3D 三维游戏引擎对资源的组织方式。

参 考 文 献

[1] 刘剑卓. Android 手机游戏开发从入门到精通. 北京：中国铁道出版社,2012.
[2] 软件开发技术联盟. Android 开发实战. 北京：清华大学出版社,2013.
[3] 吴亚峰,于复兴. Android 游戏开发大全.2 版. 北京：人民邮电出版社,2013.
[4] 佘志龙,陈昱勋. Android SDK 开发范例大全. 北京：人民邮电出版社,2011.
[5] 李刚. Android 疯狂讲义. 北京：电子工业出版社,2012.
[6] 李华明. Android 游戏编程之从零开始. 北京：清华大学出版社,2011.
[7] 蒂马尔奇奥. Android 4 游戏实战编程. 张龙,译. 北京：清华大学出版社,2013.